Biological Aspects of Circ

Biological Aspects
of
Circadian Rhythms

Edited by
J. N. Mills
Department of Physiology
University of Manchester
Manchester M13 9PT
England

℗ PLENUM PRESS · London and New York · 1973

Plenum Publishing Company Ltd.
Davis House
8 Scrubs Lane
London NW10 6SE
Telephone 01-969 4727

U.S. Edition published by
Plenum Publishing Corporation
227 West 17th Street
New York, New York 10011

ISBN: 0-306-30595-X

Library of Congress Catalog Number: 72-77044

Printed in Great Britain by
The Whitefriars Press Ltd., London and Tonbridge

Contributors

J. G. Bohlen   Research Assistant, University of Minnesota, Minneapolis, Minnesota 55455, U.S A.

F. Halberg   Department of Pathology, Medical School, University of Minnesota, Minneapolis, Minnesota 55455, U.S.A.

J. E. Harker   Girton College, Cambridge CB3 0JG, England.

F. Hawking   Medical Research Council, National Institute for Medical Research, Mill Hill, London N.W.7, England.

G. V. T. Matthews   The Wildfowl Trust, Slimbridge, Gloucester GL2 7BT, England.

J. N. Mills   Department of Physiology, Manchester University, Oxford Road, Manchester M13 9PL, England.

A. Reinberg   Laboratoire de Physiologie, Fondation Adolphe de Rothschild, 29 rue Manin 75, Paris 19, France.

H. W. Simpson   Department of Pathology, University of Glasgow, Glasgow C4, Scotland.

M. B. Wilkins   Department of Botany, University of Glasgow, Glasgow W2, Scotland.

Preface

A "biological clock" has now been inferred in so many and such diverse organisms and tissues that even a summary of the more interesting and important observations would be a tedious and encyclopaedic compilation, whose bibliography would assume a daunting size. It would also be obsolescent on the day of publication. The new titles appearing in the monthly lists are scattered through many journals, but a new journal devoted exclusively to rhythm research published its first issue in May, 1970—the *Journal of Interdisciplinary Cycle Research*—and another, *Chronobiology,* appears in 1973.

In this volume several authors have been asked to review separate aspects within their own fields of study, in the hope that thereby the reader might gain an idea of the many directions of active progress and be better placed to interrelate them than would be possible after a more exhaustive study of a limited part of the field. The outcome is a series of essays in which each contributor has exercised his individuality in ideas, style and presentation, and, at some points, in vocabulary, although the glossary includes a number of terms which have been fairly generally used.

All biological study, indeed all scientific endeavour, must begin with observation and in the study of rhythms there is still plenty of scope for such, whether of the feeding habits of insects (Ch. 6), the migratory behaviour of birds (Ch. 8), or the complex adaptive rhythms of a great range of parasites (Ch. 5). Halberg (Ch. 1) devotes some attention to means of amassing useful observations upon human subjects, and Reinberg, writing about pharmacological rhythms (Ch. 4), considers that we are not yet in a position to pass beyond observation towards a coherent explanation. There are, however, scattered observations of, for example, circadian variations in concentration of an enzyme responsible for inactivation of a drug, which point the way to further exploration; and in Ch. 2 an attempt is made to work out causal sequences between a single controlling clock and the various observable manifestations of rhythmicity. This inevitably raises the problem of whether there is such a single clock or whether the rhythms in different organs and tissues are controlled by a multiplicity of clocks, a problem which is mentioned in several chapters. Studies on plants provide the best evidence that an internal clock is a widespread property of cells, and experiments on unicellular plants (Ch. 7) offer the best hope of understanding the intimate mechanism of the clock, since both microanatomical and biochemical dissection is more practicable than with more complex organisms. Almost nothing is known of transmission processes between a clock and observable rhythmic processes in plants whereby, for example, the movements of different leaves could be coordinated; but in animals, where both nervous and hormonal means are available, such study is easier and it has been attempted in insects (Ch. 6) and in mammals (Ch. 2). In mammals difficulties arise because so many variables oscillate circadianly and since they can influence one another, often mutually, it is very difficult to exclude all external factors, rhythmic or random, that can affect any of them. There is therefore need for statistical procedures, often of some complexity, for defining rhythms, and some of these are described by Halberg (Ch. 1).

Much effort has been devoted to demonstrating that rhythms are indeed endogenous, rather than responses to rhythmic external influences. An endogenous origin is now fairly generally accepted for a wide range of rhythms, although many more are clearly simple responses to outside influences; the alternation of light and darkness for example, accounts for some of the more obvious plant rhythms, such as opening and closing of flowers. These wholly external rhythms are often neglected by students of rhythms as of little interest, although to the physiologist their functional value indicates a great diversity of evolutionary adaptions. The rhythms of parasites (Ch. 5), though largely exogenous, are of special interest for their adaptive value. When the vector, such as a night-flying mosquito, is itself periodic in its habits, it is essential that the parasite be accessible at the correct time of day. For this purpose an endogenous clock within the parasite, which would be unlikely to keep perfect time indefinitely, is less satisfactory than an appropriate response to a rhythmically varying factor within the host; the periodicity of host, parasite and vector must somehow be coordinated if the parasite is to be perpetuated. Other aspects of parasite development, such as the maturation of the sporozoites of the malarial plasmodia, need however a sufficiently accurate internal clock.

The alternation of light and darkness is the most consistent oscillation resulting from the passage of day and night. It is therefore not surprising that in most species light exerts a dominant influence upon the clock and that it operates through a great variety of light-sensitive organs and pigments, suggesting that such mechanisms have appeared many times in the course of evolution. Such mechanisms have, presumably, a genetic basis; and individual differences in rhythmic behaviour have been shown, alike in plants (Ch. 7) and in insects (Ch. 6), to be due to minor genetic differences. The alternation of light and darkness follows a very different course in different parts of the globe, from the tropics, where there is a constant LD 12:12. with little annual variation, to the poles where conditions are always LL or DD, and the annual cycle of climatic variation is dominant. The consequences of this are considered in Ch. 3. The Arctic, in particular, has been the scene of much recent investigation, some of it as yet unpublished. Responses to varying day length are prominent in species of temperate latitudes, but would be valueless in the tropics. Man has modified many of his habits far too rapidly to permit evolutionary adaption, so that the persistence of his endogenous rhythms is often disadvantageous. This is most notable in night or shift workers, or in those who fly rapidly across many time zones. Their disadvantages may be relieved when we gain a better understanding of these rhythms.

The most widespread use made of an internal clock by plants and animals alike is to time and trigger off rhythmic processes appropriate to the external meteorological oscillation. A further use, described in birds in Ch 8 but known to exist also in insects, is for navigation. Navigators who depend upon solar or sidereal observation have always needed an accurate chronometer; the same need is present for all animals navigating by the sun, and is apparently supplied by an internal circadian clock.

Rhythms with a period longer than a day are hardly mentioned here; many are known to exist, such as the monthly rhythms of intertidal animals and annual rhythms in many species. They present greater difficulties of investigation, since the student of rhythms always hopes to be able to observe several successive cycles under comparable conditions, whereas a bird over several years is likely to show changes referable to aging; and comparable conditions are more difficult to maintain for years than for days. In Ch. 8 evidence is however adduced to suggest that many birds have a timing device responsible for their annual cycles of behaviour. Whether this is ascribable to a circadian clock combined with a "demultiplication process", or whether the timing process is entirely distinct, is a matter for future study. Every chapter indeed reveals areas of ignorance wide open for further investigation.

As with any developing discipline, new technical terms have been found necessary and some former terms have needed precise definition. A glossary has thus been included, though some new terms, such as Chronobiology and Chronopharmacology, are omitted since they are self-explanatory; and where the sense is clear without their use, an attempt has been made to eschew needless technical terms.

The reader wishing access to a more general bibliography may consult a recent book, *Human Circadian Rhythms,* by R. T. W. L. Conroy and J N. Mills [Churchill, 1970], whose Introduction lists other books and published proceedings of conferences.

J.N.M.

Glossary

(N.B.—A glossary of terms used only in Chapter 5 appears at the end of that chapter)

Acrophase	The time of the peak value of the sine curve fitted to a rhythmically oscillating variable. This may be expressed as clock time, or time in relation to some other variable, or in degrees of arc.
Amplitude	One half of the total regular excursion of a wave; best restricted to mathematically fitted sine waves. Abbreviated C or A. *Range* is used in a less precise sense to describe the usual difference between maximum and minimum if no analysis has been performed.
Circadian	(of a rhythm). With a period of approximately 24 h, strictly 20 to 28 h. *Dian* is sometimes used for a more precise period of 23.9-24.1 h.
Clock	A hypothetical mechanism in the brain, or other organ, which records the passage of approximately 24 hours and transmits this information to other systems.
Clock hours	Are given according to international nomenclature in hours and minutes for 24 h (i.e., 6 a.m. = 0600; 6 p.m. and 37 min = 1837; midday = 1200; midnight = 2400 or 0000).
DD	In continuous darkness. LL—In continuous light. LD—In alternating light and darkness. LD followed by two numerals: the hours of light and darkness e.g. LD 1:1, alternating one hour each of light and darkness; LD 4:20, alternate periods of 4 h light and 20 h darkness. When it is desirable to specify the clock hours, the form L 0600-1800 D 1800-0600 is used.
Diurnal	Pertaining to the day rather than night. Contrast *nocturnal*.
Endogenous	(Of a rhythm.) Maintained from within the organism, independently of rhythmic external stimuli.
Exogenous	(Of a rhythm.) Dependent upon rhythmic stimuli from without the organism. (Loosely, of a stimulus): giving rise to a rhythm.

Free-running   (Of a rhythm.) Observed under conditions when exogenous stimuli are lacking, and endogenous rhythms are unmasked. The period usually differs slightly from 24 h.

LD, LL   See DD.

Mesor   The mean of the sine curve fitted to a rhythmic variable. This is equal to the mean of the determined values if these have been sampled at regular intervals for an integral number of cycles. The term "level" has been also used. Abbreviated Co or M.

Nocturnal   Pertaining to the night rather than day. Contrast *diurnal.*

Nychthemeral   (Of a rhythm.) Oscillating in time with the alternation of day and night. (Of habits, circumstances, etc.) Conforming to the influences imparted by the alternation of day and night.

Nychthemeron   A night and a day.

Period   The time interval occupied by a wave, q.v.

Range   See *Amplitude.*

Synchronizer   See *Zeitgeber.*

Wave   The complete pattern of a periodic variation which is repeated at regular intervals.

Zeitgeber   (Lit. "time-giver"). Some authors prefer to use synchronizer, or entraining agent. A stimulus which gives rise to an exogenous rhythm, or determines the phase or period of an endogenous rhythm. The common Zeitgeber are nychthemeral. In deference to its German origin, the plural is identical with the singular.

Contents

CHAPTER 1

Laboratory Techniques and Rhythmometry

Franz Halberg
Department of Pathology, Medical School
University of Minnesota, Minneapolis
Minnesota 55455, U.S.A.

CHRONOBIOLOGY

Today a branch of biological science, chronobiology, objectively describes an organism's or ecosystem's time structure, i.e., its non-random and thus predictable changes including rhythms [1] along with changes related to growth, development and aging. Biological time structure characterizes populations or groups as well as individual organisms or their sub-divisions: organ-systems, organs, tissues, cells and intra-cellular elements, including electron-microscopic ultrastructure. Several aspects of biological rhythms are innate, yet these rhythms are eminently adaptive: they are synchronized, modulated or at least influenced by a broad range of cyclic or other social or geophysical factors [1-4].

Chronobiology has become a scientific discipline because: (1) it specifically selects for study a ubiquitous and important biologic

Work supported by grants from the U.S. Public Health Service (5 KO6 GM13981-10), NASA (NAS 2-2738 and NGR-24-005-006) and the United States Air Force (Contract F 29600-69-C-0011)

temporal structure and the underlying factors—thus contributing (2) a rapidly broadening base of facts, some of them critical to individual health and environmental integrity, and achieving such progress by the use of (3) a special system of methods—including diverse sampling schemes with conventional as well as advanced instrumentation and electronic computer procedures for data analysis (and, in some cases for data collection as well [9]).

Herein, I wish to illustrate some of the procedures used in our laboratory for the analysis of rhythms, notably as these apply to current medical and biological practice. A review of the broader subject of rhythmometric techniques for the chronobiologist would extend far beyond the scope of a single chapter.

1.1 BACKGROUND

Many factors, technical and biological, contribute to the variability of biomedical information. Major emphasis has heretofore been placed upon improving various biophysical and biochemical techniques: many of the methods used in the not-too-distant past needed improvement, and technical variability was rightly recognized as being quite marked. Prior to the rather recent large-scale introduction of modern, highly-automated laboratory methods and quality control procedures, the effect of technical factors was usually not separated from biological variation, the latter being tacitly if not explicitly attributed to inter-individual differences due to sources of variation ranging from environmental factors over sex, age, ethnic and other background, to diet or occupation [5]. Moreover, careful investigators have recognized that in certain instances findings, say, on different ethnic groups were confounded by factors ranging from ethnic affinities over climate, body-size, basal metabolic rate and pigmentation, to affective differences [6]. In the face of an already complex problem it was certainly tempting to neglect the added possibility of intra-individual biological variability as a function of time; until proof was offered to the contrary, it seemed convenient to presume that such variability was rather small, if not negligible, for most variables and individuals except, perhaps, for certain changes associated with growth, development or senescence.

In view of this emphasis upon better measurement techniques, many laboratory procedures have markedly improved while more often than not biological sources of variation continue to be neglected, although it has become apparent to some interested individuals not only that data obtained with even the best techniques exhibit considerable variability in one and the same subject but also that part of this variability is rhythmic rather than random [3, 7-9]. A spectrum of rhythms with different frequencies [7, 10] contributes a statistically significant portion of this predictable variability, and computer methodology serves to assess the parameters of each individual rhythm.

In a number of instances, within-subject variation has now been found to be equal to, if not greater than, variation among subjects, with a considerable fraction of the variance being contributed by a number of partly innate rhythms; even though these rhythms are modified by "nurture" in its broadest sense, including aspects of our daily routines, they are nonetheless at least partly an evolutionary acquisition of our "nature". The relative contributions of nature and nurture to, say, circadian rhythms of different urinary variables can vary markedly from variable to variable.

Rather than raising the rhetorical question whether rhythms are exogenous *or* endogenous [42], one weighs the contributions of various factors to a given rhythm in a specified variable [8]. For instance, as shown in Fig. 1.1, the effects of a 21-h "day" schedule imposed in the Arctic, during the "midnight sun", upon various urinary variables in healthy individuals vary drastically, a finding suggesting gross differences in mechanisms underlying these rhythms. Circadian amplitude ratios for several of the variables here summarized [8] show indeed that a subject's

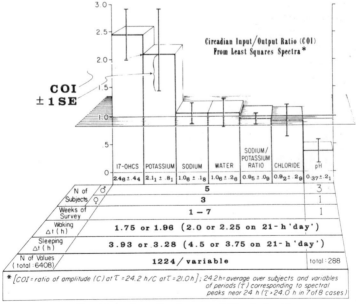

Figure 1.1. Circadian hierarchy of urinary variables in healthy man on a 21-h "day" routine during arctic summer—Spitsbergen (78°N). Courtesy of Simpson (8).

"output frequency" can be quite different from his routine of living (the 21-h periodic "input"). Second, the prominence of an "output" component desynchronized from the social 21-h routine varies from an almost negligible one (in the case of pH) over a reasonable one (for sodium, water and chloride) to one overriding the effect of the social schedule (for potassium and corticosteroids in urine). Moreover, there is no demonstrable progressive adjustment of the rhythms' frequency to 1 cycle/21 h for most variables, a point not shown in this figure [8]. It is important to realize that modern rhythm measurement—rhythmometry—utilizing properly collected and coded information, can quantitatively separate the contributions of nature and nurture, e.g., in the form of spectral amplitude ratios [8].

Apart from basic interest, this resolving power of rhythmometry awaits exploitation in the clinic as well. With this aim in mind, health care procedures should include the coding of potentially pertinent temporal aspects of an individual's "nature" as well as his "nurture". Many rhythms have now been demonstrated to be statistically signifi-cant; eventually they may become relevant to clinical practice. The time-coding of biologic data assumes even greater importance for variables exhibiting large-amplitude rhythms whether or not these are obscured by unknown or unevaluated factors and random variability, the sum total of which is herein referred to as noise. Methods for resolving rhythms in noisy time series are available and such rhythmometry is as important for many chronobiological problems as is microscopy for resolving most cells.

Against this background, we may consider the problems of specifying "normal limits" and also "clinical values" or "clinical percentiles" by focusing not only upon changes as a function of the many extrinsic and intrinsic factors mentioned earlier but also upon the many pertinent aspects of timing. The relative contributions of these factors, as yet not established, await concomitant rigorous analysis on a world-wide scale. Such analyses depend upon the standardization of data collection and upon broad scale documentation for clinical-laboratory and other biomedical purposes.

We are concerned about rhythms for either or both of the following reasons:

(1) Optimally, the study of rhythms yields new objective characteristics of health or disease.

(2) As a minimum we hope to reduce (if not eliminate) the confounding of conventional single-sample time-unqualified results by sources of variability which prevail unidentified when rhythms are not evaluated.

"Stop-gap" procedures will first be suggested to cope with at least some of the confounding sources of temporal variation. These procedures can immediately be instituted by any physician, or by technologists, nurses, orderlies or the educated lay subject himself.

Even more extensive and seemingly complex coding of many biomedical data is advocated for the purpose of resolving quantifiable endpoints of rhythms. This latter approach becomes realistic through do-it-yourself health monitoring.

1.2 MINIMAL APPROACHES

Efforts to reduce unmanageable variability by separating that attributable to rhythms should be distinguished from the description and quantification of rhythms for assessment of health as well as for detecting rhythm alteration *within* as well as beyond the so-called normal limits. Time-coded single-samples on occasion suffice for the former purpose; multiple samples, usually though not invariably [11], are needed for the latter.

1.2.1 Coding of Sample by Clock Hour and Calendar Date

This kind of coding is minimal in terms of effort but may or may not be minimal in terms of pertinent information acquired. It should be done for any sample or any one laboratory datum or for almost any other biological determination or observation. First, one must code the clock hour and minute as well as the calendar date when the sample was obtained and/or the observation made; this is a technically easy task and should be mandatory. In most individuals these data may serve for imputing the stage of a rhythm at the time of sampling and, eventually, for making a comparison between a given datum and clock-and-calendar-qualified reference standards such as statistical tolerance limits.

Whenever a majority of individuals live by clock and calendar, coding according to such schedules represents an improvement over time-unqualified data. In such relatively or mostly "synchronized" groups or populations, the importance of certain rhythms can be documented by statistics analyzed solely as a function of clock hour, without additional information; impressive statistics revealing circadian and circannual rhythms for human deaths, without even sleep-rest coding, are cases in point (Figs 1.2 and 1.3). Figure 1.2 underlines first the importance of rhythms that tip the scale between death and survival. Second, it shows that once a population is studied, even an event such as death—unique for the individual—can exhibit a rhythm. Third, again for a population, consisting of a majority of subjects adhering to a reasonably standardized mode of life, the figure demonstrates that when

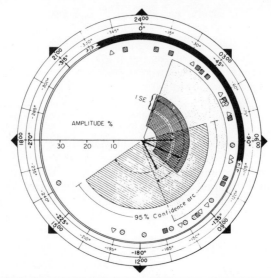

Mortality category	No. of cases	No. of (A,ϕ)'s	Symbol for acrophase (ϕ)
Paediatric	21,673	12	△ ⬤
Pulmonary	9,357	12	▨ ⬤
Cardiovascular	7,644	9	⊘ ⬤

Figure 1.2. Human circadian population rhythms in paediatric, cardiovascular, or pulmonary death as function of clock hour [47].

large samples are available clock hours can approximate the time course of rhythmic phenomena; it should be kept in mind, however, that this approach invariably ignores individuals on odd schedules, such as certain night watchmen, nurses, shift-workers, etc., and that, inter alia, it may indicate hours of changing resistance, or of changing health care quality. Figure 1.3 reveals the importance of another set of rhythmic phenomena—the human circannual system. Note the 1-year synchronization of the circannual system—in population data covering 27 years.

Nonetheless, coding solely by clock and calendar can have major shortcomings for certain individuals. For instance, without further biologic qualification, certain values such as those for plasma 17-hydroxycorticosteroids, coded only by clock and calendar time for a consistently nocturnally-active subject (e.g., a night-watchman), will differ drastically from those obtained at the same clock and calendar time for subjects on a diurnal-activity nocturnal-rest routine. Hence it is desirable—and inexpensive—to go at least one or two steps further and

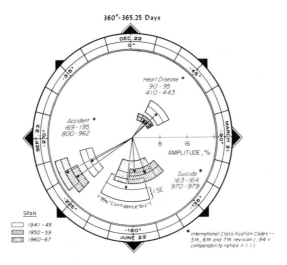

Figure 1.3. Circannual rhythms of mortality in Minnesota during 1941-1967. Note reproducibility of findings in consecutive decades.

obtain information that permits one to utilize a reference point on the sleep-wakefulness cycle (see also Fig. 1.4), or on a stable rhythm(s) with the same frequency (or frequencies) in another variable. Thereby comparison among laboratory values from individuals on different schedules may become more appropriate [11, 41].

Actually, in the case of gross differences in sleep schedules of populations even an approximate correction for an imputed sleep schedule can be quite advantageous. Thus Fig. 1.4 shows that when referred to local midnight the circadian 17-hydroxycorticosteroid acrophase differs in two populations; the difference disappears as soon as each sample's rhythm is referred to the middle of the habitual sleep span—be it in Glasgow bedrooms or equatorial rain forests.

Figure 1.5 shows plasma 17-hydroxycorticosteroid values at two times of day for about one week in two diurnally active "larks" as compared to a nocturnally-active "owl". It can be seen, first, that sampling at only two clock hours can reveal some large within-day differences—and second, that these differences themselves can drastically differ from day to day in the same subject. Third, with such two-time-point sampling, one may note the circumstance that usually, but not invariably in a given nocturnally-active subject, a blood sample at 21^{00} is lower in cortico-steroid concentration than an 05^{00} value, whereas for two diurnally active individuals the opposite seems to hold true. Fourth, for the case of blood corticosteroids, as well as for a number of other variables, a spectrum of rhythms has been demonstrated including among others

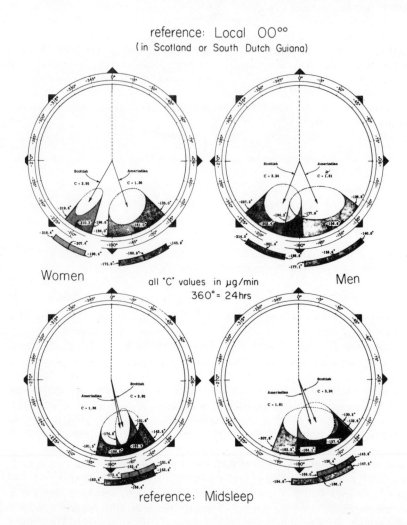

Figure 1.4. Circadian acrophase, ϕ, of 17-hydroxycorticosteroid excretion by men and women from Scotland and South Dutch Guiana (Amerindians); summaries of eight cosinors providing in circular displays an amplitude (A)-weighted ϕ and of eight ϕ-estimations, done without A-weighting, shown as arcs outside the circles: local midnight, 00^{00}, is used as phase reference for displays in the top half of the Figure; mid-sleep serves as phase reference for displays at the bottom [41].

rather prominent ultradian, circadian and yet lower-frequency changes. A single-time-point value—and even two-time-point samples that currently are withdrawn in certain clinical and biologic settings—are confounded by contributions from all of these varied frequencies. This circumstance has led to indications of a "normal range" for plasma corticosteroid extending from zero to over 20 μg per 100 ml. Since such very wide ranges are limited in information content, the possible yield of rhythmometry based upon data from at least three properly placed time points may well be tested in the clinic.

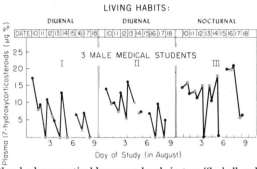

Figure 1.5. Blood adrenocortical hormone levels in two "larks" and an "owl". Comparison of samples drawn at 05^{00} (●) or 21^{00} (⊚).

1.2.2 Coding of Sleep-Regimen

For any sample such coding can readily be instituted if a brief sleep questionnaire is routinely provided for the subject at least one week prior to sampling—Scheme I (Table 1.1), prospective sleep coding. In emergencies or other cases when such as-you-go records are unavailable, the questionnaire is filled out from memory whenever samples are collected—Scheme II (Table 1.2), retrospective sleep coding. Only minimal information concerning the schedule of sleep-wakefulness is required—preferably with an indication of any insomnia. Thus, with each sample recorded in biomedical documentation, one indicates "as-one-goes" the approximate times of (1) retiring and (2) arising on each day of the week prior to sampling, or from memory at least for the 24 h preceding sampling and again as "average" retiring and arising time during the week prior to sampling [41].

Eventually this information may be recorded by self-coding, e.g., on punched cards or on a form amenable to automatic information transfer from form to punch card. For instance one may code in this order:

(a) The calendar date (Year: 4 places; Month: 2 places; Day: 2 places; Clock Hour: 2 places; and Minute: 2 places) of collection of a sample; and

Table 1.1

SCHEME I

(To be Filled Out Once a Day)

SUBJECT NAME:_____

SOCIAL SECURITY NUMBER: /_/_/_/_/_/_/_/_/_/_/_/
                    1 2 3 4 5 6 7 8 9 10 11

INTERRUPTIONS OF SLEEP (prior to "light on")   Number   Total duration

  (a) Spontaneous   /_/_/   /_/_/ (min.)
                                        12 13   14 15

  (b) Prompted by environment   /_/_/   /_/_/ (min.)
                                        16 17   18 19

LIGHT ON (or first awakening)

Year /_/_/_/_/ Month /_/_/ Day /_/_/ Hour /_/_/ Minute /_/_/
    20 21 22 23   24 25   26 27   28 29   30 31

ADDED SLEEP (after first awakening prior to last lights out)

  If "Yes": Start: Hour /_/_/ Minute /_/_/
                 32 33   34 35

        End: Hour /_/_/ Minute /_/_/
              36 37   38 39

LIGHT OUT (if coincident with retiring)

  Month /_/_/ Day /_/_/ Hour /_/_/ Minute /_/_/
      40 41   42 43   44 45   46 47

RETIRING FOR SLEEP (if not coincident with retiring)

  Month /_/_/ Day /_/_/ Hour /_/_/ Minute /_/_/
      48 49   50 51   52 53   54 55

  (b) The clock time when, on the average, the patient retired for sleep (Hour: 2 places; and Minute: 2 places) (during the week prior to study); this time is easily specifiable if the subject is required to keep records;

  (c) The average time when the patient arose during the week prior to study (Hour: 2 places; and Minute: 2 places); and also

  (d) Whether this schedule for a week prior to sampling was consistent from day to day (1 place—yes or no) and whether there was insomnia (1 place—yes or no);

  (e) Whether schedule (d) was representative of the schedule during the preceding month (1 place—yes or no).

Table 1.2

SCHEME II

RETROSPECTIVE CODING OF SLEEP-REGIMEN

SUBJECT NAME:_____

SOCIAL SECURITY NUMBER: [/ / /] – [/ /] – [/ / / /]
                              1   2   3   4   5   6   7   8   9

1. Calendar Date: Year [/ / / /] Month [/ /] Day [/ /]
                    10 11 12 13   14 15   16 17

2. Hour of Sample Collection: [/ / / /]
                            18 19 20 21

3. Did you work or rest during odd hours (e.g., work after midnight or rest late in the morning)?

   (a) during "day−1" before sampling (Yes = 1; No = 2) . . . . [/]
                                                   22

   (b) during "day−2" before sampling (Yes = 1; No = 2) . . . . [/]
                                                   23

   (c) during "day−3" before sampling (Yes = 1; No = 2) . . . . [/]
                                                   24

   (d) during "days−4 to −7" before sampling (Yes = 1; No = 2) . . . . . [/]
                                                   25

   (e) during "weeks−2 to −4" before sampling (Yes = 1; No = 2) . . . . [/]
                                                   26

If answer(s) to any of the questions in 3(a) to 3(d) is(are) "yes", fill out as best you can information required on "prospective" sleep-coding card from memory, indicating that fact on card and commenting separately in writing.

If answer(s) to any of the questions in 3(a) to 3(d) above is(are) "no", continue below, *giving preferably the "best guess" rather than no answer.*

4. Rising Time

                                          Hour   Minute

   (a) Today . . . . [/ / / /]
                                             27 28 29 30

   (b) Average, past week . . . . [/ / / /]
                                             31 32 33 34

   (c) Average, last month . . . . [/ / / /]
                                             35 36 37 38

5. Retiring Time

                                          Hour   Minute

   (a) Last night . . . . [/ / / /]
                                             39 40 41 42

   (b) Average, past week . . . . [/ / / /]
                                             43 44 45 46

   (c) Average, last month . . . . [/ / / /]
                                             47 48 49 50

If the schedule during (d) differed from one day to the next by more than one hour or if it differed, on the average, by more than two hours during (d) as compared to (e) or if there was insomnia, another card is required for coding this additional information.

1.2.3 Rhythm Coding

Ideally, any single sample or observation should be referred to the timing of a longitudinally evaluated rhythmic variable that has a favourable noise-to-signal ratio. Under many conditions circadian rhythms in body temperature or the excretion rate of potassium or corticosteroids in urine and/or a few other variables possess this quality. To obtain such a reference standard for coding, routine self-measurements and/or urine collections must be made. In a given subject one might use as zero phase the acrophase [12, 13] for oral temperature or urinary potassium in samples collected, say, about 6-9 times/day during wakefulness for two or more days prior to extensive laboratory work. Thus, one can examine in each case the relative potential merits of referring a value determined on a rhythmic variable not only to clock-hour and sleep schedule but also, say, to the acrophase, i.e., the peak of the cosine curve best approximating all data on some reference variable.

The desirability of such coding involving auxiliary rhythmometry on the more readily measured variables may depend upon the importance of possible disease-related alteration of rhythm characteristics, notably of time relations among rhythms. In any event, time-dependent changes in resistance or susceptibility not only may be significant for the onset of disease but they also matter even in determining the chances of death. Along the 24-h scale there are critical changes in the outcome of a mammal's exposure to many agents, from bacterial endotoxins to carcinogens to overdoses of cardiac and other drugs. There also are other, e.g., about-yearly (circannual) cycles in factors predisposing man to death from heart disease at one extreme or to suicide at the other, Fig. 1.3.

1.2.4 Coding Clock and Calendar Information on Sample Processing

The recording of the hour and date of the actual biochemical or other determination is an added and separate requirement for those clinical or other procedures carried out with a lag after the collection of a sample. In order to interpret the effect, if any, of varying lags between the collection of a sample and, say, a chemical determination, independent work on sample preparation and the stability of the kind of constituent studied is indispensable. Conditions of sample storage between collection and analysis should also be indicated (e.g., temperature of cold room in the case of frozen storage and the number of times and durations of any thawing of samples).

Whenever possible and especially when information on stability is incomplete, the times between sample collection and completion of biochemical or biophysical analyses will have to be kept as short as possible and similar, if not identical, in all laboratories cooperating in a standardized study.

Knowing (or if unknown, investigating) the stability of any sample during the time elapsed between its collection (e.g., withdrawal of blood) and a determination (e.g., biochemical) becomes particularly important in studies on rhythms with periods ranging from a day to a year. In such work, for a variety of reasons (among them problems in sample transmittal), variation in "time elapsed" may be unavoidable.

1.3 OPTIMAL APPROACH

First and foremost, rhythmometry yields new objective characteristics not only of health [7] but also of disease [14]. Rhythmometrically analyzed data already have indicated in a few conditions a departure from the normal time (including rhythm)–structure. Such time-structure alteration, here defined as dyschronism, includes changes in one or several of the rhythm parameters defined in 1.5.

The following conditions among others have been reported to be associated with an altered time structure revealed by rhythmometry.

(a) Transmeridian dyschronism, i.e., the rhythm alteration as well as any malaise and performance decrement following transmeridian travel (e.g. by airplane) [15, 16],

(b) Depressive emotional illness [14],

(c) Intermittent catatonia [48],

(d) Rheumatoid arthritis [17],

(e) Cancer [18],

(f) Peptic ulcer [19, 32],

(g) Chronic multiple sclerosis [20].

The chronobiological approach can be recommended on the basis of the foregoing findings and may deserve testing in the above-mentioned conditions and many others. For its implementation, self-measurement will be indicated until automatic recording and "as-you-go" analysis becomes routinely feasible for human body functions. Actually, for limited goals a statistically significant circadian rhythm (by definition one with a non-zero amplitude validated by inferential statistical means) can be sought and detected as well as quantified for at least certain healthy individuals from as few as six or nine appropriately placed values per day collected over a span of only two to five days. This point has been documented for some characteristics of circadian rhythms, though not for all of them, with respect to several important body functions [21, 22] such as temperature, heart rate, blood pressure, peak expiratory flow, grip strength and for blood and urinary variables as well.

The collection of multiple samples of urine, like the repeated sampling of biophysical variables, can be accomplished by "do-it-yourself" health monitoring. At first, such lay participation in the "preventive maintenance" of human health need involve interactions only between a given subject and a computer. The subject could well be initiated in different steps of such "autorhythmometry" at appropriate levels of his schooling [24]. Throughout life, rhythmometry can serve as a monitoring procedure until results from this technique provide an "alarm", or other (objective and/or subjective) reasons prompt a physician-contact. From then on, autorhythmometry becomes as well a medically-directed, decision-helping endeavour complemented by other conventional procedures.

1.4 COST–YIELD

So long as rhythmic variation is not ascertained and its control is not required by the practitioner, disregard of rhythms will be advocated upon economic grounds. The proper rhythm-coding of a single sample requires forethought, and it is expensive to collect and analyse multiple samples. Thus the nuisance of rhythmometric or other chronobiologic coding and the added cost for multi-sample laboratory rhythmometry often prompt one to ignore the potential benefit of these approaches. But among other factors [23] chronobiologic variation unquestionably limits the diagnostic capabilities of current laboratory procedures, notably the value of single tests [21].

Whenever unaccounted-for-variation leads to questionable or meaningless test results, the whole procedure may not only be wasteful [24] but harmful as well—if it leads to inaction in cases when preventive action could be helpful. From this viewpoint one may ask whether some laboratory tests really meet a requirement for the subject examined, just as one may question the value of a few time-unqualified blood pressure measurements [25]. Are some test results more than diagnostic "placebos" intended for meeting the "homoeostatically-trained" physician's expectation and the associated medico-legal exigencies? These questions are particularly pertinent in view of the modern laboratorians' capability to perform many diverse determinations on a single sample of tissue, blood or urine. Under the present system it would be quite innovative to do all these determinations on serial samples.

The temptation is great to exploit instead the existing (and to the consumer hardly gratuitous) opportunities for a multivariable approach on single samples, whether or not the variables investigated are pertinent to the condition under study, and to do so without reference to readily assessed rhythms in one or several pertinent variables. The attitude that, in the absence of pertinent information allowing the singling out of one or a few functions for rhythmometric evaluation, one should study all possible variables in single samples collected even without regard for rest-activity schedules seems to be potentially rather wasteful when some or perhaps all of the variables to be examined are known to be rhythmic [21]. "Multa sed non multum" indeed may be the result.

Up to now, laboratory technology has developed in such a way that in many laboratories, automatic analysers are routinely programmed for 12 to 26 kinds of determinations, whether or not they are desirable or requested. Under these circumstances, the collection and laboratory work-up of more than one sample from a given individual remains costly, quite apart from expenditures for numerical analyses. Moreover, since the automatic analysers are best started at a fixed time, we readily become slaves of instrumentation—the withdrawal time becoming a function of starting the run, even though there are precedents for testing the merits of letting the kind of information being sought rather than the logistics of analyser operation dictate the time of sampling.

The nocturnal emergence of *Wuchereria bancrofti* microfilariae into the circulating blood constitutes perhaps the oldest practical example of a need for timing the withdrawal of blood for a diagnostic purpose [26-29], see chapter 5. Sampling of blood in the evening for suspected Cushing's disease [30, 31] is yet another case in point, since the evening sample will differentiate most patients with Cushing's syndrome from other diurnally active human beings. Many other blood constituents also exhibit rhythms [33]. In these instances as well, timing will have merit, although many practitioners will lose interest in a given rhythm to the extent that its amplitude decreases. For all rhythmic variables, however, we should test the benefit from problem-aimed laboratory work, i.e., from carrying out but one or a few kinds of determinations on multiple samples from a given individual.

With the detection of rhythms with yearly and other frequencies in any one variable, new problems of course arise. These are considerations at first for research rather than for practice. Moreover, we can economize in such work substantially once we realize that under some circumstances certain kinds of pooling for clinical rhythmometry on urine greatly reduce the cost of multiple chemical determinations.

1.5 WHAT RHYTHMOMETRY YIELDS

Procedures for exploiting a single datum or so-called transverse macroscopic profiles, and a brief illustration of least squares windows [43] as well as of variance spectra [44, 45, 46] for biologists, are alluded to or illustrated elsewhere.

For any one statistically significant rhythm or for each of several rhythms of different frequencies, computer-executed rhythmometry can provide a set of the following objective numerical values, briefly a profile of rhythm endpoints (e.g., on microfilm, eventually as a "health summary" of a given subject):

(a) *mesor, M,*—corresponding to a rhythm-adjusted mean;
(b) *amplitude, A,*—measuring half the extent of predictable rhythmic change with fundamental frequency, and
(c) *acrophase, ϕ,*—measuring the timing of the predictable rhythmic change with fundamental frequency, as well as
(d) *shape,* i.e., amplitude and acrophase pairs $[(A_1,\phi_1) \ldots (A_n,\phi_n)]$ at several statistically-validated harmonics that can automatically be added vectorially at consecutive time points in order to synthesize and display the characteristic waveform, and
(e) *confidence limits* for each of the above characteristics as a first approach to more complete biologic and "clinical limits" of each rhythmic variable.

One can rapidly obtain in computer-directed plots on microfilm such endpoints as:

(f) displays in time, so-called *chronograms*, of the data themselves as original values or after transformations, e.g., into a percentage of the mean or as a "behaviour day chart", Fig. 1.6.
(g) *chronobiological windows* (W) or spectra (S)—including a display of amplitude and of an error measure as a function of trial period;
(h) *chronobiological serial sections* (SS)—including a display of rhythm characteristics as a function of time. Successive sections of (one or several) fixed length(s) [called interval(s)] of a time series are fitted with a proper (or several) cosine curve(s), chosen on the basis of independent evidence. These intervals are displaced in increments throughout the series. Characteristics thus assessed include first a probability value (P) as well as the M, A and ϕ;

In Fig. 1.7, a chronobiological serial section reveals transmeridian-flight-induced changes in circadian rhythm characteristics of human mental performance. The chronogram on top presents as dots self-administered test results obtained several times during each waking span for 55 days. Consecutive short, vertical graduations below the bottom row of the profile bracket 24 h, whereas two vertical broken lines (Flight 1 and Flight 2) each indicate the time of a transmeridian flight. The ordinate scale of the chronogram extends from the 'mean −3 standard deviations' at the bottom to the 'mean +3 standard deviations' at the top.

Each dot plotted describes the ability of a healthy man to remember random numbers correctly and to transcribe them speedily. Each dot represents performance gauged by dividing the seconds elapsed during the test (controlled by stopwatch) into the number of correctly

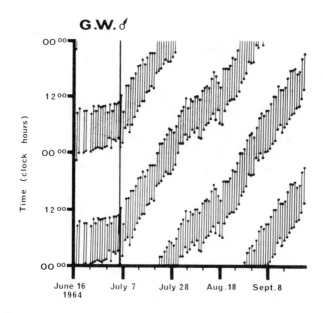

Figure 1.6. Behaviour-day chart prepared under computer control. Continuous lines bracketed by dots indicate times when a subject, alone in a cave with a watch, slept. Before the heavier vertical line he attempted to sleep at his customary hours; subsequently he went to bed and got up when he felt so inclined. The ordinate covers 48 h, each sleep span being represented twice. Original data of John N. Mills, 1964, *J. Physiol., Lond.,* 174, 217-231.

remembered numerals from a total of 70 random numbers (10 sets of 7). A considerable scatter of these performance values is seen in the top row, revealing within-day differences in quality and/or speed of mental performance recurring throughout the entire study. Also discernible is a gradual improvement in performance—a learning effect.

In the rows below the top chronogram, a so-called pergressive analysis is displayed; it involves the fit of a 24-h cosine curve (24 h ≡ 360°) to 5-day intervals displaced for analysis in 12-h increments, as indicated at the bottom of the profile. All values thus obtained are plotted at the midpoint of a given interval analysed. The number of samples for each 5-day interval analysed is shown in the bottom row; it varies from 21 to 37.

For the possible detection of any circadian rhythm, probabilities (*P* values) are imputed and shown as dots in the second row. The dashed line in this row denotes the 5% level of statistical significance. The majority of *P* values lie at the bottom of the *P* row, corresponding to statistical significance below the 1% level.

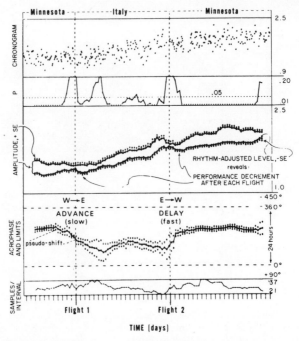

N = 323 ; Period fitted = 24h ; Interval analyzed = 5 days ; Increment = 12 hours ;
Ø reference = 00°°CST

Figure 1.7. Chronobiologic profile of human mental performance in relation to two transmeridian flights. Random number memory: ratio of correct numbers/total time. Copy of original output directly on microfilm. Original investigations and self-measurements by R. B. Sothern (c.f. also [49]).

Only P values separated by a number of increments at least equal to the interval chosen for analysis can be used for rhythm detection. Special numerical methods and test conditions are required to separate any possible effects of serial dependence from the rhythm, *per se*. The spectrum of an entire time series (not shown) or, in our case, of large sections of the series revealed as consistent by the SS serves to validate impressions from microscopic analyses on relatively short, separate intervals. In any event, imputations have been defined and quantified earlier [11].

The third row from the top shows two curves. The lower one represents roughly the performance mesor, or M—a so-called rhythm-adjusted mean; it is corrected for data collected at unequal intervals (rather frequently during waking but without interrupting the subject's sleep span for testing). One standard error of each M value is plotted below it. The distance between any point on the lower (M) curve and the corresponding point on the upper curve represents the amplitude, A. One standard error of the A is plotted as a dot above each A.

The amplitude values portray a statistically significant circadian rhythmic change in performance during most of the observation span. An effect clearly seen even in the chronogram, a steady increase in values, is now properly quantified. By the same token, a performance decrement is noticeable by a drop in M, in this case after the West-to-East flight. (Any rhythms or pararhythms with a lower-than-circadian frequency suggested by lesser fluctuations in the A and in the M line are not here considered.)

A measure of the rhythm's timing, the computative acrophase, ϕ, is shown in the bottom row. It represents the peak of the 24-h cosine function best approximating the rhythm in all data comprised within a given interval analysed. Two dashed horizontal lines in this row correspond on the ordinate to zero and $-360°$, i.e., to midnight Central Standard Time (CST) in Minnesota, which remains the ϕ reference throughout the entire profile. The ordinate actually extends by 90° below 0° (to +90°) and above 360° (to $-450°$).

In Minnesota, before the first flight (1 on the graph), the ϕ—denoting highest performance in the neighbourhood of this time—was superior to that at other times. After the first flight, the ϕ advances toward 18^{00} Central European Time, corresponding to 11^{00} CST, in this case. Actually, a shift in ϕ is apparent in the figure before the actual flight time. Since each dot represents data from a 5-day interval, some values obtained *after* the flight contribute a "pseudo-shift" effect to all dots representing a 5-day analysis interval centred at time points before flight but including post-flight data. The same applies to the second, return flight (2 on the graph). Nonetheless, ϕ adjustment to local time after the East to West flight appears to be faster than that after the West-to-East flight. This difference in the rate of phase-shift is seen in the plot of acrophases and their dispersion indices.

A performance decrement immediately following intercontinental flights, in this case and probably more generally under conditions of conventional flight schedules, comes about in two ways:

First, there is a drop in the overall rhythm-adjusted mesor of performance, after the West-to-East flight in particular; a transient circadian frequency desynchronization among the many rhythms of a given organism shifting with different speeds may contribute to this impairment.

Second, on the continent of arrival there are times of relative disadvantage (and advantage) that come about—quite apart from any change in mesor—because the circadian acrophase in performance shifts but gradually following the speedy transfer of a subject across several time zones. In other words, there is relatively low ability to perform at the clock hours when important tasks more likely need to be done, whereas at times when there may not be any important tasks to be done there is a relatively better ability to perform.

1.5.1 Computer Executed Rhythmometry Can Also Yield

Cosinor plots [11] for a group (Figs 1.4 and 1.8) or a population of healthy subjects which combine and summarize, as reference standards, the characteristics A and ϕ at each statistically significant frequency or rhythm characterizing the separate series on individuals sampled, as well as showing the average M.

In a cosinor display, ϕ is the angle formed by a vector with the phase reference shown as 0^0. The amplitude A, is usually shown by the length of a vector. In Fig. 1.4, precise values for A are given next to each vector but inter-group differences in A are ignored in order to facilitate a comparison of circadian acrophases. Radii drawn tangent to an error ellipse—representing a 95% confidence region for the A-weighted ϕ—delineate the 95% confidence arc for the acrophase. Intersects of the same ellipse with the amplitude vector and with a prolongation of this vector provide the confidence interval for amplitude. Outside the circular cosinor displays, A-unweighted ϕ estimates also are shown as lines, in the centre of the corresponding shaded 95% confidence arcs.

In Fig. 1.4, when local midnight is used as phase reference, the ϕ's of rhythm computed (with A-weighting) by cosinor or without A-weighting for Scottish and Amerindian women are different. The 95% confidence arcs (upper left) do not overlap. The A-unweighted ϕ estimations suggest an intergroup difference for groups of men as well as for women from the two populations—so long as local 00^{00} is used as ϕ reference (upper half of figure).

In the lower half of the figure, mid-sleep is used as phase reference. All acrophases, whether or not computed with A-weighting, are very similar and their 95% confidence arcs overlap. An adjustment in phase reference thus not only eliminates a spurious intergroup-difference but also provides an objective, quantitative index of a rhythm's timing in different populations. A statistically significant difference in distribution of the acrophases from the two populations was not detected by an F-test; this result does not indicate of course that no such difference existed [41].

In Fig. 1.8, elliptical confidence regions for several variables are displayed on the same graph in this computer-directed plot on microfilm of the amplitude and acrophase of certain salivary rhythms. These rhythms show differences in phase and deserve further study as potential ready substitutes for the evaluation of circadian rhythms in the corresponding variables of blood.

A-Sodium; B-Potassium; C-Chloride

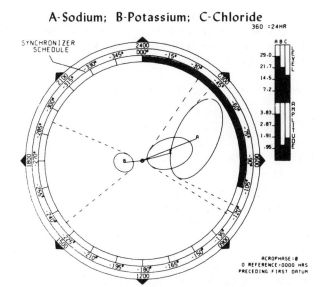

Figure 1.8. Multiple cosinor quantifies internal timing of electrolyte rhythms in human saliva; data and analyses by C. Dawes. Level = Mesor.

Cosinor plots for certain groups of patients differ with statistical significance from the cosinors of groups of healthy subjects. In any case, cosinors of properly sampled healthy individuals represent reference standards to which individual A and ϕ values can be compared in diagnosis and treatment. Endeavours to restore normality of an altered A or ϕ as well as M are justified in that we attempt to see thereby whether performance and/or health decrements are reduced or eliminated.

Indeed, the computer-capability for resolving rhythms may be applied to test chronopathological hypotheses in the clinic as follows:

(1) whenever rhythm(s) are consistently detected and quantified in health; and
(2) quantifiable dyschronism (rhythm alteration) consistently characterizes the afflicted organism(s); and it can be shown further that
(3) dyschronism is at least a contributor to disease or a critical determinant thereof.

   It becomes reasonable, indeed highly desirable, to search for means to
(4) correct and preferably prevent dyschronism and consequent human chronopathology.

In the long run, an assessment of normality or abnormality may include the rhythmometric scrutiny of laboratory data collected at several stages in the healthy subject's lifetime. Indeed, the reference standards for rhythm-coded laboratory data or for rhythm-endpoints may be more than some broad "clinical limits" for a given population. An individual's "health summary" on microfilm may eventually be appraised in the context of background information on his rhythmometrically assessed earlier physical and mental functions and performance, while allowing for changes expected to occur in these parameters with age for his sex and as a result of other factors to which he specifically might have been exposed.

1.6 THE ELECTRONIC COMPUTER— A "HIGH-POWER MICROSCOPE" [34]

In aiming toward a higher resolution of rhythms for a better understanding of both physiology and pathology, modern electronic computer methods resolve rhythms as body "constituents" in time—constituents comparable to the cells as spatial aspects of organisms. To such valid biomedical computer applications as the literature search or the monitoring of critically ill patients has now been added the spectral analysis of a set of biologic rhythms with several frequencies in one and the same physiologic variable—with objective estimates of amplitude, acrophase, mesor, etc., for each frequency component present in the data and—what is equally important— providing appropriate displays comparable in terms of readability to the stained tissue section. These displays of rhythms are more objectively interpreted than is the histologic section, in that confidence intervals also are routinely provided for each parameter estimated—whereas much microscopic interpretation is currently restricted to somewhat subjective statements.

The number of tasks already known to be required for a spectral analysis of rhythms, quite apart from the developmental work yet to be done, also needs to be clarified. Just as histology requires more than a microscope for the resolution and interpretation of findings on cells as the constituents of our spatial structure, so the resolution and interpretation of findings on rhythms, even at this early stage, already requires displays from several different kinds of electronic analyses especially designed for the study of biological rhythms-displays that took years to develop. At best, perhaps, the extent of this developmental work can be compared to that necessary for the very early steps that had to be taken not only to obtain from a tissue block, as starting material, the appropriate and representative number of mounted and properly stained sections, but also—what is at least equally important—to acquire experience in interpreting the sections, e.g., in distinguishing artifact from structure.

Thus, the time-consuming developments corresponding to the design of procedures and instruments for the cutting, fixing, embedding, sectioning, mounting and staining of tissues have their rough counterparts in the numbers and kinds of computer programs that are applied in order to obtain biologically informative displays; and the training of a "fledgling" pathologist already involves at least the equivalent of examining the pertinent sections from a few hundred autopsies. Only by virtue of such developments—and only after the time required for gaining experience in the discipline under discussion—can one derive and utilize the necessary resolving power of either kind, i.e., for resolving biologic rhythms in time series or for the now classical resolution of cells in tissues.

In order to better define health and disease, we are now able to evaluate temporal as well as spatial structure at several levels of resolution. Whether one deals with the microscopic result on histologic structure or with the computer resolution of the spectral (rhythm) structure, both the "macroscopic" and the "microscopic" (3) findings can be integrated to obtain more complete information. The important point remains that this can be done without necessarily equating the findings themselves when they are made at two different levels of resolution. The naked eye serves one important purpose and the microscopic or computer method serves another complementary aim—the "microscopic" result often being more immediately pertinent to certain problems in classical morphologic pathology and, possibly, in a chronopathology as well.

1.7 SUMMARY AND SUGGESTIONS

For a quantitative approach to medical and broader socio-ecologic goals, the chronobiologist gathers numerical objective reference standards for rhythmic biophysical, biochemical and behavioural variables. These biological reference standards can be derived by specialized computer analyses of largely self-measured (until eventually automatically recorded) time series (autorhythmometry). Objective numerical values for individual and population parameters of reproductive cycles can be obtained concomitantly with characteristics of about-yearly (circannual), about-daily (circadian) and other rhythms.

Mesors, amplitudes and acrophases of rhythms may also represent multipurpose sentinel endpoints for

(a) detecting biologic effects from air (and, perhaps, water) pollution [35, 36],
(b) signalling and reducing if not eliminating—by appropriate scheduling—performance decrements associated with ill-timed shift work or transmeridian flights; and more importantly,

(c) monitoring healthy productivity by computer-linked auto-rhythmometry for health assessment [49], as well as

(d) examining by rhythm assessment and treating—with timing according to rhythms—the "unawarely sick", "not-so-sick" [37], "early ill" [38, 39] or "severely-ill" (as a "third approach" [37] complementing "multitest" but single-time-point screening and other conventional public health and hospital measures).

As a minimum, when a single sample is used for say, a biochemical "pattern definition of normal limits" on a set of simultaneously measured variables—a promising new approach [50]—one may also exploit benefit derived from "knowing the sampling time", preferably as the stage of an easily measured rhythm such as that in body temperature. Moreover, a pattern definition can be carried out on the rhythm parameters themselves. Thus, a yet newer and more complete approach to health and disease becomes feasible. Any complexity from such an endeavor will be likely compensated by the potential value of rhythm pattern definition for medicine and ecology more generally.

REFERENCES

1. J. N. Mills, Human circadian rhythms. *Phys. Rev.*, **46**, 128-171 (1966).
2. R. T. W. L. Conroy and J. N. Mills, *Human Circadian Rhythms*, pp. 236. J. and A. Churchill, London (1970).
3. F. Halberg, Chronobiology. *A. Rev. Physiol.*, **31**, 675-725 (1969).
4. A. Reinberg and F. Halberg, Circadian chronopharmacology. *Ann. Rev. of Pharmacology*, **2**, 455-492 (1971).
5. Documenta Geigy. Scientific Tables. Sixth Edition (Konrad Diem, ed.). Published by Geigy Pharmaceuticals, Ardsley, N.Y. (1962).
6. H. C. Friedmann, 17-ketosteroid in Indian males. *Lancet*, **2**, 262-266 (1954).
7. F. Halberg, M. Engeli, C. Hamburger and D. Hillman. Spectral resolution of low-frequency, small-amplitude rhythms in excreted ketosteroid; probable androgen-induced circaseptan desynchronization. *Acta endocr. Copenh. Supplement*, **103**, 54 (1965).
8. H. W. Simpson, M. C. Lobban and F. Halberg, Near 24-hour rhythms in subjects living on a 21-hour routine in the arctic summer at $78°N$—revealed by circadian amplitude ratios. *Arctic Anthropology*, **7**, 144-164 (1970).
9. F. Halberg, Frequency spectra and cosinor for evaluating circadian rhythms in rodent data and in man during Gemini and Vostok flights. *COSPAR Life Sciences and Space Research*, **8**, 188-214 (1970).
10. E. Haus and F. Halberg, Circannual rhythm in level and timing of serum corticosterone in standardized inbred mature C-mice. *Env. Research*, **3**, 81-106 (1970).
11. F. Halberg, Y. L. Tong and E. A. Johnson, Circadian system phase—an aspect of temporal morphology; procedures and illustrative examples. Proc. International Congress of Anatomists. In: *The Cellular Aspects of Biorhythms, Symposium on Biorhythms*, pp. 20-48. Springer-Verlag (1967).
12. F. Halberg and A. Reinberg, Rhythmes circadiens et rhythmes de basses frequencies en physiologie humaine. *J. Physiol., Paris*, **59**, 117-200 (1967).
13. E. Haus, D. Lakatua and F. Halberg, The internal timing of several circadian rhythms in the blinded mouse. *Expl. Med. Surg.*, **25**, 7-45 (1967).

14. F. Halberg, Physiologic considerations underlying rhythmometry, with special reference to emotional illness. In: J. de Ajuriaguerra (ed.), *Symposium on Biological Cycles and Psychiatry, Symposium Bel-Air III*, pp. 73-126. Masson et Cie, Geneve (1968).
15. F. Halberg, W. Nelson, R. Doe, F. C. Bartter and A. Reinberg, Chronobiologie. *J. Eur. Toxicol.*, 6, 311-318 (1969).
16. K. E. Klein, H. Brüner, H. Holtmann, H. Rehme, J. Stolze, W. D. Steinhoff and H. M. Wegmann, Circadian rhythm of pilots' efficiency and effects of multiple time zone travel. *Aerospace Med.*, 41, 125-131 (1970).
17. E. Haus and F. Halberg, Circadian acrophases of human eosinophil rhythm in patients with progressive or remitting rheumatoid arthritis, as compared to patients with osteoarthritis and healthy subjects. *Rass. Neur. Veg.*, 21, 227-234 (1967).
18. M. Garcia-Sainz and F. Halberg, Mitotic rhythms in human cancer, re-evaluated by electronic computer programs—evidence for temporal pathology. *J. Natl. Cancer Inst.*, 37, 279-292 (1966).
19. B. Tarquini, M. della Corte and R. Orzalesi, Circadian studies on plasma cortisol in subjects with peptic ulcer. *J. Endocrinology*, 38, 475-476 (1967).
20. N. Montalbetti, P. A. Bonini, S. Marforio, F. Halberg and J. Reinhardt, Transverse circadian rhythmometry of 17-hydroxycorticosteroid excretion in patients with multiple sclerosis. *Rass. Neur. Veg.*, 22, 101-118 (1968).
21. F. Halberg, J. Reinhardt, F. Bartter, C. Delea, R. Gordon, A. Reinberg, J. Ghata, H. Hofmann, M. Halhuber, R. Günther, E. Knapp, J. C. Pena and M. Garcia-Sainz, Agreement in endpoints from circadian rhythmometry on healthy human beings living on different continents. *Experientia*, 25, 107-112 (1969).
22. R. Günther, E. Knapp and F. Halberg, Referenznormen der Rhythmometrie: Circadiane Acrophasen von zwanzig Körperfunktionen. *Z. ang. Bäder-und Klimaheilkunde*, 16, 123-153 (1969).
23. D. Stamm, Ringversuche in der Klinischen Chemie. *Schweiz. med. Wschr.*, 101, 429-437 (1971).
24. F. Halberg, Education, biologic rhythms and the computer, in: *Engineering, Computers and the Future of Man*. Proceedings of Conference on Science and the International Man: The Computer, Chania, Crete, June 29-July 3, 1970. International Science Foundation, Paris. In press.
25. H. Levine and F. Halberg, Circadian rhythms of the circulatory system. Literature Review: Computerized case study of transmission flight and medication effects on a mildly hypertensive subject. SAM-TR-72-3, pp. 64 (1972).
26. R. Engel, F. Halberg, W. L. P. Dassanayake and J. DeSilva, Adrenal effects on time relations between rhythms of microfilariae and eosinophils in the blood. *Am. J. tropical Med. Hygiene*, 11, 653-663 (1962).
27. P. Manson, Additional notes on filaria sanguinis hominis and filaria disease. *Med. Rep. Shanghai. Spec. Series*, 18th issue, 31-51 (1880).
28. R. Engel, F. Halberg, W. L. P. Dassanayake and J. DeSilva, 24-hour rhythms in blood eosinophils and *Wucheria bancrofti* microfilariae before and after $\triangle_1$9αfluorocortisol. *Nature*, 181, 1135-1136 (1958).
29. S. MacKenzie, A case of filarial haematochyluria. *Trans. path. Soc. London*, 33, 394-410 (1882).
30. R. P. Doe, J. A. Vennes and E. G. Flink, Diurnal variation of 17-hydroxycorticosteroids, Na, K, Mg and creatinine in normal subjects and in cases of treated adrenal insufficiency and Cushing's syndrome. *J. clin. Endocr. Metab.*, 20, 253-265 (1960).
31. M. W. Knapp, P. M. Keane, J. G. Wright, Circadian rhythm of plasma 11-hydroxycorticosteroids in depressive illness, congestive heart failure and Cushing's syndrome. *Br. med. J.*, 2, 27-30 (1967).

32. A. Lunedei and M. Cagnoni, Le sindromi diencefaliche. Problemi in discussione. 68th Congresso della Societa Italiana di Medicina Interna, pp. 413. Edizioni Luigi Pozzi, Roma (1967).

33. F. Halberg, Circadian temporal organization and experimental pathology. VII Conferenza Internazionale della Societa per lo Studio dei Ritmi Biologici, Siena (September, 1960), pp. 20.

34. F. Halberg, Resolving power of electronic computers in chronopathology—an analogy to microscopy. *Scientia*, 101, 412-419 (1966). (Fr. translation 172-179 in Supplement.)

35. A. Reinberg, P. Gervais, J. Frambourg, C. Abulker, F. Halberg, D. Vignaud and J. Dupont, Rythmes circadiens de fonctions respiratoires et de la temperature chez des asthmatiques sejournant en milieu hypoallergenique. *Press méd.*, 78, 1817-1821 (1970).

36. F. Halberg, Francine Halberg and A. Giese, Estimation of objective parameters for circannual rhythms in marine invertebrates. *Rass. Neur. Veg.*, 23, 173-186 (1969).

37. V. Ramalingaswami, Unfulfilled expectations and the third approach. *Brit. J. med. Education*, 2, 246-248 (1968).

38. S. R. Garfield, The delivery of medical care. *Scientific American*, 222, 15-23 (1970).

39. M. F. Collen and L. F. Davis, The multitest laboratory in health care. *Occup. Med.*, 11, 355-360 (1969).

40. F. Halberg, W. Nelson, W. J. Runge, O. H. Schmitt, G. C. Pitts, John Tremor and O. E. Reynolds, Plans for orbital study of rat biorhythms. Results of interest beyond the Biosatellite program. *Space Life Sciences*, 2, 437-471

41. F. Halberg and H. Simpson, Circadian acrophases of human 17-hydroxycortico-steroid excretion referred to midsleep rather than midnight. *Human Biology*, 39, 405-413 (1967).

42. F. Halberg, Physiologic 24-hour periodicity; general and procedural consider-ations with reference to the adrenal cycle. *Z. Vitam. Horm. Fermentforsch.*, 10, 225-296 (1959).

43. F. Halberg, Some aspects of biologic data analysis; longitudinal and transverse profiles of rhythms, in: *Circadian Clocks*, pp. 13-22 (J. Aschoff, ed.). Proceedings of the Feldafing Summer School, Sept. 1964. North-Holland, Amsterdam (1965).

44. F. Halberg and H. Panofsky, I.—Thermo-variance spectra; method and clinical illustrations. *Exp. Med. Surg.*, 19, 284-309 (1961).

45. H. Panofsky and F. Halberg, II.—Thermo-variance spectra; simplified computational example and other methodology. *Exp. Med. Surg.*, 19, 323-338 (1961).

46. D. M. Mercer, Analytical methods for the study of periodic phenomena obscured by random fluctuations. *Cold Spring Harbor Symp. Quant. Biol.*, 25, 73-86 (1960).

47. M. Smolensky, F. Halberg and F. Sargent II, Chronobiology of the life sequence. In: Advances in Climatic Physiology. S. Itoh, K. Ogata and H. Yoshimura (eds.), pp. 281-318. Igaku Shoin Ltd., Tokyo (1972).

48. H. W. Simpson, L. Gjessings, F. Halberg and A. Fleck, A phase analysis of the somatic and mental variables in Gjessings Case 2484 of periodic catatonia. In: *Chronobiology*. Proceedings of the International Society for the Study of Biological Rhythms, Little Rock, Ark., in the Press. L. E. Schering, F. Halberg and J. E. Pauly (eds.) Igaku Shoin, Tokyo (1973).

49. F. Halberg, E. A. Johnson, W. Nelson, W. Runge and R. Sothern, Autorhyth-mometry—Procedures for physiologic self-measurements and their analysis. *Physiology Teacher*, 1, 1-11 (1972).

50. R. R. Grams, E. A. Johnson and E. S. Benson, Laboratory data analysis system. *Am. J. Clin. Path.*, 58, 177-219 (1972).

CHAPTER 2

Transmission Processes Between Clock and Manifestations

J. N. Mills

Department of Physiology,
University of Manchester, Manchester, U.K.

2.1 INTRODUCTION

Of the innumerable variables which have been shown to wax and wane over the course of the 24 h, a more modest but still considerable number have been shown to be endogenous by one or more of the criteria enumerated elsewhere [1]. The mechanism of the circadian timekeeper has been investigated mainly in unicellular organisms such as *Gonyaulax,* as described in Chapter 7. In more complex organisms, whether animal or plant, the rhythms are usually coordinated and consequently synchronous; all the leaves or petals of a plant move at the same time, whether their movements represent a response to light and darkness, or whether they are free-running in constant illumination. In animals such as man and the rat, in which a large number of different rhythmic manifestations can be observed, these are commonly in phase with one another, or at least show some regular phase relationship. The concept has thus arisen of a "clock", usually supposed to reside in the brain, which controls the other rhythms by one or another form of mediation. Individual cells, tissues or organs may have their own inherent rhythmicity, which they may continue to manifest after removal from the body [2, 3] but they appear in most circumstances to have their period set, and to be locked in phase, by some master clock.

This chapter is concerned with intermediation between the clock and the external manifestations, which one may liken to the "hands" of the clock. Discussion will be largely confined to mammals since our understanding, though scanty, is less woefully inadequate here than in other groups of organisms; likewise, only those rhythms will be considered which we may suppose to be endogenous, and for whose transmission from the clock we have some evidence: in this way the potential bulk of the chapter will be grossly reduced.

2.1.1 Pitfalls in Interpretation

Some misconceptions must first be cleared away. As has been discussed by Aschoff [4] a number of influences, or Zeitgeber, set the period of the rhythms to a fairly precise period of 24 h; among the most important of these are light, temperature, and social activity. The following argument suggests that these exert their effect at, or very near to, the clock. We know from free-running experiments [5-16] that the clock does not keep time very precisely, running on a period of around 25 h. Supposing that under normal nychthemeral conditions the clock was not entrained to a precise 24-h rhythm, then the various other influences, following an exact 24-h period, would become progressively out of phase with the clock, so that any process which was affected by both would wax and wane with a period of 25 days. This is not a generally recorded phenomenon, so it is presumably the clock, rather than the final manifestation, which has been entrained to a precise 24-h cycle. However, many nychthemeral influences can also operate lower down in the causal sequence upon the transmission mechanism or upon the external manifestations. Thus, the urine flow is greater by day than by night both because we drink more by day, and because of an inherent circadian rhythm; if the clock were disconnected from the kidney and we continued our usual nychthemeral habits, we should probably still continue to produce more urine by day than by night. Figure 2.1 shows the great variety of possible factors exerting a rhythmic influence upon urine flow.

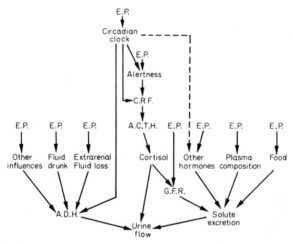

Figure 2.1. Diagrammatic representation of some factors possibly conducive to circadian periodicity in urine flow. E.P. indicates external periodic influences, including environment and habit [1]. (Courtesy Physiological Reviews.)

It is commonly and correctly argued that if the adrenals formed an essential part of the transmission between clock and renal excretion, then after adrenalectomy the clock could no longer affect renal excretion. Observations of continued renal rhythmicity after adrenalectomy, or in human cases of adrenal insufficiency [17], do not however disprove the adrenal involvement in the transmission process unless this rhythm in the adrenalectomized animal is demonstrated by the same stringent criteria that are used initially to demonstrate that the renal rhythm is endogenous. Observations of continued rhythmicity in human sufferers from Addison's disease, or in those who have undergone surgical adrenalectomy, and who are following in any respect nychthemeral existence, are therefore little evidence against the role of the adrenals.

A simple diagrammatic explanation is shown in Fig. 2.2, where the arrows indicate influences, waxing and waning circadianly, upon a series of physiological processes indicated by different letters. In a system as simple as this, if it were possible to maintain process D at a constant level throughout the 24 h, then processes H, N and O would lose their circadian rhythmicity, while the other processes would remain rhythmic.

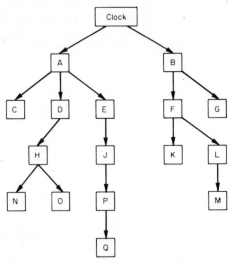

Figure 2.2. Hypothetical branching chain of causation between an endogenous clock, and a series of rhythmic manifestations designated A to Q.

In reality the causal nexus is likely to be much more comples, both through cross connexions, and through external influences which may merely contribute "noise" to the system, or may themselves be circadian. Figure 2.3 represents such a system, derived from a series of plausible links, in which for simplicity only three exogenous circadian influences have been included, and known negative feedback loops such as that from the adrenals to the hypothalamus have been omitted. Clearly, the persistence of a rhythm in alertness in the absence of rhythmic stimulation from the adrenals is no evidence against the role of the adrenals in the transmission process, unless all other rhythmic influences such as social contact have been rigorously excluded.

2.1.2 Criteria for Intermediary Links

A series of criteria has been laid down to demonstrate that, of two processes oscillating circadianly, A is the sole and sufficient cause of

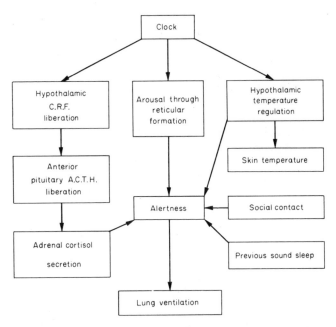

Figure 2.3. Part of a web of known or plausible causal connexions between an endogenous clock and physiological variables. For simplicity, only two exogenous influences are included, and known negative feedback pathways are omitted.

B [1] : 1. A must be capable of inducing B at any time in the 24 h; 2. B must cease to be rhythmic if A is maintained at a constant level; 3. a phase shift of A must be followed immediately, without any transients, by an equal phase shift of B; 4. the variations of A which occur spontaneously over the course of the nychthemeron must be of appropriate size to induce the circadian variations in B. These criteria are only strictly applicable to a simple causal system such as that of Fig. 2.2, though when multiple causation is suspected, as in Fig. 2.3, they can be used to assess the relative importance of different contributory causes. One cannot always, however, apply these criteria rigidly, since with many circadian variables the physiological background to any deliberately introduced influence must vary. An example may be given from a rhythm of much shorter period, the heart beat; no one doubts that the sole and sufficient cause of this is the rhythmic initiation of impulses in the sino-atrial node; yet there is a time, the refractory period, during which an impulse arising here will be without effect. In like manner, an organ responding to the rhythmic secretion of cortisol might be unable to respond at night, without a sufficient recovery period after its previous response. Such failure to respond at the wrong time is not irrefutable evidence against the morning secretion of cortisol being the mediator between the clock and a customary morning response.

2.1.3 A Single Clock?

It has been thus far presumed that there is a single clock, but in view of the claim that isolated animal as well as plant [18, 19] tissues can exhibit circadian rhythmicity this presumption needs justification. When exogenous and endogenous influences have a slightly different period, as in subjects living on a "day" of slightly over or under 24 h, both periods are usually seen in the final manifestations: one may for example find two urinary components of which one follows a 24-h cycle and the other a cycle whose period is equal to the artificial "day", or the same function may follow two periods simultaneously [20] leading to "beats". We appear here to have an internal clock and an exogenous rhythm out of synchrony and competing, rather than two separate clocks. The soundest ground for postulating two clocks would be the demonstration of two distinct functions oscillating with rhythms of somewhat different periods, neither of which corresponded to the period of any exogenous influence. Such a phenomenon would best be sought among people or animals screened from exogenous circadian influences.

A few instances of such apparent dissociation have been reported. Thus, an occasional individual, isolated from all time clues, has shown a free-running rhythm of near 24 h in temperature and an activity rhythm of around 36 h. The claim that this implies the existence of two independent clocks is advanced with caution by Aschoff [15] who writes: "Both rhythms possibly have to be looked upon as separate oscillators". The weakness of the claim lies in the assumption that habits of waking and sleeping necessarily indicate the operation of an internal clock. There is little doubt that in most circumstances an endogenous rhythm of wakefulness and sleepiness influences the actual habits of waking and sleeping, and in lower animals may entirely account for the activity rhythms. In people following ordinary nychthemeral habits such a rhythm is in phase with exogenous influences: the alternation of light and darkness, and the habits of the societies in which most men live. Even in such everyday circumstances, however, the voluntary act of going to bed is subject to other influences, interest or boredom, and the demands of work and recreation, to name but a few. When subjects are living in isolation, the strongest evidence that their sleeping habits are governed by an internal clock is the emergence of a circadian rhythm from superficially irregular habits [12] or the tendency for the period elapsing from one bedtime to the next to be either around 24 h, or about double this, with intermediate times rarely represented [13]. It is clear that the time of retiring or rising is not solely determined by an internal clock; and it would therefore be rash to conclude that when an exceptional man lives for a time on a "day" of 36 or 48 h he must have an unusual clock. It is conceivable that his unusual habits result from some one or more of the numerous other influences that postpone the time one goes to bed, such as lack of fatigue, or interest in an absorbing book, or other continuously interesting occupation.

Observations on isolated organs and tissues, and upon lowly organisms, suggest the reasonable possibility that timing devices are a common characteristic of living cells, even though in a complex animal such as man they are largely or wholly subordinate to a central clock. Dissociation between different rhythmic manifestations is most commonly seen in subjects living on abnormal time schedules, which are discussed below, since they offer evidence about transmission processes.

2.1.4 Exogenous Influences on the Clock

It has been pointed out that many exogenous influences, such as the taking of food, fluid and exercise, can affect rhythmic variables such as urine flow or body temperature, without directly interfering with the clock or its transmission processes. Although some external Zeitgeber clearly constrain the clock to keep precise 24-h time, it is difficult to demonstrate that an exogenous influence, deliberately introduced for experimental purposes, can influence the clock directly.

The critical evidence for the existence of a clock is that it continues in the absence of external Zeitgeber: thus, if the clock has been reset it will continue in the absence of the influence which reset it. When subjects in the Arctic have lived on days of abnormal length, and urinary rhythms have followed the artificial day of, for example, 21 h, these rhythms have reverted immediately to normal 24-h timing as soon as the subjects returned to a day of this length [21]. The artificial day was thus influencing the kidney at some level lower than the clock, which had apparently remained unaffected by many months of abnormal habits. The only instance known to the writer where a clock appeared to be entrained to a day of abnormal length is the record of Kleitman [22] that, after a month in the Mammoth Cave of Kentucky, living on a day of 28 h, the rhythm of sleepiness and alertness stuck to a 28-h period after return to a normal existence.

By contrast, an external phase shift seems to affect the clock and not merely the manifestations. People who go abroad and stay for long enough to be fully adapted to a phase shift of, perhaps, 6 h, and who subsequently return weeks or years later, must again go through a period of readaptation. Documented observations come from Sharp's experiments in the Arctic, in which subjects adapted to a phase reversal (a 12-h phase shift) and on returning to normal time took as long to re-adjust [23].

2.1.5 Origin of the Clock

It is probable also that the development of rhythmicity in the first year or two of life [24] involves an influence of the external Zeitgeber upon the clock itself. The distinction has indeed been drawn between the time at which a function in the human infant oscillates with a 24-h period, and the somewhat later stage in development at which that rhythm becomes endogenous, persisting during the temporary suspension of environmental rhythmicity.

The absence of rhythmicity in Eskimos, by contrast with its persistence in those who, though brought up in temperate latitudes, have lived for some years in the Arctic, further suggests that in man the clock develops under the influence of external Zeitgeber (Chapter 3). Even in the most northerly regions which man inhabits, there is a season around the equinoxes when day and night succeed in the customary 24-h pattern, although in winter there is continual night and in summer continual day. It seems that more than a few weeks' exposure to a 24-h alternation of day and night is needed to develop the clock.

This is in sharp contrast to the situation in mice, which have been reared in conditions of constant illumination for as many as six generations and which still show circadian rhythmicity in their activity [25]. It would seem that in this species the clock is genetically determined, needing an external Zeitgeber only to adjust it precisely to a 24-h period.

Two questions of great interest await investigation. First, what is the situation in a wider range of species? and secondly, would a baby brought up on a non-24-h time schedule develop a clock of appropriate period, or would he show the same resistance as do adults to the imposition of a long or a short day? It could well be that 24-h timing is genetically built into the human clock, as in rats and mice, although some external rhythm is needed to develop the genetic potential. The existence of critical stages of development for exogenous influences on other central nervous function, such as testosterone upon hypothalamic rhythmicity, suggests another possibility: that exposure to a non-24-h rhythm at a critical developmental stage might induce a clock with a rhythm of the same period. This would put the possessor at such a disadvantage in human society that the experiment would be hard to justify ethically; nevertheless, a man whose physiological rhythms have difficulty in following a 24-h period has recently been described [26].

It is a mistake to suppose that the foetus *in utero* is shielded from all nychthemeral influences. The high level of plasma oestriol in women towards the end of pregnancy is believed to be of foetal origin, and it shows a clear circadian rhythm [27]. When samples were collected from 25 women on two successive days, there was between 0830 and 1630 h a mean fall of 48.3 ± 5.8 μg/litre plasma, and between 1630 and 2130 a rise of 28.6 ± 3.6 μg/litre plasma. It is hardly conceivable that endogenous rhythms in 25 foetuses could be so well synchronized, so it is a fair presumption that the foetuses were responding to rhythmic influences from their mothers, such as their activity cycles.

It may be added that even if the postulate of an internal clock were false, and physiological rhythms resulted from some undetected geophysical rhythm [28], the arguments of this chapter would still apply. Whether rhythmicity starts in a clock, or in an exogenous rhythm detected at an unknown site, there must still be some more or less complex train of transmission to the effector processes that are measured and recorded. Transmission processes here discussed are mainly those involving transfer of information from one cell or tissue to another. The rhythmic processes within a cell are also presumably linked by a causal sequence, involving for example induction of an enzyme and consequent depletion of a substrate and accumulation of a product which may in turn influence other reactions. The best studied of such sequences is in the pineal, which is considered below.

2.2 TEMPERATURE; PSYCHOMETRIC PERFORMANCE

Although the rhythm in body temperature has been described as largely environmental [20], there is adequate evidence [1, 30] for an endogenous rhythm. When in an Arctic summer subjects changed their time by 12 h, sleeping by day, the temperature rhythm took a few days to adapt [31]. This was a genuine adaptation of the clock rather than a direct response to exogenous influences, since a similar lag of a few days was needed to re-adapt on reverting to normal time. Even after the time shift of one hour involved in daylight saving, it is claimed that several days are needed for the adaptation of the temperature rhythm [22].

This most familiar of bodily rhythms [29] should be one of the easiest to interpret, since deep temperature is determined by the balance between two variables only, heat production and heat loss. It is also relatively easy to measure accurately and frequently, and can therefore usefully be studied as a mediator of other rhythms.

Owing to the high heat capacity of the body there may be a considerable lag between a change in thermal balance and the consequent change in deep temperature; an alteration of 10% in heat production or loss at rest would have to persist for some 5 h before the body temperature had changed by $1^\circ C$. Moreover, deep temperature can be conserved in the face of heat loss by decreasing the volume of the core whose temperature is maintained.

2.2.1 Metabolic Rate

When we consider rhythmic variations in metabolic rate, we must bear in mind that they can both cause a change in body temperature, and result from it by the familiar effect of temperature on metabolic processes. In any unrestrained condition, metabolic rate varies widely through alterations in the level of muscular activity, and to a lesser extent from the ingestion of food, and since all species of animal investigated show a circadian rhythm in activity, it is not surprising that their rhythm in metabolic rate has too large an amplitude to result from their body temperature rhythm [33]. Conversely, quite large changes in the level of muscular activity have a relatively small influence on body temperature, owing to the operation of thermo-regulatory mechanisms.

In man it is reasonably practicable to maintain constant inactivity throughout the 24 h and thus to study variations in metabolism due to causes other than exercise. Voigt and Engel [34] in observations on six men and nine women observed what they refer to as the "familiar (bekannt) minimum of oxygen consumption between 0000 and 0400 h" in a brief note in which they also record a similar rhythm in a subject working at a fixed speed on an ergometer. Further carefully controlled observations are recorded by Bornstein and Völker [35] and reproduced in Fig. 2.4. They made six or seven measurements of oxygen consumption during the 24 h on a subject fasting for two days and at strict bed rest. To avoid suspicion that the low rate at night was a result of sleep, they made the subject sleep at different times on four different occasions, though they fail to record how well he slept at unfamiliar hours. The low metabolic rate persisted at night even when he was awake, although when he went to sleep at 0500 h the usual morning rise in oxygen consumption was postponed. In many experiments the exogenous influences have been less well controlled. Thus, a slow fall of oxygen consumption in the course of the night in nine young soldiers [36] may well, as the authors suggest, be due to the declining influence of the last meal.

More recent experiments upon four males, almost fasting and remaining recumbent and awake in a room at uniform temperature [37], with oxygen consumption recorded during the first 20 min of each hour, have failed to confirm any circadian variation. The same was true of two women in the first half of the menstrual cycle, but two observed in the second half of the cycle had a metabolic rate which rose during the night; their deep temperature fell much less during the course of the night than did that of the other subjects, and it is suggested that the high

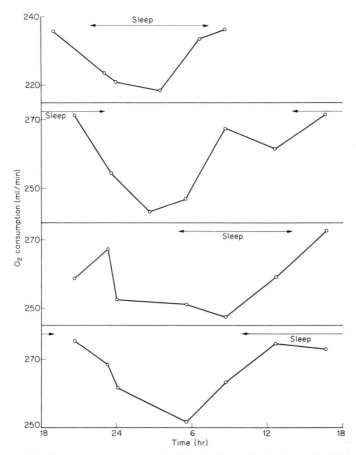

Figure 2.4. Oxygen consumption of subject fasting and recumbent, but sleeping at different hours on different occasions. Note that the low values during the hours after midnight are independent of sleep. After Bornstein and Völker [35].

morning temperature during the luteal phase may result from this high nocturnal metabolism.

Apart from this luteal influence, changes in metabolic rate appear to play little, if any, part in the circadian variations of deep body temperature; much larger variations in metabolic rate occur with the alternation of sleep and activity than any recorded as occurring spontaneously in resting subjects; and the temperature rhythm may remain circadian in subjects whose activity cycle is of 12 h, Fig. 2.5, or 36 h [15]. Even if variation in heat production is not responsible for the rhythm in deep temperature, there is in both rats and birds [38] a

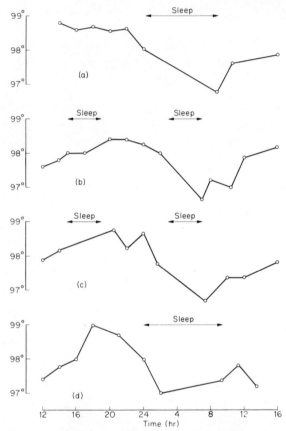

Figure 2.5. Body temperature during four consecutive 24-h periods. (a) and (d), nychthemeral existence; (b) and (c), sleeping for 4 h in every 12.

circadian variation of metabolic rate quite apart from that induced by the rest-activity cycle. This is easily demonstrable in birds which, when fasting and in the dark, remain inactive for 24 h. In rats exposed to LD 12 : 12 the metabolic rhythm, with high values in the dark period, is unimpaired by removal of thyroid, adrenals or pituitary [38a], but it presumably results from an activity rhythm. Similar endocrine ablations on birds which were subsequently kept fasting in the dark might reveal the cause of their rhythms in metabolic rate.

2.2.2 Heat Loss

The temperature rhythm is much more closely paralleled by variations in heat loss. Most workers who have recorded skin temperature over the

24 h have observed, in the extremities, the appropriate changes [37, 39, 40, 41], a rise of skin temperature, which will increase heat loss, in the late evening before the fall of body temperature, and a fall in the morning, conserving heat and conducing to the morning rise in deep temperature. This rhythm is not observed on the forehead, neck, trunk, or proximal part of the limbs [37, 39, 40], which seem to be less involved in the control of heat loss. An increase in heat loss from the hand to a calorimeter has also been demonstrated in the evening [42]. At variance with these reports, and without obvious explanation, is an account of random and inconsistent behaviour of finger and toe temperature in the course of the day [43].

The changes in heat flow through the skin [38], consequent upon these variations in skin temperature, are mainly attributable to changes in cutaneous blood flow. Observations upon seven male subjects during the waking hours, from 0600 to 1200 or 2400 h, have however revealed some discrepancies between heat flow and blood flow. The two will only be consistently associated when blood remains at core temperature, and the possibility thus arises that there are circadian variations in the operation of the counter-current heat exchange between arteries and veins [44], whereby arterial blood is cooled before reaching the extremities.

Sweating is another mechanism for the dissipation of heat, and has been found to occur [41] around the same time as the rise in temperature of the feet. It has been further claimed in a preliminary communication that the threshold temperature at which sweating begins follows a similar circadian rhythm to that of deep temperature [45].

2.2.3 Cause of Temperature Rhythm

In summary, it would seem that the circadian rhythm in body temperature cannot be ascribed to rhythmicity in any one of the factors responsible for its regulation, but rather reflects a rhythm in the "set" of the controlling mechanism in the hypothalamus.

Since the hypothalamus has also been suggested as a site for the clock, it would seem natural that the heat-regulating centres here have direct nervous connexions with the clock. The pyrogenic action of etiocholanolone and other steroids, some of which are precursors or metabolites of adrenal steroids [46], suggests a more complex mechanism: that perhaps the clock operates upon the adrenal cortex, one or more of whose steroids is conveyed back to the hypothalamus to elevate the body temperature. The action of a single injection of an appropriate steroid develops and subsides in around 24 h, so the liberation would have to be severely restricted in time, but this is not implausible, since it is probably true of the liberation of cortisol in the early morning (see p. 54).

The similar pyrogenic action of progesterone metabolites [47] may likewise be responsible, rather than progesterone itself, for the high temperature in the second half of the menstrual cycle.

A further complication is that other endocrine secretions, such as the thyroid, determine the proportion of steroids metabolized by the etiocholanolone pathway [48]. Whether such pyrogenic steroids are produced with a circadian rhythm, and whether they contribute to the temperature rhythm, are subjects needing further investigation.

2.2.4 Consequences of the Temperature Rhythm

The next question which arises is how far the changes in body temperature may be responsible for other circadian rhythms.

The alleged rhythm in metabolic rate has been discussed above. In the quiescent, wakeful, fasting human subject the variations in oxygen consumption from the lowest to the highest (Fig. 2.4) represent an increase of 8 to 12½%, which could result directly from a rise in temperature of $1°C$, a fairly usual range for the temperature rhythm.

2.2.5 Psychometric Performance

Performance in a number of psychometric tests has been alleged to vary circadianly, and to reflect the temperature rhythm. Thus, Kleitman [22], p. 152 presents curves of a number of test scores which over the working day follow a course very similar to that of body temperature. Apart from the obvious objection that parallelism does not imply casual connexion, we have here the further difficulty that we do not know what is being measured. When we collect and analyse a sample of urine we can interpret out test result, subject to ascertainable errors, in terms of a specifiable renal function such as the rate of sodium excretion. The interpretation of a psychometric test score is more dubious, and only with the simplest tests, such as reaction time, can we ascribe performance to a limited number of specifiable functions: receptor delay, conduction time along nerve fibres and across synapses, and an unknown process whereby incoming sensory information is transferred to an outgoing motor tract. Differential reaction time involves a longer central delay over a process of which we know even less. All psychometric tests involve three parts, and may be planned so that the delay, and hence any variation in delay, is mainly due to one component: the recognition of the appropriate sensory signal, the transfer to a motor tract, and the effector action. The central process may be regarded as a simple "switching", as in reaction time, or may be signified by a number of terms subjectively meaningful but incapable of physiological definition: problem-solving, decision-making, and so forth.

Many assessments are intrinsically non-numerical but can be converted into an arbitrary numerical score. The subject may rate his alertness, sleepiness, or appetite, on a linear or digital scale, or his answers to a questionary may be scored. We can thus talk about the influence of measurable physiological factors upon psychometric test scores, but not upon psychological functions, except insofar as we can give a name to a function derived from the well-correlated results of a series of tests. Only when many test scores show similar circadian rhythms does it become profitable to postulate a common underlying process, and to search for a cause among physiological rhythms. A tentative approach on these lines is to suggest [49] that various test scores may reflect the state of arousal, a term which has physiological as well as psychological meaning, and which is said—purely conjecturally—to be related to body temperature [50].

Another difficulty is that it is much harder to exclude interfering factors in psychometric than in physiological measurements. Fatigue, boredom, motivation, practice, post-prandial changes, may all have large effects on test scores, making it difficult to uncover circadian rhythmic changes and even more difficult to assess their intermediate causes.

2.2.6 Influence of Temperature upon Performance

Many workers since Kleitman have commented upon the parallelism between performance and temperature in subjects living nychthemerally. This has been observed, for example, with an adding test [51], as well as with simple reaction time [52], and with speed of manipulation [53].

Evidence suggestive of a closer association between temperature and performance comes from a series of papers [54-57] in which industrial workers on a variety of different shift systems were tested for vigilance and quick arithmetic. Occasions when the shift was changed provided an opportunity to study the speed of adjustment of the different rhythms, and in general those men whose temperature rhythms adjusted soon to a new shift showed performance rhythms which likewise adapted rapidly. The correlation between temperature and performance was not always as good as one might wish, since in individual subjects temperature adapted sometimes before, sometimes after, performance [57]. This may however reflect only the uncontrolled variability of psychometric tests and their vulnerability to such factors as fatigue and inadequate motivation.

The simplest hypothesis of causation would be that the brain works better when its temperature is higher, at least over the physiological range. There is very little experimental work to test this hypothesis directly, by examining performance after a deliberate elevation of deep body temperature, apart from a single paper [58]. The higher temperatures here used were outside the physiological range, and of the two main tests vigilance declined over the 12 days of the experiment whilst speed of addition improved, reflecting presumably the two common intruding influences, boredom and practice. There is however some evidence in this study that a rise of temperature from "normal" (unspecified) to $37.3°$C was followed by an improvement in signal detection and speed of arithmetic (Fig. 2.6). If temperature does affect performance, it could do so through action on the arousal mechanism in the reticular formation, or by action upon some other region in the nervous system involved in the performance measured.

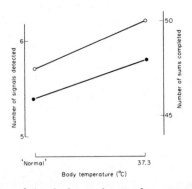

Figure 2.6. ○, number of signals detected, out of ten presented, and ●, number of multiplication sums completed in a fixed time, at "normal" temperature and at $37°$C. Mean of results on twelve subjects. Data of Wilkinson *et al* [58].

A much simpler function, which is said to vary circadianly [59, 8], is the ability to estimate short intervals of time, from around ½ to 2 mins. Some attempt this with the aid of a motor or mental activity, such as tapping or counting seconds, while others attempt to assess "empty" time, but both methods of estimation are presumed to depend upon some nervous activity which might speed up when brain temperature is higher, causing the subject to produce a shorter interval than requested, or to overestimate a presented interval. Many workers have recorded diurnal rhythms in time estimation which correspond to the observed [59-61], or expected [63, 64, 108], temperature rhythm, though contrary reports have appeared [50, 55].

A rhythm in time estimation has also been observed in two subjects isolated underground, in whom this and other measured rhythms followed a free-running period slightly longer than 24 h [8]; though in a third subject under similar conditions the expected inverse relationship between estimation and production of time intervals was not seen [65].

Deliberate Alteration of Temperature

If time estimation is thus dependent upon a thermolabile clock, it should be possible to alter a subject's assessment of time in the appropriate direction by deliberately altering his temperature. Appropriate changes have been claimed when temperature was raised [66, 67] or lowered [68] deliberately or in fever [69]. Others have however found less convincing alterations [70, 71], suggesting that factors other than a simple thermolabile clock are involved, though conflation of results from several groups of workers suggests a clear relationship, with a 10% speeding up of time sense for a 1° F rise in body temperature [68].

2.2.7 Adrenal Influence upon Performance

Since so much discussion of performance rhythms has inevitably followed a consideration of the influence of temperature, it is convenient here to consider other factors which may contribute to the rhythm in performance.

An association between ability and adrenaline excretion in the course of the day has been claimed [72]. Out of a group of 186 students, 11 were selected as extreme "morning" types, for comparison with a group of 11 who claimed to be extreme "evening" types. Their self-assessment was not borne out by their performance on tests of arithmetic and rapid colour-naming at different times of day, but did correspond to their diurnal change in differential reaction time, which improved throughout the day in the "evening" but not in the "morning" types. Adrenaline excretion likewise declined progressively in the "morning" but not in the "evening" types. As others have claimed, the "morning" types were more introverted, the "evening" types more extraverted. The same synchrony between a circadian rhythm of alertness and of adrenaline excretion has been claimed in a group of subjects kept awake for three days and practising rifleshooting at regular intervals [73].

Though their interest is obvious, these observations offer no indication of whether adrenaline improves performance, or whether some common antecedent factor such as alertness is responsible for both. The same applies to cortisol, whose early morning secretion, before the time of waking, has been described as an alerting mechanism [74, 75], and may be responsible for a morning peak in adrenaline secretion through its effect upon the methylation of noradrenaline [76].

In conclusion, while the association between vigilance, alertness and other measurable aspects of performance with adrenal cortical and medullary activity is suggestive of some causal connexion, we have no evidence that the adrenals contribute to performance rhythms. There is however suggestive, if as yet inconclusive, evidence that the temperature rhythms may form a direct link between the "clock" and rhythms in performance.

2.2.8 Other Rhythms Dependent upon Temperature

By contrast with this wide uncertainty, there is one rhythm which seems to be closely controlled by body temperature, that of some parasitic microfilariae, discussed in Chapter 5.

Another rhythm which may depend on temperature is that of pulse rate, which is as easy to measure; a large number of recorded values [77] show a striking parallelism between the two leading to the suggestion that the rhythm in pulse rate results from the temperature rhythm, with an increase of 10 to 20 beats per min for a temperature increase of $1°F$. An increase of this order has been demonstrated when a human subject with pharmacologically denervated heart is deliberately heated [78]. The occasional dissociation between these rhythms [79] does not disprove a causal connexion, as explained below, since when they are dissociated pulse rate is following an obvious exogenous rhythm of rest and activity. The pulse rate rhythm has indeed been held to be wholly environmental in origin [20], and the evidence of an endogenous rhythm is slight: a slow pulse rate in airline pilots at the time when they would normally be asleep [80] is one fragment of evidence. If an endogenous rhythm in pulse rate exists, then it is at least plausible that it results from the temperature rhythm.

The fairly regular association between temperature and activity rhythms when both are free-running in the absence of exogenous influences is open to several interpretations; the rise of temperature could result from activity, or could impart wakefulness, or both could result from the operation of an internal clock. Conversely, the dissociations which are frequently observed in free-running experiments suggest that the temperature rhythm is responsible for the rhythm of neither the adrenals nor the kidney; both have in fact remained rhythmic when temperature had lost its rhythm in a subject in a space capsule simulator [81].

If one considers the direct action of temperature upon chemical reactions, and consequently upon a variety of bodily processes [82], it is remarkable how little evidence there is that the temperature rhythm is responsible for the circadian rhythm in other functions. Perhaps this is a consequence of the rather small amplitude of the temperature swings in homoeotherms.

2.3 ADRENAL CORTEX AND OTHER ENDOCRINES

The adrenal cortex has often been suggested as a mediator of other circadian rhythms [75, 3] but the evidence for its mediation is seldom conclusive and sometimes purely speculative, resting merely upon the similar time course of the adrenal rhythm and that of, for instance, bronchial resistance [83]. The evidence is best for its implication in the eosinophil rhythm, suggestive for mitotic rhythms and some aspects of urinary rhythms, and frankly speculative for other rhythms less specifically assignable to a single organ, such as susceptibility to noxious agents, and indeed death itself.

2.3.1 Eosinophils

The morning fall in eosinophil count seems to have been first noted in 1931 by von Domarus [84]; it is also seen in dogs [85], while in rats and mice, which are active by night, low values are seen in the early night [86, 87]. Pincus's observation of a morning steroid peak [88] suggested a possible connexion between the two rhythms [89, 90] (Fig. 2.7), though a similar inverse relationship with the rhythm in body temperature was also noted [91]. Rather surprisingly, in view of both previous and subsequent findings, the observation that eosinophil count was greatly reduced by ACTH given at 0800 h [92] was accompanied by the statement that there was no significant change in control subjects given no injection. The importance of the adrenals in the eosinophil rhythm was apparent when it was found that the normal rhythm was absent in patients with Addison's disease, or after adrenalectomy [75,

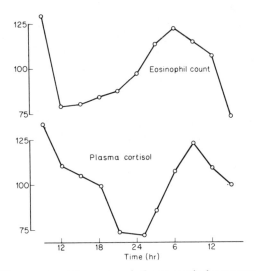

Figure 2.7. Mean eosinophil count and plasma cortisol concentration in samples collected every 3 h from a group of eight healthy males, following ordinary nychthemeral habits. Values are expressed as percentage of 30-h mean. After Gordon, Spinks, Dulmanis, Hudson, Halberg and Bartter [90].

93-96] (Fig. 2.8); and the possibility that this merely reflected a permissive action was ruled out when it was found that the rhythm could be restored in Addisonians by giving a single dose of 25 mg of cortisone by mouth at 0600, but not by dividend doses distributed throughout the day [96]. Though cortisone is not the natural hormone of the adrenals, this is by no means a pharmacological dose; and a similar substantial fall in eosinophil count has been induced by 20 mg of the natural hormone, cortisol [92]. Furthermore, estimation of blood steroid levels after administration of eosinopenic doses of steroids [97] suggests an effect with physiological levels of hormone. It thus appears firmly established that the morning secretion of cortisol plays an essential intermediate role in the eosinophil rhythm.

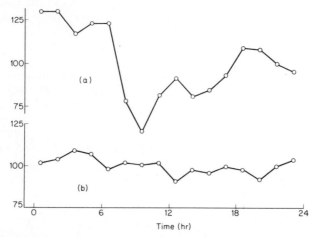

Figure 2.8. Mean eosinophil counts on samples collected every 90 min. A, healthy subjects; B, adrenal insufficiency without replacement therapy, limited activity. Values are expressed as percentage of 24-h mean. After Halberg and Kaiser [95].

Observations upon the adaptation of eosinophil rhythms to time shifts [98], and upon the influence of light upon eosinophil count [99], are probably to be interpreted as an effect upon the adrenals. Whether this is exerted by resetting the clock, or by a direct hypothalamic-pituitary mechanism, is not known.

2.3.2 Enzymes

Evidence is beginning to accumulate that the adrenal cortex is responsible for induction of some enzymes whose concentrations in the liver show circadian variations. The difficulty in such studies, discussed elsewhere [30], is that for the methods at present in use each animal provides only one estimate, since it is killed to obtain liver samples; the group of animals used must therefore first be synchronized by some powerful exogenous Zeitgeber. In a recent account, which includes references to other work [100], the rhythm of phosphoenolpyruvate carboxykinase in mouse liver was studied by killing groups of animals at 4-hourly intervals from 0800 until 2400; minimal enzyme activity was observed at midday and maximal at 2000. Exogenous cortisone given 4 h before death raised the low morning concentrations substantially but had very little effect on the high afternoon values. After adrenalectomy the usual rhythm was replaced by a modest progressive rise in concentration from morning till night, and cortisone now elevated the level equally at all hours. The simplest interpretation is that the rhythm is largely due to the familiar evening secretion of glucocorticoid, but that in the absence of the adrenals an exogenous influence, perhaps due to feeding habits, maintains some rhythmicity. It is thus plausible, though not firmly established, that an endogenous rhythm in liver enzyme concentration is wholly due to the adrenocortical rhythm. Further evidence might be obtained by the use of a larger animal in which repeated liver biopsy might permit the demonstration of a rhythm in a single individual.

For another liver enzyme, tyrosine transaminase, some evidence [100a] suggests that its rhythm in the rat is due to vagal stimulation of hepatic cells at the onset of the evening activity period; but this enzyme can also be induced by other circadianly varying agencies, such as glucocorticoids and dietary protein. It therefore offers a good example of multiple causation, such as is indicated in Fig. 2.3 and complicates attempts to disentangle causal sequences.

2.3.3 Mitosis

A circadian variation in the number of cells in mitosis was observed over 30 years ago in human foreskin removed at circumcision [101, 102], but a quotation will show that the interest aroused by the finding was of a different nature from that usually shown today:

> "It appears that during the day, the emphasis lies on work, digestion ... and that during the night, when the need for these functions is diminished, attention is turned towards repairing run-down tissues and building new ones."

In one of these investigations [102] no samples were collected between 0100 and 0700 h, and the maximum could only be defined as between 2000 and 0100 h; but in the other investigation [101] samples were collected at all hours of the 24 except 1800, and a fitted sine curve indicates a maximal rate between 1900 and 2000 h. Biopsy of the skin

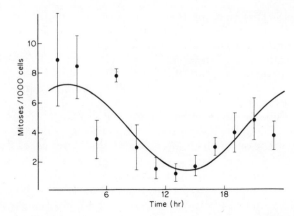

Figure 2.9. Mitotic rate in human skin biopsy samples. Each point is the mean of 2-14 samples collected at that hour, with S.E. of mean. Best-fitting sine curve has been inserted. After Fisher [104].

has shown that similar rhythms are present in the adult [103, 104] (Fig. 2.9). In the bone marrow the myeloid tissue shows a rhythm of much smaller amplitude and the erythroid tissue no discernible rhythm [105].

Observations upon small rodents have enabled us to proceed beyond naive teleological speculation to an attempt to specify the direct influence upon potentially dividing cells. In mice and rats, as might be expected in a nocturnal animal, the rhythm is out of phase with that in man, with a low mitotic rate at night and a high rate in the morning. This has been shown in several tissues, including the skin of the pinna, the oral mucosa, the connective tissue of the periodontal membrane, the thyroid, and the bone marrow. In the adrenals, in contrast, mitotic rate is higher by night [87, 105a, 106]. In the early experiments counts were only made at two times of day, when mitoses were expected to be frequent or scarce. The "internal timing", that is, the temporal relationship with other circadian rhythms, needs a finer definition of phase, which has been attained by 4-hourly sampling on mice [107, 108]. It was found that a high level of corticosterone in the serum and in the adrenal glands occurred about the same time, and preceded a sharp fall in mitotic rate in skin and liver (Fig. 2.10). Taken together, these observations suggested that an increased adrenal function at the beginning of the nocturnal activity period is manifested by an increased rate of adrenal mitosis, and by an increased synthesis and release of corticosterone, which might in some way diminish the rate of mitosis in other tissues.

The earlier work is fully reviewed by Bullough [109], who concludes that the inception of mitosis is normally critically dependent upon

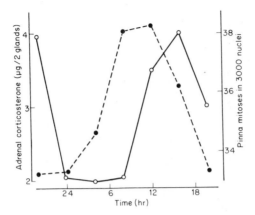

Figure 2.10. Mice, fasting. o———o, adrenal corticosterone, ●– – – –●, mitotic rate in pinna. After Halberg *et al* [107].

energy supply from the tricarboxylic acid cycle. He observed [110] that mitosis may be inhibited by adrenaline and its oxidation products, and by cortisone but not by deoxycorticosterone except in excessive dosage. The inhibition was exerted whether the hormones were injected into an animal, or were added to a culture medium with fragments of pinna. More recent work [111] suggests that the inhibition is not direct, but that, in mouse epidermis, adrenaline enhances the effect of the epidermal chalone, whilst cortisol prolongs the effect, probably by slowing the rate of loss of adrenaline from the cells.

There is thus suggestive evidence that the adrenocortical rhythm may mediate the rhythm in mitosis in the epidermis, and perhaps in other tissues. It is said that mitotic rhythms are lost after adrenalectomy [75, 112], though other workers [113], who killed groups of rats every two hours and examined the mitotic activity of the gastric mucosa, claimed that a circadian rhythm was still present after adrenalectomy.

Further doubt is cast upon this simple hypothesis by another consideration. Mitosis involves replication of nuclear material, which can be assessed by the increased incorporation of tritiated thymidine or isotopic phosphorus into DNA. This precedes mitosis by some 6-12 h; the best estimate, obtained from sampling every 2 h [114], is 10 h, whilst less accurate estimates are derived when sampling has been only every 4 [115] or 6 [105] h. The difference in timing between the rise of plasma cortisol and the fall in mitotic activity is shorter than this; if we envisage mitosis as one event in a long biochemical process, beginning many hours earlier with DNA replication and ending perhaps an hour or two later with the reconstitution of two independent cells, then the rise in plasma corticosteroid concentration comes too late to slow down the process at the beginning of the activity period.

Although the correct temporal relations are again difficult to establish, the anterior pituitary may also be involved in the control of the mitotic rhythm. Growth hormone has been shown to increase in human plasma early in sleep [116], and in the liver of immature mice it increases the frequency of liver mitoses [117], whilst hypophysectomy in larvae of the Urodele *Amblystoma maculatum* leaves the mitotic rate in corneal epithelium low and non-rhythmic [118]. On this interpretation the variable effect of hypophysectomy upon mitotic activity in different tissues [119] would result from the secretion by the anterior pituitary of two hormones, growth hormone and ACTH, which affect mitosis in opposite directions.

Further evidence could be obtained, though tediously, by experiments on abnormal time schedules, since the mitotic rhythms in mice can be altered by changing the routine of light and darkness [120]. The adrenal rhythm can be similarly entrained, after a sufficient lapse of time, whereas the available evidence [116] suggests that the growth hormone rhythm would be immediately entrained on reversal of the sleeping habits. If these two hormones and mitotic activity could be studied together on an altered time schedule, it might be possible to observe which were more closely associated with one another.

Whatever be the Zeitgeber, and however it may mediate its effect, malignant tumours can escape from its influence and assume a mitotic rhythm of their own, with a period less than 24 h [121].

2.3.4 Urinary Potassium and Sodium

Disturbances of the sodium excretory rhythm have been reported in patients with adrenal disease, but such patients, whether they suffer from excess [17, 122, 123] or deficiency [124] of steroids, are unusually susceptible to postural effects on the kidney whereby recumbency increases excretion of chloride and sodium. If the patients are more strictly recumbent at night than by day, this postural effect may swamp the usual nocturnal sodium retention; such patients therefore offer no clear evidence on whether the adrenals are involved in the urinary rhythm. While studies on patients in whom posture and meals have not been carefully controlled may afford information of interest to the clinician, they do little to enlarge our understanding of endogenous rhythms. When patients with adrenal deficiency have been kept strictly recumbent, with meals and any maintenance therapy uniformly distributed over the 24 h, normal excretory rhythms for sodium and chloride have been observed [122, 125-127]. Indeed there is little evidence to implicate the adrenals directly in either of these electrolyte rhythms.

The adrenals can more readily be implicated in the excretion of potassium [128] and phosphate. Cortisol administration causes a fall in plasma phosphate concentration [129-131], due in part at least to increased uptake by muscle [132], with a fall in its excretion [133-135], and a rise in plasma concentration [136] and in excretion of potassium [133, 134, 136, 137] which may be associated with a fall in hydrion excretion [136, 137].

These urinary changes are observed in the morning, with about the right phase relation to the cortisol increase [90], and phosphate excretion rises again as plasma cortisol falls [135]. Evidence against the involvement of the adrenals comes from a report that the potassium, as well as the sodium, excretory rhythm persisted in a patient recumbent, fed every 6 h, and given steroids every 12 h [122, 126]; the major exogenous influences were here excluded, but the patient was living in a nychthemeral society. Evidence for the involvement of the adrenals comes from observations upon three healthy subjects who remained recumbent and fasting on waking in the morning, and in whom the usual morning rise in potassium excretion was prevented either by inhibiting cortisol synthesis with an inhibitor of the 11-β-hydroxylase enzyme, or by blocking ACTH production with regular doses of prednisone [138]. The potassium excretory and the steroid rhythms are said to be the most persistent if a subject alters the phase or the period of his habits, by subjecting himself to an abrupt time shift [23, 139, 140] or by living on a day of abnormal length [20, 141, 142].

Of two subjects who both reversed their sleep and activity habits, and lived in continuous light [143], one showed good adaptation of plasma steroid and of potassium excretory rhythms. The authors state that in the other subject the steroid rhythm was well adapted but the potassium excretory rhythm was bimodal, reflecting competing influences of habit and an endogenous rhythm; inspection of the data however shows a distinct appearance of bimodality in the steroid rhythm of this second subject, and potassium excretion was thus following a course similar to that of his plasma steroids, which could well have determined the renal behaviour. Against this, however, it has been claimed that, in subjects living on a 21-h day, these two rhythms can become dissociated [21, 144]. A similar dissociation has been noted in night workers, many of whom showed excretory rhythms for potassium as well as sodium which were well adapted to night work despite failure of their plasma steroid rhythm to adapt [145] (Fig. 2.11). Furthermore, administration by slow intravenous drip of amounts of cortisol around midnight comparable with the amount normally secreted in the morning, though raising the excretion of potassium, have failed to lift it to the level normally occurring spontaneously after waking [146]. In a single reported instance a steadily rising steroid level failed to disturb the usual pattern

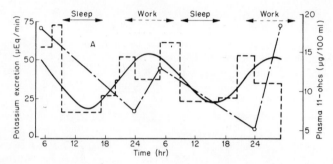

Figure 2.11. Night worker: urinary potassium following a nocturnal, and plasma corticosteroid concentration, o — · — o a diurnal, pattern. Best-fitting sine curves on the urinary data are fitted separately to each 24 h.

of potassium excretion [147]; this may not be of much significance, since the high values of plasma steroid were apparently due to stress in a subject who disliked venepuncture, and may therefore have represented very brief bursts of secretion at the time when blood samples were collected; this is unlikely to have the same effect upon the kidney as a more sustained high plasma level.

2.3.5 Urinary Phosphate

The evidence that the adrenals are responsible for the morning decline in phosphate excretion is weaker, although the time relationships are entirely appropriate, and when sampling continues during the night plasma cortisol rises and phosphate excretion falls. However, it is generally observed [143, 148] that phosphate excretion adapts immediately or very rapidly to any change of habit, falling during the first few hours after waking even the first day after a sudden time shift [149] (Fig. 2.12). By contrast, the steroid rhythm, assessed by its excretory pattern which follows the plasma level with a lag of about 3 h, is often one of the slowest to adapt [23, 139, 140]. It is disappointing how seldom these two rhythms have been studied together.

The suspected dissociation would be better investigated in subjects on a day of abnormal length, since adaptation here is likely to be much slower. We have a few observations upon night workers [145, Figs 1 and 6], in which phosphate excretion fell sometimes when the subject rose, even though his plasma cortisol was not rising, and at other times immediately after a rise in plasma cortisol, at some time other than when he arose from sleep. It would seem then that there may be two influences, one from the adrenals, and one associated in some other way with rising from sleep, both of which coincide in a usual nychthemeral

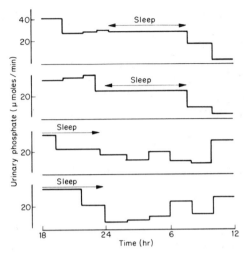

Figure 2.12. Phosphate excretion of subject in isolation unit during four consecutive days. After the second day the clock was advanced 8 h equivalent to an eastward flight, but excretion is plotted on real time. Note that phosphate excretion falls for some hours on waking, and that this behaviour adapts immediately to the time change.

existence. In the experiments with adrenal inhibitors mentioned above [138] the usual morning fall of phosphate excretion persisted. This confirms the presence of an extra-adrenal influence, and further indicates that it is associated with waking, even though the subject remains recumbent and fasting. It is not known whether in these conditions the plasma phosphate also fell.

2.3.6 Origin of Adrenal Rhythms

The origin of the corticosteroid rhythm has been extensively studied and reviewed [150]. The main control over glucocorticoids, cortisol and corticosterone, is by the corticotrophic hormone (ACTH) secreted into the bloodstream by the anterior pituitary, and this in turn is controlled by the releasing factor (CRF) produced in the hypothalamus and descending in the hypothalamo-hypophyseal portal vessels. If the clock resides in the region of the hypothalamus, it has thus a known route of access to the adrenal cortex, and the higher nervous control offers an obvious means whereby environmental factors, social or climatic, could influence its activity.

It is generally assumed that the varying levels of corticosteroids in the blood reflect varying output by the adrenals, rather than a varying metabolic loss superposed upon a constant production, and experimental study of metabolic loss has confirmed this [151]. It is now known that release from the gland does not follow a simple circadian rhythm. Secretion has been described as confined to a period of 6 h or less in the morning [151]; but more detailed definition has been obtained by sampling every 30 min in four subjects [151a] or every 20 min in six subjects, of whom one was examined on two occasions separated by two months [152, 153]. It appears that cortisol is secreted in a series of 7 to 13 episodes in the course of the 24 h, and the exponential fall of concentration after each peak suggests that secretion largely or completely ceases between the episodes. Plasma concentration may remain around zero for some hours before and after bedtime; and a high density of secretory episodes just before and after the time of rising accounts for the high concentrations regularly recorded in the morning. Although secretion is most abundant when REM sleep is most frequent, the individual secretory episodes do not correspond to times of REM sleep.

Plasma concentration of ACTH oscillates circadianly [154, 155] and presumably imparts the rhythm to the adrenals, since the variations in concentration are adequate to account for the rhythm in plasma cortisol: a similar relationship between the concentrations of the two hormones is observed when the adrenals are stimulated by continuous infusion of ACTH [156] (Fig. 2.13). The rhythm in ACTH is not dependent upon

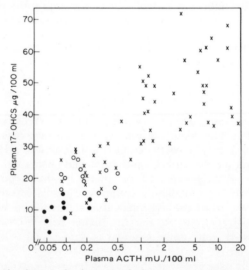

Figure 2.13. Simultaneously determined plasma concentrations of cortisol, and of ACTH (log scale). ○ 0600 h; ● 1800 h. X subjects infused with ACTH for 24 h. After Ney *et al.* [156].

any feedback from the steroids produced under its influence, since the ACTH rhythm persists in Addison's disease [157], and after adrenalectomy in rats [158]. It must be remembered that rats, which are active nocturnally, have all their physiological rhythms reversed in phase when compared with a diurnal species such as man; their maximum corticosteroid production is thus in the early evening. The ACTH content of their pituitaries is also maximal in the afternoon [158], showing that the pituitary rhythm is one of synthetic activity rather than of release of hormone from a preformed store.

The use of small rodents has permitted a much deeper exploration into the mechanisms of endocrine rhythms by *in vitro* studies on the glands. Hamster adrenals cultured *in vitro* for five days are said to show a circadian rhythm in their steroid synthesis [3], suggesting that they have an inherent rhythmicity, apart from that imparted by the pituitary. The published figures are however somewhat erratic, and possible errors in interpreting such experiments have been indicated [159]. A more extensive series of *in vitro* experiments has been reviewed [160] in which adrenals, pituitaries or hypothalami were removed from rats sacrificed at 4-hourly intervals throughout a complete nychthemeron. The hypothalamic content of CRF, the pituitary content of ACTH, and the synthetic ability of the adrenals were all shown to vary rhythmically, with peak values in the afternoon, appropriate to the early evening rise in plasma corticosterone. The only observations seriously at variance with this simple picture, of a circadian hypothalamic rhythm determining a pituitary rhythm which in turn imparts rhythmicity to the adrenal cortex, are upon varying sensitivity of the target organs. Mouse adrenal glands excised at 0400 are maximally sensitive to ACTH, as assessed from the log-dose-response curve [161]. This maximal sensitivity coincides with the time of minimal activity and vice versa. An explanation suggested by recent work [162] is that an adrenal exposed to a certain concentration of ACTH, such as prevails in the early afternoon, is less sensitive to an incremental dose than is an adrenal in the morning without ACTH influence. The varying adrenal sensitivity may not therefore be intrinsic but may be a function of the ACTH rhythm, although the search for an internal oscillating system in the adrenal has not yet been dropped.

The sensitivity of the human pituitary also seems to vary circadianly, since a standard dose of a synthetic ACTH-releasing factor caused a bigger increase of plasma cortisol at 1700 and at 2400 h than at 0800 h [163]. If however the adrenals secrete at a constant rate, varying only the number of minutes during any hour in which they are actively secreting [153], it is inevitable that a greater increase in activity can be induced at a time of day when they spend a larger proportion of each hour in idleness.

Ablation experiments in mice suggest that the hypothalamus is the only region of the brain essential for the adrenal rhythm [160]. Naturally occurring brain damage in man [164, 165] is usually too diffuse to permit precise localization, though it seems that the circadian rhythm in adrenal activity can be lost while other adrenal control mechanisms are preserved, and there is some evidence for specific association between hypothalamic disease and abnormality in human corticosteroid rhythms [166].

2.3.7 The Pineal

Circadian rhythms in the pineal gland have been extensively studied in the rat, but also in the pigeon [234] and in other birds [235]. The unique product of this gland is melatonin, formed by acetylation and subsequent methylation of 5-hydroxytryptamine. The circadian variation in content of melatonin in the gland is largely determined by variation in the activity of the acetylating enzyme. This rhythmically depletes the cells of its substrate, 5-hydroxytryptamine, and by providing a varying supply of substrate to the methylating enzyme is responsible for the varying production of melatonin [236]. The varying activity of the acetylating enzyme is under sympathetic control, which accounts for circadian variation in the noradrenaline content of the gland [237]. Though melatonin has a variety of actions [238], its normal function in the mammal is largely unknown. Its very striking biochemical rhythms, in part endogenous and in part dependent upon light [239], have led to speculation about its possible function in control of other rhythms. In the rat, however, where it has been most extensively studied, circadian rhythm in locomotor activity persists in the absence of the pineal [240], although pinealectomized rats adapt their activity rhythm to a phase shift in the external Zeitgeber much more readily than do intact animals [241]. It is probably more closely involved in activity rhythms in birds. [v. Ch. 8].

2.3.8 Other Hormones

There are many claims that other hormones show a nychthemeral rhythm, but in rather few has any attempt been made to exclude even the grosser exogenous influences. A rhythm in follicle-stimulating and luteinizing hormones in subjects following ordinary nychthemeral habits, and in growth hormone in subjects either fasting or with diet strictly controlled [168], could result from a hypothalamic rhythm, but evidence is lacking. The rhythms in gonadotrophins are independent of ovarian feedback, since they persist after oophorectomy [167].

The growth hormone rhythm is unusual in that, although concentration is on average higher by night than by day, there are large oscillations with a period of around 6 h. It is said to be reversed at once if a subject sleeps by day, so it may be wholly exogenous [116].

Growth hormone production in response to a given level of hypo-glycaemia is much greater in the morning than at night [169]. The growth hormone rhythm could, as mentioned below, contribute to the urinary sodium rhythm, and has been tentatively implicated in the morning fall of plasma phosphate [170].

There is some evidence that the relationship between the rhythms of the thyroid and the anterior pituitary thyrotrophic hormone (TSH) is different from those so far described. In plasma samples collected every 4 h, TSH concentration was minimal at 1600 h and maximal at 0400 h while free thyroxine showed precisely the reverse [171]; this suggests that the thyroid rhythm was primary and the pituitary rhythm resulted from negative feedback. The thyroid rhythm in turn may have been wholly exogenous, since thyroxine loss from the blood is closely dependent upon food intake [172].

The secretory activity of the adrenal medulla is so dependent upon stress and a variety of emotional stimuli [173] that most reported rhythms are likely to be wholly exogenous. The most striking instance of an adrenal medullary rhythm of apparently endogenous origin [73] derives from observations on Swedish military personnel who spent 72 h in simulated stressful activity, continuously awake, and with no breaks longer than 15 min, and with no indication of the alternation of day and night. The urinary excretion of adrenaline showed a very pronounced rhythm with a period of around 24 h. Since the formation of the enzyme which catalyses the methylation of noradrenaline to form adrenaline is stimulated by the adrenal cortex [76], this adrenaline rhythm may arise from the corticosteroid rhythm already discussed. The evidence for a rhythmic production of noradrenaline [174; 30, p. 40] is less clearly established and has no known origin, but, like that of adrenaline, it persists despite continuous rest in bed with 4-hourly meals [175].

The earlier decline of adrenaline excretion in introverted, "morning" types than in extraverted, "evening" types [72] has already been mentioned, but we are largely ignorant of the causal nexus involved.

2.4 URINARY RHYTHMS

Although a nychthemeral rhythm has been detected in the excretion of water and of most urinary constituents, we are almost entirely ignorant how these various rhythms are controlled. Even if any of the cells composing the kidney has an intrinsic rhythmicity, excretion is normally constrained to a 24-h period; and since the kidney has no known information channel from the outside world it must be subject to some rhythmic influence from within the body. Since we do not yet understand the intimate nature of the signal to the kidney in many homoeostatic or other adaptive renal responses, it is however particularly difficult to identify the circadian signal.

Some urinary rhythms are, so far as we know, wholly exogenous; urea excretion probably simply follows the pattern of protein intake, and may be determined by the plasma urea concentration, and there is little evidence for endogenous rhythms in uric acid excretion. Excretion of calcium and magnesium are normally dependent on the pattern of meals, though an endogenous rhythm can be demonstrated if other potent influences are excluded. At the other extreme, excretion of potassium and acid are not greatly affected by the routine of living and mainly reflect a circadian endogenous influence, while both exogenous and endogenous factors substantially affect excretion of sodium, chloride, and phosphate.

2.4.1 Water Excretion

Water excretion is the resultant of several distinct renal processes, each susceptible to so many internal and external influences that it will not be separately considered. The variety of these influences is indicated in Fig. 2.1. In people on a usual nychthemeral existence it is very dependent upon the pattern of fluid ingestion; but if fluid intake is modest, and grosser exogenous circadian influences are excluded, urine flow appears to be osmotically determined by the osmotic load, mainly sodium and potassium and associated anions, which in their passage down the collecting duct retain water in an amount varying with the total solute load (Fig. 2.14). This is to be expected so long as there is no substantial change in the composition of the renal medullary tissue around the collecting duct.

2.4.2 Adrenal Influence

The influence of cortisol upon urinary rhythms has already been considered. Aldosterone, by contrast, appears to be of no importance; a careful study of its rate of secretion over successive 3-h periods, by dilution of isotopically labelled hormone [128], has shown that it is secreted in greatest amount in the morning, when sodium excretion is high: since aldosterone causes sodium retention, it cannot therefore contribute to the usual circadian rhythm. This morning secretion of aldosterone is probably dependent upon posture, for in subjects remaining recumbent voluntarily [176], or as a result of paraplegia [177, 178], aldosterone excretion shows no circadian rhythm.

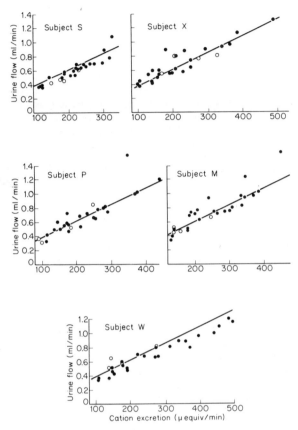

Figure 2.14. Dependence of urine flow upon electrolyte (Na + K) excretion in five subjects living on a 12 h day, ● awake, ○ asleep. The regression line is calculated from the pooled data from all five subjects [148].

2.4.3 Plasma Concentration

The most obvious signal to the kidney would be the plasma concentration of the substance being excreted, and indeed if the clearance of a substance is reasonably constant, its excretion will be directly proportional to plasma concentration. This is approximately true of urea, and it has been suggested above that the rhythm in its excretion results solely from variation in plasma concentration. Creatinine clearance is also supposed to be fairly constant, but its plasma concentration is very low, and difficult to measure reliably; the relatively minor oscillations between a low night and higher diurnal excretion

[179, 7] are commonly held to reflect a rhythm in glomerular filtration rate, which has been observed by several authors [128, 176, 180-182].

Phosphate

Another urinary constituent whose excretion is directly affected by varying plasma concentration is phosphate, which is filtered at the glomerulus and then reabsorbed up to a maximal tubular capacity [183], which may however vary with glomerular filtration rate [184] and probably with endocrine influences. As has been mentioned above, excretion of phosphate usually declines after waking [124, 180, 185] (Fig. 2.12), at a time when glomerular filtration is rising, so the variations in glomerular filtration are unlikely to contribute to the phosphate excretory rhythm, which in any case is easily disturbed by altered habits of waking and sleeping. When subjects remain under reasonably constant conditions and show a well-defined excretory rhythm, this is closely paralleled by the rhythm in plasma phosphate concentration [128; 30, p. 57] (Fig. 2.15), which appears to be the immediate signal to the kidney.

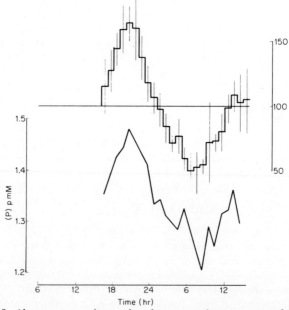

Figure 2.15. Above, mean urinary phosphate excretion, per cent of 24-h mean, and S.D., below, mean plasma phosphate concentration, in subjects awake and active and with identical hourly food and fluid intake [1]. (Courtesy Physiological Reviews.)

The cause of the morning fall in plasma phosphate concentration is unknown, since very few plasma estimations have been made in circumstances when the usual rhythms might be disturbed. As has been already mentioned, cortisol depresses plasma phosphate, and the phase relations between the cortisol and urinary phosphate rhythms are appropriate for the adrenals to determine the phosphate excretory behaviour. Evidence against an intermediary role for cortisol is that in a variety of circumstances the behaviour of plasma cortisol and urinary phosphate are dissociated; but with the paucity of determinations of plasma phosphate it is impossible to say whether the dissociation is between cortisol and plasma concentration of phosphate, or between plasma phosphate and the renal behaviour.

Potassium

Plasma potassium concentration is not generally considered to be a major influence upon renal excretion of potassium [186], partly because very large changes in potassium excretion can be observed with negligible change in plasma concentration, and partly because urinary potassium is derived from the cells of the kidney rather than from the plasma [187]. Over short periods however urinary potassium has been shown to follow plasma concentration so closely as to suggest a causal connexion [188] and this supposition is borne out by the unilateral increase in excretion when potassium is infused into one renal artery [189]. Again, some authors have failed to observe appropriate circadian variation in plasma concentration of potassium, but the variations to be sought are small, and readily obscured by traces of haemolysis; and there are accounts of circadian variation in plasma potassium concentration of sufficient magnitude to account for the urinary changes [128, 190a]. A recent report [190] describes similar changes in serum potassium concentration, but these are not held by the authors to be responsible for the urinary rhythm. The statistical methods employed admittedly fail to show that a serum potassium rhythm *does* underlie the urinary rhythm, but this is not the same as showing that it does not.

Another line of approach is to consider whether reported circadian variations in plasma potassium are large enough to account for the urinary rhythm. Figure 2.16 shows examples of the spontaneous variation in plasma concentration and urinary excretion of potassium over 24 h, and the changes induced when a large dose of potassium is taken in the early afternoon. While the absolute levels are very different, the slopes of the regression lines are sufficiently similar to suggest that change in plasma concentration may be a determining factor in short-term variations of potassium excretion. It is thus perfectly possible that

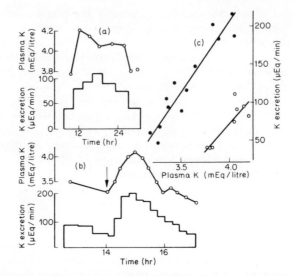

Figure 2.16. Plasma concentration and rate of urinary excretion of potassium. (a), control data over 24 h, after Wesson [128]. (b), own data [188]; at arrow 60 mEq KHCO₃ was ingested. (c), urinary excretion plotted against plasma concentration, with linear regression inserted. ○, data from (a); ●, data from (b). Note the different scales used in the three plots.

plasma potassium concentration is an obligate intermediary between the clock and the kidney, and conceivable that cortisol mediates between the clock and plasma potassium concentration.

Acid

There is no good evidence that plasma composition plays a part in controlling the rhythm of acid excretion, which almost all workers find to be higher by night than by day. The carbon dioxide tension and hence the acidity of the blood increase during sleep [191], and an increased CO_2 tension can increase the urinary acid [192]; but any rise of CO_2 in a sleepless night is trifling [193] and quite inadequate to explain the urinary acidity [194]. There is moreover an alternative explanation: potassium and hydrion are believed to compete with one another for exchange with sodium in the distal convoluted tubule [195] and since when disturbing influences are minimized the circadian excretory rhythms are perfect mirror images of one another, it is suggested that they are casually related by their competition for sodium [196]. If the potassium change is primary, and determined by its plasma concentration, we can thus account for the acid excretory rhythm.

2.4.4 Cause of Urinary Sodium Rhythm

Our greatest ignorance concerns the cause of the sodium excretory rhythm, though this merely reflects our ignorance of the means whereby sodium excretion is regulated in other circumstances. This rhythm is commonly of very large amplitude [180] (Fig. 2.17), but it is disturbed much more readily than is that of potassium by posture [176, 197], exercise [182] and meals [198].

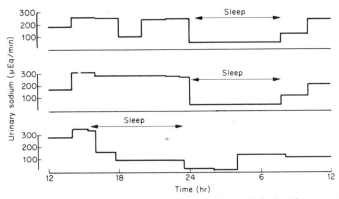

Figure 2.17. Sodium excretion of subject in isolation unit during three consecutive days. After the second day the clock was advanced 8 h, equivalent to an eastward flight, but excretion is plotted on real time. Note the very large variations in sodium excretion, which is only partially adapted to the time change.

Plasma concentration of sodium shows no circadian rhythm [128, 180, 190], nor is it generally considered to be a major influence on sodium excretion. Glomerular filtration rate is believed by some to be an important influence in the circadian rhythm [181, 182]. Contrary evidence is that the circadian rhythms reported for glomerular filtration rate have sometimes been of very large and sometimes of negligible amplitude; and it has been claimed both that deliberately induced changes in GFR are without significant effect on sodium excretion [199] and that large changes in sodium excretion can be induced by other means, despite reverse changes in GFR [200]. It remains possible that GFR changes, though of no importance in many adaptive alterations of sodium excretion, still contribute substantially to its circadian variation; but there are many other influences to be explored, of which some have recently been reviewed [201]. Increased hydrostatic pressure in the peritubular blood vessels has been shown by direct micropuncture to inhibit reabsorption of sodium, and this may explain, at least in part, the effects on sodium excretion of vasoactive peptides such as bradykinin [202, 203] and angiotensin [204, 205], as well as the effects of experimental vascular manipulations [206]. Many groups of workers have claimed to demonstrate a natriuretic substance in the plasma of animals infused with saline [200, 207, 208], and it has been specifically suggested that this is a prostaglandin [209], but a recent careful series of investigations in three independent laboratories [210] has failed to reveal such a substance. It has been argued already that the salt-retaining action of aldosterone cannot play a part in the circadian rhythm; this action of aldosterone is however, both in hypophysectomized rats [211] and in isolated perfused kidneys [212], very dependent upon the presence of adequate amounts of anterior pituitary growth hormone. Another sodium-retaining steroid is contained in cardiac muscle, and released in variable amounts [213-217], and a substance which may well be identical is released by skeletal muscle in exercise [217]. Another sodium-retaining substance is claimed to be released by perfused lungs [214, 218]. Any of these may be the unknown sodium-retaining factor postulated by other workers [219].

Little is known about the respective role and importance of these various influences upon sodium excretion, and even less about any circadian variation. Growth hormone alone has been shown to vary circadianly in its concentration in the plasma, and since it is secreted mainly at night it has the right phase relationship to contribute to the sodium excretory rhythm. Its close dependence upon sleep, meals and fasting, which do not grossly affect sodium excretion, cast doubt however upon its role. We are indeed in no position to speculate usefully upon the immediate signal to the kidney to excrete more sodium by day and less by night; and even if it were concluded that GFR played a part, we still do not know the origin of the circadian variations in GFR.

Chloride excretion follows a course similar to that of sodium in its circadian rhythm as in other circumstances, and it is likely that whatever determines the sodium rhythm also determines that of chloride; an occasional lag of the morning urinary rise in sodium behind that of chloride may reflect the influence of the morning secretion of aldosterone in the ambulant subject [128].

Calcium, magnesium and strontium are all said to show nychthemeral variation in their excretion, but they are so largely affected by activity, meals and other circumstances that no clear endogenous rhythm can be defined. It would thus be premature to speculate upon any influence of the clock upon these electrolytes, and the interested reader is referred elsewhere [30].

2.4.5 Miscellaneous Influences

Two obvious influences upon renal function have not yet been considered, the nervous system and the posterior pituitary antidiuretic hormone (ADH). Those who have estimated ADH in the urine [220, 221] have not demonstrated any circadian rhythm, but those who have assayed the apparent ADH content of plasma, by injection into diuretic rats, have found higher values by night than during the day [222, 223]. There must always be some doubt about the specificity of a biological assay, although in the more recent of these papers it was demonstrated that the material was inactivated by treatment with thioglycolate; these findings obviously need confirmation by the better techniques now available. In both investigations the subjects, both healthy and hospital patients, were exposed to the usual nychthemeral influences; but if the findings are substantiated then ADH secretion at night could obviously contribute to the usual oliguria.

There is some confusion as to the part played by the nervous system in the control of renal function. The most reliably denervated human kidneys consist of transplants; and while it has been claimed [224] that transplanted human kidneys show the usual rhythm, other workers have found considerable disturbance or even reversal of some rhythms [225, 226]. Patients with renal transplants are usually receiving substantial doses of drugs and renal function may be in other respects imperfect, so these observations contribute little to our knowledge of the causation of the usual urinary rhythms of healthy subjects.

Clues as to the origin of urinary rhythms may also be sought in any dissociation of components in subjects living on abnormal time schedules. Such studies have been usually concerned with other variables beyond the renal ones, so the conclusions to be drawn from them will be considered together.

2.5 ABNORMAL TIME SCHEDULES

The study of subjects living on abnormal time schedules is useful in establishing the endogeneity of rhythms, but can only offer negative evidence about transmission processes. In such circumstances, when some rhythms are disturbed, any two rhythms which preserve their usual temporal relationship may, or may not, have some simple or direct causal connexion; two rhythms which are dissociated cannot be directly related as cause and effect.

2.5.1 Days Longer or Shorter than 24 hours

Subjects have been exposed, in the Arctic, in an underground laboratory, or elsewhere, to artificial days of 18 h [79], 21 h [20, 21, 22, 142], 22 h [141, 227, 229], 22.2/3 and 26.2/3 h [15, 229], and 28 h [20, 22, 79, 142], whilst paraplegics have been submitted to a routine cycle of

19 h [198]. With the exception of a single report [22] on a subject living on a 28-h day in the Mammoth Cave of Kentucky, it appears that the clock never adjusts its period to an abnormal imposed length, since rhythms revert immediately to a 24-h period on resumption of normal habits.

While people are living on these abnormal time schedules they are subjected to two competing influences, an internal clock with a 24-h period, and the sum of all exogenous influences with a different period; many functions continue to show the operation of both influences, since study of their rhythmicity reveals two distinct periods [228]. This may lead to a striking appearance of "beats", the measured rhythms, such as urinary flow [20], having a large amplitude when the two influences are in phase, and a low amplitude when they are out of phase. In general, the clock influence wanes and the exogenous influences become increasingly prominent; and the very different time-course of adaptation of different components often suggests relative independence of one another.

The temperature rhythm often fails to conform to the activity rhythm [9, 15, 229], as has also been observed in subjects trying to live on an abnormal day-length in a nychthemeral society; it is clear that the temperature rhythm is not a simple result of the alternation of activity and rest. Others however have observed the temperature rhythm responding promptly to an abnormal pattern of life [20]. When the rhythm of sleepiness has failed to adapt to an imposed abnormal day, the temperature rhythm has been in phase with the spontaneous rhythm of wakefulness rather than with the imposed habits [79], suggesting that the temperature rhythm may be a cause of the rhythm in sleepiness and wakefulness, rather than a consequence of the activity rhythm.

As mentioned above (p. 44) pulse rate has been found to adapt immediately to a day of artificial length, when temperature failed to do so [79], and one is tempted to suppose that body temperature is not involved in transmission of rhythmicity to the pulse rate. This is however a logical fallacy: in these experiments the rhythm in pulse rate was apparently wholly exogenous, and one can rightly infer that body temperature is not involved in the transmission of an external rhythmicity to the pulse rate. When however exogenous influences on the pulse rate are minimized, and an endogenous rhythm emerges, temperature may well mediate the influence from the clock.

It is possible to generalize from this example. If, on a day of abnormal length, a rhythm A maintains a 24-h period and another rhythm B adopts the imposed artificial period, then B cannot mediate between the clock and A, but A may mediate between the clock and B. This is indicated diagrammatically in Fig. 2.18. On an artificial day, if the exogenous influences resulting from the abnormal periodicity of sleep, waking, meals and other habits affect B more powerfully than A, they

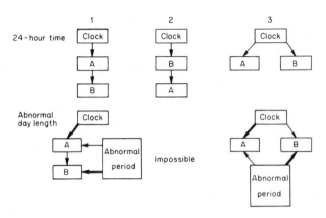

Figure 2.18. Three possible causal connexions between an endogenous clock and two processes, A and B. If, on a day of abnormal length, A maintains a 24-h rhythm and B adopts the abnormal imposed rhythm, 1 and 3 remain possible, but 2 is excluded.
—— dominant influence;   —— weak influence.

may swamp the clock influence upon B while A maintain as 24-h rhythm (Fig. 2.18, 1). If however A is constrained into an abnormal period, it cannot transmit a 24-h rhythm from the clock to B. In the original example A represents temperature, which may maintain 24-h time on an abnormal day length, and B represents pulse rate which is very readily affected by exogenous rhythms of rest and exercise, but, when such influences are eliminated, may respond to the rhythm in body temperature. This hypothesis would be at once excluded if on an abnormal time schedule temperature picked up the artificially imposed rhythm while pulse rate adhered to 24-h time. The third possible causal relationship, of a common mediator between the clock and A and B (Fig. 2.18, 3) similarly remains possible even if A or B assumes an artificially imposed rhythm while the other adheres to 24-h time.

The rhythmic excretion of potassium adapts only slowly, and in some subjects not at all, to a day of abnormal length, while the rhythm in sodium and chloride adapts more readily [20, 21]. In part, this presumably represents the greater susceptibility of sodium and chloride excretion to a variety of exogenous influences; in subjects whose habits and diet are not carefully controlled, potassium excretion follows a much more regular rhythmic pattern than does that of sodium or chloride. It is therefore possible that the clock affects all three electrolytes by the same intermediate link, and, following the same logic as before, their dissociation shows only that the rhythm in potassium excretion is not a result of that in sodium or chloride. Likewise, neither can result directly from the rhythm in body temperature, which adapts more rapidly than the urinary rhythms.

Excretion of potassium and of corticosteroids are both slow to adapt, and as has been considered above there are many grounds for implicating the adrenal cortex as a mediator between the clock and the excretion of potassium; but in subjects who have lived for a long time on a 21-h day there are sufficient instances of dissociation between excretion of potassium and of corticosteroids to throw some doubt upon the hypothesis of close causal connexion [21, 144].

2.5.2 Abrupt Time Shifts

Dissociation between the behaviour of different rhythmic variables may also be studied in subjects who abruptly change the phase of their habit rhythms, either by deliberately altering their clock time when living in isolation in an Arctic summer [140] or underground, or by flying into a different time zone, or by changing over from day to night work or vice versa [149]. This is a field of study which is being actively pursued, and many results are as yet unpublished or published only in preliminary form, but in general it supports the findings from experiments on days of abnormal length, that steroid secretion and urinary potassium adapt at a similar rate [149a]. In a recent study on night workers mentioned above [145], the potassium excretory rhythms of many men were adapted to their pattern of work while the rhythm in plasma 11-hydroxycortico-steroids was not (Fig. 2.11). The logical position is very similar to that described above for days of abnormal period, and it will be seen that this is not conclusive evidence against the mediation of the adrenals between clock and potassium excretion, since the urinary potassium rhythm may have been exogenous. Though logically possible, this is however unlikely, since potassium excretion is not very susceptible to exogenous influences.

2.5.3 Subjects Isolated from Time Clues

Another source of information about the connexion or dissociation between different rhythmic variables comes from subjects screened from all circadian rhythmic influences, either in underground caves or specially constructed sound-proof and light-proof chambers. Most subjects in such circumstances follow a sleep-activity cycle of somewhat over 24 h, with the other measured rhythms following a closely similar period. This has been observed by one or more groups of workers for temperature, pulse rate, excretion of electrolytes and of corticosteroids, and ability to estimate short intervals of time [5, 7, 8, 10, 11, 13, 15]. Aschoff's group report that of 50 subjects thus studied in isolation, 36 showed complete synchrony between all measured variables. It thus appears that either the clock, which is "free-running" and not adhering precisely to 24-h time, is exerting its influence over all the measured variables, or that they are all influenced by the subject's habits of sleep and activity. The latter is improbable, since in subjects living on days of abnormal length their activity pattern does not seem to be a very potent influence upon their rhythms; but in any case the continued association between different rhythms leaves these experiments of only modest value in uncovering transmission processes. The altered phase relations sometimes observed must eventually be incorporated into our picture, but they are not all consistent between one subject and another; for example, two subjects isolated in caves [16] showed peaks of rectal temperature and of urinary 17-hydroxycorticosteroids earlier in relation to the sleep-activity cycle than when they were living on normal time; but whereas in one, the peak temperature and steroid excretory peaks in isolation were close together, in the other they were over 6 h apart. Others have also reported a temperature peak earlier in the activity cycle when subjects are living in isolation [230]. The maximum and minimum were evenly spaced, whereas under more normal conditions, in the presence of Zeitgeber, the time from maximum to minimum was less than the time from minimum to maximum.

A particularly interesting finding from such isolation experiments is that two variables can occasionally oscillate with a different period, both of them departing from 24 h. The best authenticated dissociations of this sort, which have already been discussed, are between the temperature and activity cycles. A subject whose temperature oscillated with a period of 24.8 h and whose habits of waking and sleeping followed a rhythm of 33.2 h [11] would appear to be influenced by two completely independent clocks, if we could be certain that the actual habits of waking and sleeping necessarily indicate an endogenous rhythm of wakefulness and sleepiness. Dissociation has also been observed in simulated orbital space flight [81]. Under such conditions and in constant light one subject lost all temperature rhythm, but over a period of six days his steroid excretion maintained a 24-h period corresponding to his activity cycle, and his potassium excretory rhythm followed a period of 29-31 h (Fig. 2.19). This is further evidence that the adrenals are not responsible for the potassium excretory rhythm. As with so much work on rhythms, we could come to much safer conclusions if we could measure many more variables simultaneously: when steroid and urinary potassium rhythms are dissociated, what happens to plasma potassium concentration?

Little has yet been published about the physiological rhythms of human subjects in orbit or in space. There is however an account of a monkey in an orbiting satellite [231] whose sleeping habits followed a

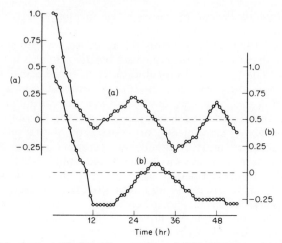

Figure 2.19. Autocorrelation of excretion of (a), 17-OH corticosteroids, (b), potassium, in a subject exposed to constant light for six days in a space flight simulator. Body temperature was non-rhythmic. Good correlation with values 24 and 48 h later, as in (a), indicates a repeating pattern; correlation with values 30 h later, as in (b), similarly indicates a repeating 30-h pattern. After Rummel, Sallin and Lipscomb [81].

24-h cycle, while several other functions, including body temperature and pulse rate, followed a cycle of over 24 h. This supports the suggestion made above that the habits of sleep and activity are not the main cause of other rhythms; it is also in conformity with the suggestion that the temperature rhythm may determine the rhythm in pulse rate, but fails to support the further suggestion that the temperature rhythm also determines that of sleepiness.

2.6 CONCLUSION

Of the very large number of variables which oscillate nychthemerally [30] a large number, though a small proportion, has been shown to have an endogenous rhythm. It is therefore disappointing that to so few can we ascribe an immediate cause, the signal from the clock. Figure 2.20 shows those few links that are reasonably established, more for which there is some evidence, and others which are largely speculative. The hypothalamus has been taken as a starting point not through conviction that this is the site of the clock, but because this region controls, directly or indirectly, so many functions that vary circadianly. The ascription of causal links to one or other category, and the complete omission of many more, is very largely a matter of the author's individual, perhaps

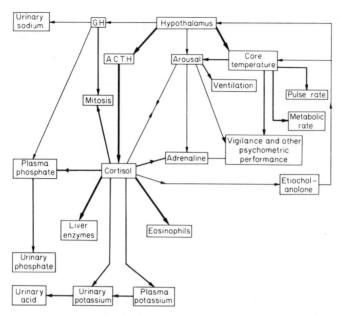

Figure 2.20. Summary of connexions considered in this chapter between the hypothalamus and circadianly rhythmic functions. ▬▬▶▬ connexions reasonably established. ──▶── connexions based upon evidence which is incomplete or conflicting. ──▶── speculative connexions. A connexion, and the arrow indicating its direction, indicates not only a causal sequence, but that rhythmic variations in the one function may be the cause of rhythmic variations in other; the negative feedback of cortisol upon ACTH is thus omitted since it appears to play no part in the control of rhythmicity. Some possible connexions, e.g. upon urine flow and sodium, have been omitted through their complexity.

arbitrary, judgment; but however one assesses the evidence, it is clear that our ignorance vastly exceeds our understanding. At least this should be a spur to further work: or are we chasing a chimaera in searching for transmission from a universal controlling clock?

Most physiological functions are involved in homoeostasis and hence are controlled by negative feedback mechanisms. It is a natural property of such a system to oscillate, so it is superficially tempting, for example, to ascribe the rhythmicity of the adrenals to the negative feedback of cortisol upon the anterior pituitary production of ACTH, or the rhythm of urinary acidity to an overcorrection of a low plasma pH. Where this "pendulum" hypothesis has been tested, it has been clearly disproved: the production of ACTH remains rhythmic even when the response of the adrenals is absent; and when the kidney is constrained to put out an alkaline urine rich in potassium for three successive night, so soon as this constraint is removed the usual rhythms reappear in their normal phase [232]. It is hard to avoid the conclusion that a controlling clock was running continuously, even when it was unable to affect the kidney. It is moreover hard to see how a series of oscillations resulting from a large number of distinct feedback systems could, in normal circumstances, be constrained to a synchronous rhythm of 24 h without some central controlling mechanism. One thing is quite clear: the subjective awareness of time [233] is quite distinct from the time-recording system responsible for circadian rhythms, and the two may be very far apart in their estimates [6, 13].

REFERENCES

1. J. N. Mills, Human circadian rhythms. *Physiol. Rev.*, 46, 128-171 (1966).
2. E. Bünning, Das Weiterlaufen der "physiologischen Uhr" im Säugerdarm ohne zentrale Steuerung. *Naturwissenschaften*, 45, 68 (1958).
3. R. V. Andrews and G. E. Folk, Circadian metabolic patterns in cultured hamster adrenal glands. *Comp. Biochem. Physiol.*, 11, 393-409 (1964).
4. J. Aschoff, Tierische Periodik unter dem Einfluss von Zeitgebern. *Z. Tierpsychol.*, 15, 1-30 (1958).
5. J. Aschoff and R. Wever, Spontanperiodik des Menschen bei Ausschluss aller Zeitgeber. *Naturwissenschaften*, 49, 337-342 (1962).
6. M. Siffre, *Hors du Temps* (Julliard, Paris, 1963). [*Beyond Time* (Chatto and Windus, London, 1965)].
7. J. N. Mills, Circadian rhythms during and after three months in solitude underground. *J. Physiol. (Lond.)*, 174, 217-231 (1964).
8. F. Halberg, M. Siffre, M. Engeli, D. Hillman and A. Reinberg, Etude en libre-cours des rhythmes circadiens du pouls, de l'alternance veille-sommeil et de l'estimation du temps pendant les deux mois de séjour souterrain d'un homme adulte jeune. *C.R. Acad. Sci. Paris*, 260, 1259-1262 (1965).
9. B. R. Clegg and K. E. Schaefer, Studies of circadian cycles in human subjects during prolonged isolation in a constant environment using 8-channel telemetry systems, S.M.R.L. Report No. 66-4 U.S.N. Submarine Med. Cent., Groton, Connecticut.
10. M. Siffre, A. Reinberg, F. Halberg, J. Ghata, G. Perdriel and R. Slind, L'isolement souterrain prolongé—étude de deux sujets adultes sains avant, pendant et après cet isolement. *Presse Méd.*, 74, 915-919 (1966).
11. J. Aschoff, U. Gerecke and R. Wever, Desynchronization of human circadian rhythms. *Jap. J. Physiol.*, 17, 450-457 (1967).
12. J. N. Mills, Keeping in step—away from it all. *New Scientist*, 33, 350-351 (1967).
13. J. Colin, J. Timbal, C. Boutelier, Y. Houdas and M. Siffre, Rhythm of the rectal temperature during a 6-month free-running experiment. *J. Appl. Physiol.*, 25, 170-176 (1968).
14. M. Apfelbaum, A. Reinberg, P. Nillus and F. Halberg, Rhythmes circadiens de l'alternance veille-sommeil pendant l'isolement souterrain de sept jeunes femmes. *Presse Méd.*, 77, 879-882 (1969).
15. J. Aschoff, Desynchronization and resynchronization of human circadian rhythms. *Aerospace Med.*, 40, 844-849 (1969).

16. J. Ghata, F. Halberg, A. Reinberg and M. Siffre, Rhythmes circadiens désynchronisés du cycle social (17-hydroxycorticostéroides, température rectale, veille-sommeil) chez deux sujets adultes sains. *Ann. Endocr. Paris*, 30, 245-260 (1969).

17. R. P. Doe, J. A. Vennes and E. B. Flink, Diurnal variation of 17-hydroxycorticosteroids, sodium, potassium, magnesium and creatinine in normal subjects and in cases of treated adrenal insufficiency and Cushing's syndrome. *J. clin. Endocr.*, 20, 253-265 (1960).

18. W. Enderle, Tagesperiodische Wachstums- und Turgorschwankungen an Gewebekulturen. *Planta*, 39, 570-588 (1951).

19. T. van den Driessche, Circadian rhythms in Acetabularia: photosynthetic capacity and chloroplast shape. *Exp. Cell Res.*, 42, 18-30 (1966).

20. P. R. Lewis and M. C. Lobban, Dissociation of diurnal rhythms in human subjects living on abnormal time routines. *Quart. J. exp. Physiol.*, 42, 371-386 (1957).

21. H. W. Simpson and M. C. Lobban, Effect of a 21-hour day on the human circadian excretory rhythms of 17-hydroxycorticosteroids and electrolytes. *Aerospace Med.*, 38, 1205-1213 (1967).

22. N. Kleitman, *Sleep and Wakefulness*, 2nd Ed. (University of Chicago Press, Chicago and London) (1963).

23. G. W. G. Sharp, S. A. Slorach and H. J. Vipond, Diurnal rhythms of keto- and ketogenic steroid excretion and the adaptation to changes of the activity-sleep routine. *J. Endocr.*, 22, 377-385 (1961).

24. T. Hellbrügge, The development of circadian rhythms in infants. *Cold Spr. Harb. Symp. quant. Biol.*, 25, 311-323 (1960).

25. J. Aschoff, Exogenous and endogenous components in circadian rhythms. *Cold Spr. Harb. Symp. quant. Biol.*, 25, 11-28 (1960).

26. A. L. Elliott, J. N. Mills and J. M. Waterhouse, A man with too long a day. *J. Physiol. (Lond.)*, 212, 30-31P (1971).

27. M. Selinger and M. Levitz, Diurnal variation of total plasma estriol levels in late pregnancy. *J. clin. Endocr.*, 29, 995-997 (1969).

28. F. A. Brown, Jr., Living clocks. *Science*, 130, 1535-1544 (1959).

29. J. Aschoff, Der Tagesgang der Körpertemperatur beim Menschen. *Klin. Wschr.*, 33, 545-551 (1955).

30. R. T. W. L. Conroy and J. N. Mills, *Human Circadian Rhythms* (Churchill, London, 1970).

31. G. W. G. Sharp, Reversal of diurnal temperature rhythms in man. *Nature*, 190, 146-148 (1961).

32. G. Felton, Effect of time cycle change on blood pressure and temperature in young women. *Nurs. Res.*, 19, 48-58 (1970).

33. C. Kayser and A. A. Heusner, Le rhythme nychthéméral de la dépense d'énergie. *J. Physiol. Paris*, 49, Suppl. 3-116 (1967).

34. E. D. Voigt and P. Engel, Tagesrhythmische Schwankungen des Energieverbrauches bei Arbeitsbelastung. *Pflügers Arch. ges. Physiol.*, 307, R 89-90 (1969).

35. A. Bornstein and H. Völker, Uber die Schwankungen des Grundumsatzes. *Z. ges. exp. Med.*, 53, 439-450 (1926).

36. M. B. Kreider, E. R. Buskirk and D. E. Bass, Oxygen consumption and body temperatures during the night. *J. appl. Physiol.*, 12, 361-366 (1958).

37. K. F. Koe, W. Höfler and K. Lüders, Mittlere Hauttemperatur und peripherer Extremitätentemperaturen bei den tagesperiodischen Anderungen der Wärmeabgabe. *Arch. Phys. Ther. (Leipzig)*, 20, 221-226 (1968).

38. J. Aschoff and H. Pohl, Rhythmic variations in energy metabolism. *Fed. Proc.*, 29, 1541-1552 (1970).

38a. C. Kayser, J. L. Imbs and J. Karacz, Rythme circadien de la consommation d'oxygène du Rat et glandes endocrines. *C.R. Soc. Biol.*, 164, 1372-1374 (1970).

39. F. Heiser and L. H. Cohen, Diurnal variations of skin temperature. *J. industr. Hyg.*, 15, 243-254 (1933).

40. G. Hildebrandt and P. Engelbertz, Bedeutung der Tagesrhythmik für die physikalische Therapie. *Arch. phys. Ther. (Leipzig)*, 5, 160-170 (1953).

41. E. H. Geschickter, P. A. Andrews and R. W. Bullard, Nocturnal body temperature regulation in man: a rationale for sweating in sleep. *J. Appl. Physiol.*, 21, 623-630 (1966).

42. J. Aschoff, Einige allgemeine Gesetzmässigkeiten physikalischer Temperaturregulation. *Pflügers Arch. ges. Physiol.*, 249, 125-136 (1947).

43. K. Honma, K. Sekine, E. Harada and K. Kimura, The diurnal patterns of the temperature in human toes, fingers and ears. *Z. Biol.*, 115, 299-310 (1966).

44. R. E. Smith, Circadian variations in human thermoregulatory responses. *J. appl. Physiol.*, 26, 554-560 (1969).

45. G. W. Crockford, C. T. M. Davies and J. S. Weiner, Circadian changes in sweating threshold. *J. Physiol. (Lond.)*, 207, 26-27P (1970).

46. A. Kappas, W. Soybel, D. K. Fukushima and T. F. Gallagher, Studies on pyrogenic steroids in man. *Trans. Ass. Amer. Phycns.*, 72, 54-61 (1959).

47. A. Kappas, W. Soybel, P. Glickman and D. K. Fukushima, Fever-producing steroids of endogenous origin in man. *Arch. intern. Med.*, 105, 701-708 (1960).

48. L. Hellman, H. L. Bradlow, B. Zumoff, D. K. Fukushima and T. F. Gallagher, Thyroid-androgen interrelations and the hypocholesteremic effect of androsterone. *J. clin. Endocr.*, 19, 936-948 (1959).

49. J. J. Sherwood, A relation between arousal and performance. *Amer. J. Psychol.*, 78, 461-465 (1965).

50. M. J. F. Blake, Time of day effects on performance in a range of tasks. *Psychon. Sci.*, 9, 349-350 (1967).

51. N. T. Loveland and H. L. Williams, Adding, sleep loss and body temperature. *Percept. Mot. Skills*, 16, 923-929 (1963).

52. K. E. Klein, H. M. Wegmann and H. Brüner, Circadian rhythm in indices of human performance, physical fitness and stress resistance. *Aerospace Med.*, 39, 512-518 (1968).

53. D. S. P. Schubert, Simple task rate as a direct function of diurnal sympathetic nervous system predominance. *J. comp. physiol. Psychol.*, 68, 434-436 (1969).

54. M. J. F. Blake and W. P. Colquhoun, Experimental studies of shift work. *Ergonomics*, 8, 376 (1965).

55. W. P. Colquhoun, M. J. F. Blake and R. S. Edwards, Experimental studies of shift-work. I: A comparison of "rotating" and "stabilized" 4-hour shift systems. *Ergonomics*, 11, 437-453 (1968).

56. W. P. Colquhoun, M. J. F. Blake and R. S. Edwards, Experimental studies of shift-work. II: Stabilized 8-hour shift systems. *Ergonomics*, 11, 527-546 (1968).

57. W. P. Colquhoun, M. J. F. Blake and R. S. Edwards, Experimental studies of shift work. III: Stabilized 12-hour shift systems. *Ergonomics*, 12, 865-882 (1969).

58. R. T. Wilkinson, R. H. Fox, F. R. Goldsmith, I. F. G. Hampton and H. E. Lewis, Psychological and physiological responses to raised body temperature. *J. appl. Physiol.*, 19, 287-291 (1964).

59. G. J. Stephens and F. Halberg, Human time estimation. *Nurs. Res.*, 14, 310-317 (1965).
60. F. Halberg, The 24-hour scale: a time dimension of adaptive functional organization. *Perspect. Biol. Med.*, 3, 491-527 (1960).
61. D. Pfaff, Effects of temperature and time of day on time judgments. *J. exp. Psychol.*, 76, 419-422 (1968).
62. R. Günther, E. Knapp and F. Halberg, Circadiane Rhythmometrie mittels elektronischer Rechner zur Beurteilung von Kurwirkungen, in: *Kurverlaufs- und Kurerfolgsbeurteilung*, W. Teichmann (ed.). Synposion II (Sanitas-Verlag, Bad Wörishofen, 1968), pp. 106-111.
63. R. Günther, E. Knapp and F. Halberg, *Referenznormen der Rhythmometrie: Circadiane Acrophasen von zwanzig Körperfunktionen.* (Sonderband der Z. für Bäder- und Klimaheilkunde) (Schattauer Verlag, Stuttgart, 1969).
64. D. H. Thor, Diurnal variability in time estimation. *Percept. Mot. Skills*, 15, 451-454 (1962).
65. P. Fraisse, M. Siffre, G. Oleron and N. Zuili, Le rhythme veille-sommeil et l'estimation du temps, in: *Cycles Biologiques et Psychiatrie*, J. de Ajuriaguerra (ed.), Symposium Bel-Air III, Geneva, September 1967 (Georg & Co. and Masson & Co., Geneva and Paris, 1969), pp. 257-265.
66. H. Piéron, *The sensations: their functions, processes and mechanisms* (Yale University Press, New Haven, 1952), pp. 293-294.
67. R. S. Kleber, W. T. Lhamon and S. Goldstone, Hyperthermia, hyperthyroidism, and time judgment, *J. Comp. Physiol. Psychol.*, 56, 362-365 (1963).
68. A. D. Baddeley, Time estimation at reduced body temperature. *Amer. J. Psychol.*, 79, 475-479 (1966).
69. H. Hoagland, The physiological control of judgments of duration: evidence for a chemical clock. *J. gen. Psychol.*, 9, 267-287 (1933).
70. C. R. Bell and K. A. Provins, Relation between physiological responses to environmental heat and time judgments. *J. exp. Psychol.*, 66, 572-579 (1963).
71. C. R. Bell, Time estimation and increases in body temperature. *J. exp. Psychol.*, 70, 232-234 (1965).
72. P. Pátkai, Interindividual differences in diurnal variations in alertness, performance and adrenaline excretion. *Acta physiol. scand.*, 81, 35-46 (1971).
73. L. Levi, Physical and mental stress reactions during experimental conditions simulating combat. *Försvarsmedicin*, 2, 3-8 (1966).
74. F. Halberg, Some physiological and clinical aspects of 24-hour periodicity. *J.-Lancet (Minneapolis)*, 73, 20-32 (1953).
75. F. Halberg, E. Halberg, C. P. Barnum and J. J. Bittner, in: *Photoperiodism and Related Phenomena in Plants and Animals*, R. B. Withrow (ed.) (American Association for the Advancement of Science, Washington), Publ. 55 (1959), pp. 803-878.
76. R. J. Wurtman and J. Axelrod, Control of enzymatic synthesis of adrenaline in the adrenal medulla by adrenal cortical steroids. *J. biol. Chem.*, 241, 2301-2305 (1966).
77. N. Kleitman and A. Ramsaroop, Periodicity in body temperature and heart rate. *Endocrinology*, 43, 1-20 (1948).
78. T. O. Nunan and J. D. Sheehan, The direct effect of temperature on heart rate in exercise. *Irish J. med. Sci.*, 3, 436 (1948).
79. N. Kleitman and E. Kleitman, Effect of non-24-hour routines of living on oral temperature and heart rate. *J. appl. Physiol.*, 6, 283-291 (1953).

76   J. N. MILLS

80. J. S. Howitt, J. S. Balkwill, T. C. D. Whiteside and P. D. G. Whittingham, A preliminary study of flight deck work loads in civil air transport aircraft, U.K. Ministry of Defence (Air Force Dept.), *FPRC,* 1240 (1966).
81. J. Rummel, E. Sallin and H. Lipscomb, Circadian rhythms in simulated and manned orbital space flight. *Rass. Neurol. Veg.,* 21, 41-56 (1967).
82. W. J. Crozier, The distribution of temperature characteristics for biological processes; critical increments for heart rates. *J. gen. Physiol.,* 9, 531-546 (1926).
83. P. Wylicil and J. M. Weber, Zirkadianrhythmus des Bronchialwiderstandes. *Med. Welt (Berlin),* 40, 2183-2187 (1969).
84. A. von Domarus, Die Bedeutung der Kammerzählung der Eosinophilen für die Klinik. *Dtsch. Arch. Klin. Med.,* 171, 333-358 (1931).
85. F. Halberg, E. Halberg, D. C. Wargo and M. B. Visscher, Eosinophil levels in dogs with surgically established arteriovenous anastomoses. *Amer. J. Physiol.,* 174, 313-315 (1953).
86. F. Halberg, M. B. Visscher and J. J. Bittner, Eosinophil rhythm in mice: range of occurrence; effects of illumination, feeding and adrenalectomy. *Amer. J. Physiol.,* 174, 109-122 (1953).
87. F. Halberg, H. A. Zander, M. W. Houglum and H. R. Mühlemann, Daily variations in tissue mitoses, blood eosinophils and rectal temperatures of rats. *Amer. J. Physiol.,* 177, 361-366 (1954).
88. G. Pincus, *Recent Progress in Hormone Research,* Vol. I, pp. 123-141 (Academic Press, New York, 1947).
89. R. P. Doe, E. B. Flink and M. G. Goodsell, Relationship of diurnal variation in 17-hydroxycorticosteroid levels in blood and urine to eosinophils and electrolyte excretion. *J. clin. Endocr.,* 16, 196-206 (1956).
90. R. D. Gordon, J. Spinks, A. Dulmanis, B. Hudson, F. Halberg and F. C. Bartter, Amplitude and phase relations of several circadian rhythms in human plasma and urine: demonstration of rhythm for tetrahydrocortisol and tetrahydrocorticosterone. *Clin. Sci.,* 35, 307-324 (1968).
91. F. Halberg, E. Halberg and R. J. Gully, Effect of modifications of the daily routine in healthy subjects and in patients with convulsive disorder. *Epilepsia,* Ser. 3, 2, 150 (1953).
92. A. G. Hills, P. H. Forsham and C. A. Finch, Changes in circulating leucocytes induced by the administration of pituitary adrenocorticotrophic hormone (ACTH) in man. *Blood,* 3, 755-768 (1948).
93. F. Halberg, S. L. Cohen and E. B. Flink, Two new tools for the diagnosis of adrenal cortical dysfunction. *J. Lab. clin. Med.,* 38, 817 (1951).
94. F. Halberg, M. B. Visscher, E. B. Flink, K. Berge and F. Bock, Diurnal rhythmic changes in blood eosinophil levels in health and in certain diseases. *J.-Lancet,* 71, 312-319 (1951).
95. F. Halberg and I. H. Kaiser, Lack of physiologic eosinophil rhythm during advanced pregnancy of a patient with Addison's Disease. *Acta endocr. (Kbh.),* 16, 227-232 (1954).
96. H. D. Kaine, H. S. Seltzer and J. W. Conn, Mechanism of diurnal eosinophil rhythm in man. *J. Lab. clin. Med.,* 45, 247-252 (1955).
97. D. H. Nelson, A. A. Sandberg, J. G. Palmer and F. H. Tyler, Blood levels of 17-hydroxycorticosteroids following the administration of adrenal steroids and their relation to levels of circulating leucocytes. *J. Clin. Invest.,* 31, 843-849 (1952).
98. G. W. G. Sharp, Reversal of diurnal leucocyte variations in man. *J. Endocr.,* 21, 107-114 (1960).

99. G. W. G. Sharp, The effect of light on diurnal leucocyte variations. *J. Endocr.*, 21, 213-218 (1960).
100. L. J. Phillips and L. J. Berry, Hormonal control of mouse liver phosphenolpyruvate carboxykinase rhythm. *Amer. J. Physiol.*, 219, 697-701 (1970).
100a. Ira B. Black and D. J. Reis, Cholinergic regulation of hepatic tyrosine transaminase activity. *J. Physiol.*, 213, 421-433 (1971).
101. Z. K. Cooper, Mitotic rhythm in human epidermis. *J. Invest. Derm.*, 2, 289-300 (1939).
102. A. C. Broders and W. B. Dublin, Rhythmicity of mitosis in epidermis of human beings. *Proc. Mayo Clin.*, 14, 423-425 (1939).
103. L. E. Scheving, Mitotic activity in human epidermis. *Anat. Rec.*, 127, 363 (1957).
104. L. B. Fisher, The diurnal mitotic rhythm in the human epidermis. *Brit. J. Derm.*, 80, 75-80 (1968).
105. A. M. Mauer, Diurnal variation of proliferative activity in the human bone marrow. *Blood*, 26, 1-7 (1965).
105a. Y. A. Romanov and V. P. Rybakov, Duration of mitosis and diurnal rhythm of mitotic activity. *Bull. exp. Biol. Med. U.S.S.R.*, 70, 934-936 (1970).
106. H. R. Mühlemann, T. M. Marthaler and P. Loustalot, Daily variations in mitotic activity of adrenal cortex, thyroid and oral epithelium of the rat. *Proc. Soc. Exp. Biol. (N.Y.)*, 90, 467-468 (1955).
107. F. Halberg, J. H. Galicich, F. Ungar and L. A. French, Circadian rhythmic pituitary adrenocorticotropic activity, renal temperature and pinnal mitosis of starving, dehydrated C mice. *Proc. Soc. Exp. Biol. (N.Y.)*, 118, 414-419 (1965).
108. F. Halberg, Y. L. Tong and E. A. Johnson, Circadian system phase—an aspect of temporal morphology; procedures and illustrative examples, in: *The Cellular Aspects of Biorhythms*, H. von Mayersbach (ed.), Symposium on Rhythmic Research, VIIIth International Cong. Anat., Wiesbaden, 1965 (Springer-Verlag, Berlin, 1967).
109. W. S. Bullough, The energy relations of mitotic activity. *Biol. Rev.*, 27, 133-168 (1952).
110. W. S. Bullough, Stress and epidermal mitotic activity. I. The effects of the adrenal hormones. *J. Endocr.*, 8, 265-274 (1952).
111. W. S. Bullough and E. B. Laurence, The role of glucocorticoid hormones in the control of epidermal mitosis. *Cell Tiss. Kinet.*, 1, 5-10 (1968).
112. F. Halberg and R. B. Howard, 24-hour periodicity and experimental medicine. *Postgrad. Med.*, 24, 349-358 (1958).
113. R. H. Clark and B. L. Baker, Effect of adrenalectomy on mitotic proliferation of gastric epithelium. *Proc. Soc. exp. Biol., N.Y.*, 111, 311-315 (1962).
114. J. M. Brown and R. J. Berry, The relationship between diurnal variation of the number of cells in mitosis and of the number of cells synthesizing DNA in the epithelium of the hamster cheek pouch. *Cell Tiss. Kinet.*, 1, 23-33 (1968).
115. F. Halberg, Periodicity analysis: a potential tool for biometeorologists. *Int. J. Biomet.*, 7, 167-191 (1963).
116. J. F. Sassin, D. C. Parker, J. W. Mace, R. W. Gotlin, L. C. Johnson and L. G. Rossman, Human growth hormone release: relation to slow-wave sleep and sleep-waking cycles. *Science*, 165, 513-515 (1969).
117. T. Litman, F. Halberg, Ellis and J. J. Bittner, Pituitary growth hormone and mitoses in immature mouse liver, *Endocrinology*, 62, 361-365 (1958).

118. L. E. Scheving and J. J. Chiakulas, Effect of hypophysectomy on the 24-hour mitotic rhythm of corneal epithelium in Urodele larvae. *J. exp. Zool.*, 149, 39-43 (1962).

119. L. K. Romanova, Labor and diurnal rhythm of mitotic activity of the cells of the interalveolar septa of the lungs in hypophysectomized rats. *Bull. exp. Biol. Med.*, 62, 1040-1042 (1966).

120. F. Halberg and C. P. Barnum, Continuous light or darkness and circadian periodic mitosis and metabolism in C and D_3 mice. *Amer. J. Physiol.*, 201, 227-230 (1961).

121. M. Garcia-Sainz and F. Halberg, Mitotic rhythms in human cancer, revaluated by electronic computer programs—evidence for chronopathology. *J. Nat. Cancer Inst.*, 37, No. 3, 279-292 (1966).

122. J. D. Rosenbaum, B. C. Ferguson, R. K. Davis and E. C. Rossmeisl, The influence of cortisone upon the diurnal rhythm of renal excretory function. *J. Clin. Invest.*, 31, 507-520 (1952).

123. L. A. de Vries, S. P. ten Holt, J. J. van Daatselaar, A. Mulder and J. G. G. Borst, Characteristic renal excretion patterns in response to physiological, pathological and pharmacological stimuli. *Clin. Chim. Acta*, 5, 915-937 (1960).

124. F. C. Bartter, C. S. Delea and F. Halberg, A map of blood and urinary changes related to circadian variations in adrenal cortical function in normal subjects. *Ann. N.Y. Acad. Sci.*, 98, 969-983 (1962).

125. O. Garrod and R. A. Burston, The diuretic response to ingested water in Addison's disease and panhypopituitarism and the effect of cortisone thereon. *Clin. Sci.*, 11, 113-128 (1952).

126. J. D. Rosenbaum, S. Papper and M. M. Ashley, Variations in renal excretion of sodium independent of change in adrenocortical hormone dosage in patients with Addison's disease. *J. Clin. Endocr.*, 15, 1459-1474 (1955).

127. G. W. Liddle, Analysis of circadian rhythms in human adrenocortical secretory activity. *Arch. intern. Med.*, 117, 739-743 (1966).

128. L. G. Wesson, Electrolyte excretion in relation to diurnal cycles of renal function. *Medicine (Baltimore)*, 43, 547-592 (1964).

129. T. F. Frawley, The role of the adrenal cortex in glucose and pyruvic acid metabolism in man, including the use of intravenous hydrocortisone in acute hypoglycaemia. *Ann. N.Y. Acad. Sci.*, 61, 464-493 (1955).

130. S. H. Ingbar, E. H. Kass, C. H. Burnett, A. S. Relman, B. A. Burrows and J. H. Sisson, The effects of ACTH and cortisone on the renal tubular transport of uric acid, phosphorus and electrolytes in patients with normal renal and adrenal function. *J. lab. clin. Med.*, 38, 533-541 (1951).

131. J. N. Mills and S. Thomas, The acute effect of adrenal hormones and carbohydrate metabolism upon plasma phosphate and potassium concentrations in man. *J. Endocr.*, 16, 164-179 (1957).

132. J. N. Mills and S. Thomas, The influence of adrenal corticoids on phosphate and glucose exchange in muscle and liver in man. *J. Physiol. (Lond.)*, 148, 227-239 (1959).

133. J. C. Laidlaw, J. F. Dingman, W. C. Arom, J. T. Finkenstaedt and G. W. Thorn, Comparison of the metabolic effects of cortisone and hydrocortisone in man. *Ann. N.Y. Acad. Sci.*, 61, 315-323 (1955).

134. J. N. Mills and S. Thomas, The acute effects of cortisone and cortisol upon renal function in man. *J. Endocr.*, 17, 41-53 (1958).

135. R. S. Goldsmith, A. W. Siemsen, A. D. Mason, Jr. and M. Forland, Primary role of plasma hydrocortisone concentration in the regulation of the normal forenoon pattern of urinary phosphate excretion. *J. Clin. Endocr.*, 25, 1649-1659 (1965).

136. F. Bartter and P. Fourman, The different effects of aldosterone-like steroids and hydrocortisone-like steroids on urinary excretion of potassium and acid. *Metabolism*, 11, 6-20 (1962).
137. J. N. Mills, S. Thomas and K. S. Williamson, The effects of intravenous aldosterone and hydrocortisone on the urinary electrolytes of the recumbent human subject. *J. Physiol. (Lond.)*, 156, 415-423 (1961).
138. M. J. Imrie, J. N. Mills and K. S. Williamson, Circadian variations in renal and adrenal function: are they connected? *Mem. Soc. Endocr.*, 13, 3-13 (1963).
139. E. B. Flink and R. P. Doe, Effect of sudden time displacement by air travel on synchronization of adrenal function. *Proc. Soc. Exp. Biol. N.Y.*, 100, 498-501 (1959).
140. P. J. Martel, G. W. G. Sharp, S. A. Slorach and H. J. Vipond, A study of the roles of adrenocortical steroids and glomerular filtration rate in the mechanism of the diurnal rhythm of water and electrolyte excretion. *J. Endocr.*, 24, 159-169 (1962).
141. P. R. Lewis and M. C. Lobban, Patterns of electrolyte excretion in human subjects during a prolonged period of life on a 22-hour day. *J. Physiol. (Lond.)*, 133, 670-680 (1956).
142. P. R. Lewis and M. C. Lobban, The effects of prolonged periods of life on abnormal time routines upon excretory rhythms in human subjects. *Quart. J. exp. Physiol.*, 42, 356-371 (1957).
143. D. T. Krieger, J. Kreuzer and F. A. Rizzo, Constant light: effect on circadian pattern and phase reversal of steroid and electrolyte levels in man. *J. Clin. Endocr.*, 29, 1634-1638 (1969).
144. H. W. Simpson, Studies on the daily rhythm of the adrenal cortex, Ph.D. Thesis, University of Glasgow (1965).
145. R. T. W. L. Conroy, A. L. Elliott and J. N. Mills, Circadian excretory rhythms in night workers. *Brit. J. Industr. Med.*, 27, 356-362 (1970).
146. M. J. Imrie, J. N. Mills and K. S. Williamson, The renal action of small doses of cortisol at night. *J. Endocr.*, 27, 289-292 (1963).
147. R. I. S. Bayliss, Ciba Foundation Colloquia on Endocrinology, 8, 649 (discussion) (1955).
148. J. N. Mills and S. W. Stanbury, Persistent 24-hour renal excretory rhythm on a 12-hour cycle of activity. *J. Physiol. (Lond.)*, 117, 22-37 (1952).
149. J. N. Mills and S. Thomas, Diurnal excretory rhythms in a subject changing from night to day work. *J. Physiol. (Lond.)*, 137, 65-66P (1957).
149a. Ann L. Elliott, J. N. Mills, D. S. Minors and J. M. Waterhouse, The effect of real and simulated time zone shifts upon the circadian rhythms of body temperature, plasma 11-hydroxycorticosteroids and renal excretion in human subjects. *J. Physiol. (Lond.)*, 221, 227-257 (1972).
150. C. T. Nichols and F. H. Tyler, Diurnal variation in adrenal cortical function. *Ann. Rev. Med.*, 18, 313-324 (1967).
151. F. Ceresa, A. Angeli, G. Boccuzzi and G. Molino, Once-a-day neurally stimulated and basal ACTH secretion phases in man and their response to corticoid inhibition. *J. clin. Endocr.*, 29, 1074-1082 (1969).
151a. D. T. Krieger, W. Allen, F. Rizzo and H. P. Krieger, Characterization of the normal temporal pattern of plasma corticosteroid levels. *J. Clin. Endocrinol.*, 32, 266-284 (1971).
152. L. Hellman, F. Nakada, J. Curti, E. D. Weitzman, J. Kream, H. Roffwarg, S. Ellman, D. K. Fukushima and T. F. Gallagher, Cortisol is secreted episodically by normal man. *J. clin. Endocr.*, 30, 411-422 (1970).
153. E. D. Weitzman, D. Fukushima, C. Nogeire, H. Roffwarg, T. F. Gallagher and L. Hellman, Twenty-four hour pattern of the episodic secretion of cortisol in normal subjects. *J. clin. Endocr.*, 33, 14-22 (1971).

154. H. Demura, C. D. West, C. A. Nugent, K. Nakagawa and F. H. Tyler, A sensitive radioimmunoassay for plasma ACTH levels. *J. clin. Endocr.*, 26, 1297-1302 (1966).
155. S. A. Berson and R. S. Yalow, Radioimmunoassay of ACTH in plasma. *J. clin. Invest.*, 47, 2725-2751 (1968).
156. R. L. Ney, N. Shimizu, W. E. Nicholson, D. P. Island and G. W. Liddle, Correlation of plasma ACTH concentration with adrenocortical response in normal human subjects, surgical patients, and patients with Cushing's Disease. *J. clin. Invest.*, 42, 1669-1677 (1963).
157. G. M. Besser, D. R. Cullen, W. J. Irvine and J. Landon, Plasma corticotrophin levels in primary and secondary adrenocortical insufficiency. *J. Endocr.*, 43, x-xi (1969).
158. P. Cheifetz, N. Gaffud and J. F. Dingman, Effects of bilateral adrenalectomy and continuous light on circadian rhythm of corticotropin in female rats. *Endocrinology*, 82, 1117-1124 (1968).
159. A. Heusner, Sources of error in the study of diurnal rhythm in energy metabolism, in: *Circadian Clocks*, J. Aschoff (ed.), Proc. Feldafing Summer School, Sept. 1964 (North-Holland Publishing Co., Amsterdam, 1965).
160. F. Ungar, *In vitro* studies of circadian rhythms in hypothalamic-pituitary-adrenal systems. *Rass. Neurol. Veg.*, 21, 57-70 (1967).
161. F. Ungar and F. Halberg, Circadian rhythm in *in vitro* response of mouse adrenal to ACTH. *Science*, 137, 1058-1060 (1962).
162. F. Ungar, personal communication (1970).
163. G. W. Clayton, L. Librik, R. L. Gardner and R. Guillemin, Studies on the circadian rhythm of pituitary adrenocorticotropic release in man. *J. clin. Endocr.*, 23, 975-980 (1963).
164. M. G. White, N. W. Carter, F. C. Rector, D. W. Seldin, S. J. Drewry, J. P. Sandford, J. P. Luby, R. H. Unger, N. M. Caplan, W. Shapiro and S. Eisenerg, Pathophysiology of epidemic St. Louis encephalitis. I. Inappropriate secretion of anti-diuretic hormone. II. Pituitary-adrenal function. III. Cerebral blood flow and metabolism. *Ann. Intern. Med.*, 71, 691-702 (1969).
165. B. Rüedi, J. Wertheimer, J.-P. Felber and A. Vannotti, Disturbances of plasma cortisol circadian rhythm and of cortisol-ACTH feedback mechanism in central nervous system diseases. *Rass. Neurol. Veg.*, 21, 199-215 (1967).
166. D. T. Krieger, S. Glick, A. Silverberg and H. P. Krieger, A comparative study of endocrine tests in hypothalamic disease. Circadian periodicity of plasma 11-OHCS levels, plasma 11-OHCS and growth hormone response to insulin hypoglycaemia and metyrapone responsiveness. *J. clin. Endocr.*, 28, 1589-1598 (1968).
167. B. B. Saxena, G. Leyendecker, W. Chen, H. M. Gandy and R. E. Peterson, Radioimmunoassay of follicle-stimulating (FSH) and luteinizing (LH) hormones by chromatoelectrophoresis. *Acta endocr. (Kbh.)*, Suppl. 142, 185-206 (1969).
168. S. M. Glicks and S. Goldsmith, The physiology of growth hormone secretion, in: *Growth Hormone*, Int. Cong. Series, No. 158, 84-88, Excerpta Medica Foundation, Amsterdam (1967).
169. K. Takebe, H. Kunita, S. Sawano, Y. Horiuchi and K. Mashimo, Circadian rhythms of plasma growth hormone and cortisol after insulin. *J. clin. Endocr.*, 29, 1630-1633 (1969).
170. D. E. McMillan, J. J. Deller, G. M. Grodsky and P. M. Forsham, Evaluation of clinical activity of acromegaly by observation of the diurnal variation of serum inorganic phosphate. *Metabolism*, 17, 966-976 (1968).

171. T. Lemerchand-Béraud and A. Vannotti, Relationships between blood thyro-trophin level, protein bound iodine and free thyroxine concentration in man under normal physiological conditions. *Acta endocr. (Kbh.)*, 60, 315-326 (1969).
172. P. W. Nathanielsz, A circadian rhythm in the disappearance of thyroxine from the blood in the calf and the thyroidectomized rat. *J. Physiol. (Lond.)*, 204, 79-90 (1969).
173. L. Levi, Sympatho-adrenomedullary and related biochemical reactions during experimentally induced emotional stress, in: *Endocrinology and Human Behaviour*, R. P. Michael (ed.), Oxford University Press: London, 1968, pp. 200-219.
174. A. Cession-Fossion, R. Vandermeulen, P. Lefebvre and J. J. Legros, Variations nychthémérales de la catécholaminémie chez l'homme normal au repos. *Rev. Méd. Liège*, 22, 285-286 (1967).
175. A. Reinberg, J. Ghata, F. Halberg, P. Gervais, Ch. Abulker, J. Dupont and Cl. Gaudeau, Rhythmes circadiens du pouls, de la pression artérielle, des excrétions urinaires en 17-hydroxycorticostéroides catécholamines et potassium chez l'homme adulte sain, actif et au repos. *Ann. Endocr., Paris*, 31, 277-287 (1970).
176. A. H. Vagnucci, A. P. Shapiro and R. H. McDonald, Effects of upright posture on renal electrolyte cycles. *J. appl. Physiol.*, 26, 720-731 (1969).
177. J. L. Claus-Walker, R. E. Carter, H. S. Lipscomb and C. Vallbona, Daily rhythms of electrolytes and aldosterone excretion in men with cervical spinal cord section. *J. clin. Endocr.*, 29, 300-301 (1969).
178. J. L. Claus-Walker, R. E. Carter, H. S. Lipscomb and C. Vallbona, Analysis of daily rhythms of adrenal function in men with quadriplegia due to spinal cord section. *Paraplegia*, 6, 195-207 (1969).
179. H. Sirota, D. S. Baldwin and H. Villarreal, Diurnal variations of renal function in man. *J. clin. Invest.*, 29, 187-192 (1950).
180. S. W. Stanbury and A. E. Thomson, Diurnal variations in electrolyte excretion. *Clin. Sci.*, 10, 267-293 (1951).
181. L. G. Wesson and D. P. Lauler, Diurnal cycle of glomerular filtration rate and sodium and chloride excretion during responses to altered salt and water balance in man. *J. clin. Invest.*, 40, 1967-1977 (1961).
182. M. Toor, S. Massry, A. I. Katz and J. Agmon, Diurnal variations in the composition of blood and urine of man living in hot climate. *Nephron*, 2, 334-354 (1965).
183. J. Anderson, A method for estimating Tm for phosphate in Man. *J. Physiol. (Lond.)*, 130, 268-277 (1955).
184. D. Longson, J. N. Mills, S. Thomas and P. A. Yates, Handling of phosphate by the human kidney at high plasma concentrations. *J. Physiol. (Lond.)*, 131, 555-571 (1956).
185. H. K. Min, J. E. Jones and E. B. Flink, Circadian variations in renal excretion of magnesium, calcium, phosphorus, sodium, and potassium during frequent feeding and fasting. *Fed. Proc.*, 25, 917-921 (1966).
186. R. W. Berliner, Renal mechanisms for potassium excretion. *Harvey Lect.*, 55, 141-171 (1960).
187. D. A. K. Black, H. E. F. Davies, E. W. Emery and E. G. Wade, Renal handling of radioactive potassium in man. *Clin. Sci.*, 15, 277-283 (1956).
188. J. N. Mills, The acute response to potassium ingestion. *J. Physiol. (Lond.)*, 128, 47P (1955).
189. F. A. Harrison, I. R. McDonald and K. Olsson, Unilateral renal excretory responses to close arterial infusions in conscious sheep. *J. Physiol. (Lond.)*, 210, 125-127P (1970).

190. M. Buchsbaum and E. K. Harris, Diurnal variation in serum and urine electrolytes. *J. appl. Physiol.*, 30, 27-35 (1971).

190a. T. Morimoto and K. Shiraki, Circadian variation in circulating blood volume. *Jap. J. Physiol.*, 20, 550-559 (1970).

191. S. J. Rune and N. A. Lassen, Diurnal variations in the acid-base balance of blood. *Scand. J. Clin. Lab. Invest.*, 22, 151-156 (1968).

192. E. S. Barker, R. B. Singer, J. R. Elkington and J. K. Clark, The renal response in man to acute experimental respiratory alkalosis and acidosis. *J. clin. Invest.*, 36, 515-529 (1957).

193. J. N. Mills, Changes in alveolar carbon dioxide tension by night and during sleep. *J. Physiol. (Lond.)*, 122, 66-80 (1953).

194. D. Longson and J. N. Mills, The failure of the kidney to respond to respiratory acidosis. *J. Physiol. (Lond.)*, 122, 81-92 (1953).

195. R. W. Berliner, T. J. Kennedy and J. Orloff, Relationship between acidification of the urine and potassium metabolism. *Amer. J. Med.*, 11, 274-282 (1951).

196. J. N. Mills and S. W. Stanbury, A reciprocal relationship between K^+ and H^+ excretion in the diurnal excretory rhythm in man. *Clin. Sci.*, 13, 177-186 (1954).

197. S. Thomas, Effects of change of posture on the diurnal renal excretory rhythms. *J. Physiol. (Lond.)*, 148, 489-506 (1959).

198. O. Lindan, W. R. Baker, Jr., R. M. Greenway, P. H. King, J. M. Piazza and J. B. Reswick, Metabolic rhythms of the quadriplegic patient. 1. Effect of rhythmic and random feeding and body turning schedule on the hourly excretion pattern of urinary metabolites. *Arch. phys. Med.*, 46, 79-88 (1965).

199. M. D. Lindheimer, R. C. Lalone and N. G. Levinsky, Evidence that acute increase in glomerular filtration has little effect on sodium excretion in dog unless extracellular volume is expanded. *J. clin. Invest.*, 46, 256-265 (1967).

200. C. I. Johnston and J. O. Davis, Evidence from cross circulation studies for a humoral mechanism in the natriuresis of saline loading. *Proc. Soc. exp. Biol., N.Y.*, 121, 1058-1062 (1966).

201. H. E. de Wardener, Control of sodium reabsorption. *Brit. Med. J.*, 3, 611-616 and 676-683 (1969).

202. M. A. Barraclough and I. H. Mills, Effect of bradykinin on renal function. *Clin. Sci.*, 28, 69-74 (1965).

203. J. R. Gill, Jr., K. L. Melmon, L. Gillespie, Jr. and F. C. Bartter, Bradykinin and renal function in normal man: efects of andrenergic blockade. *Amer. J. Physiol.*, 209, 844-848 (1965).

204. M. A. Barraclough, N. F. Jones and C. D. Marsden, Effect of angiotensin on renal function in rat. *Amer. J. Physiol.*, 212, 1153-1158 (1967).

205. M. F. Lockett, Effects of salt loading and haemodilution on the responses of perfused cat kidneys to angiotensin. *J. Physiol. (Lond)*, 193, 639-647 (1967).

206. E. de Bono and I. H. Mills, Intrarenal monitoring of cardiac output in the regulation of sodium excretion. *Lancet, ii*, 1027-1032 (1965).

207. F. C. Rector, Jr., G. Van Giesen, F. Kiil and D. W. Seldin, Influence of expansion of extracellular volume on tubular reabsorption of sodium independent of changes in glomerular filtration rate and aldosterone activity. *J. clin. Invest.*, 43, 341-348 (1964).

208. R. Lichardus and J. W. Pearce, Evidence for a humoral natriuretic factor released by blood volume expansion. *Nature, Lond.*, 209, 407-409 (1966).

209. J. B. Lee and J. F. Ferguson, Prostaglandins and natriuresis: effect of renal prostaglandins on PAH uptake by kidney cortex. *Nature, Lond.*, 222, 1185-1186 (1969).
210. F. S. Wright, B. M. Brenner, C. M. Bennett, R. I. Keimowitz, R. W. Berliner, R. W. Schrier, P. J. Verroust, H. E. de Wardener and H. Holzgreve, Failure to demonstrate a hormonal inhibitor of proximal sodium reabsorption. *J. clin. Invest.*, 48, 1107-1113 (1969).
211. M. F. Lockett and C. N. Roberts, Hormonal factors affecting sodium excretion in the rat. *J. Physiol. (Lond.)*, 167, 581-590 (1963).
212. M. F. Lockett and C. N. Roberts, Some actions of growth hormone on the perfused cat kidney. *J. Physiol. (Lond.)*, 169, 879-888 (1963).
213. K. F. Ilett and M. F. Lockett, A renally active substance from heart muscle and from blood. *J. Physiol. (Lond.)*, 196, 101-109 (1968).
214. M. F. Lockett, Hormonal actions of the heart and of lungs on the isolated kidney. *J. Physiol. (Lond.)*, 193, 661-669 (1967).
215. M. F. Lockett and R. W. Retallack, The influence of heart rate on the secretion of a substance closely resembling the 18-monoacetate of D-aldosterone by the hearts of cats under chloralose anaesthesia. *J. Physiol. (Lond.)*, 208, 21-32 (1970).
216. M. F. Lockett and R. S. Retallack, The release of a renally active substance by perfused rat hearts. *J. Physiol. (Lond.)*, 212, 733-738 (1971).
217. M. F. Lockett and R. W. Retallack, The isolation of a substance very closely resembling the 18-mono-acetate of D-aldosterone from the venous blood of activated muscle and from contracting muscle. *J. Physiol. (Lond.)*, 204, 435-442 (1969).
218. M. F. Lockett, The separation of renal activity from lung and from the venous effluent from perfused lungs. *J. Physiol. (Lond.)*, 212, 719-731 (1971).
219. J. O. Davis, J. E. Holman, C. C. J. Carpenter, J. Urquhart and J. T. Higgins, An extra-adrenal factor essential for chronic renal sodium retention in presence of increased sodium-retaining hormone. *Circulat. Res.*, 14, 17-31 (1964).
220. R. Goldman and E. Luchsinger, Relationship between diurnal variations in urinary volume and the excretion of antidiuretic substance. *J. clin. Endocr.*, 16, 28-34 (1956).
221. J. C. Goodwin, F. A. Jenner, S. E. Slater, The diurnal pattern of excretion of antidiuretic hormone. *J. Physiol. (Lond.)*, 196, 112-113P (1968).
222. T. Zsótér and S. Sebök, Daily variation of antidiuretic substance in the blood serum. *Acta Med. Scand.*, 152, 47-52 (1955).
223. E. Szczepanska, J. Preibisz, K. Drzewiecki and S. Kozlowski, Studies on the circadian rhythm of variations of the blood antidiuretic hormone in humans. *Pol. med. J.*, 7, 517-523 (1968).
224. H. E. Gunn, Jr., A. L. Unger, D. M. Hume and J. A. Schilling, Human renal transplantation: an investigation of the functional status of the denervated kidney after successful homotransplantation in identical twins. *J. Lab. clin. Med.*, 56, 1-13 (1960).
225. G. M. Berlyne, N. P. Mallick, Y. K. Seedat, E. C. Edwards, R. Harris and W. McN. Orr, Abnormal urinary rhythm after renal transplantation in man. *Lancet, ii*, 435-436 (1968).
226. R. McMillan, D. B. Evans and J. G. Lines, Sodium excretion in functioning renal allografts. *Brit. J. Surg.*, 55, 863 (1968).
227. P. R. Lewis, M. C. Lobban and R. I. Shaw, Patterns of urine flow in human subjects during a prolonged period of life on a 22-hour day. *J. Physiol. (Lond.)*, 133, 659-669 (1956).

228. F. A. Jenner, J. C. Goodwin, M. Sheridan, I. J. Tauber and M. C. Lobban, The effect of an altered time regime on biological rhythms in a 48-hour periodic psychosis. *Brit. J. Psychiat.*, 114, 215-224 (1968).

229. J. Aschoff, E. Pöppel and R. Wever, Circadiane Periodik des Menschen unter dem Einfluss von Licht-Dunkel-Wechseln unterschiedlicher Periode, *Pflügers Arch. ges. Physiol.*, 306, 58-70 (1969).

230. J. Aschoff, U. Gerecke and R. Wever, Phase relationship between the circadian period of activity and the body temperature of humans. *Pflügers. Arch. ges. Physiol.*, 295, 173-183 (1967).

231. P. M. Hahn, T. Hoshizak and W. R. Adey, Biosatellite-III results - circadian rhythms of Macaca nemestrina monkey in biosatellite III. *Aerospace Med.*, 42, 295-304 (1971).

232. J. N. Mills, S. Thomas and P. A. Yates, Reappearance of renal excretory rhythm after forced disruption. *J. Physiol. (Lond.)*, 125, 466-474 (1954).

233. R. E. Ornstein, On the experience of time (Penguin, Harmondsworth, 1969).

234. W. B. Quay, Rhythmic and light induced changes in levels of pineal 5-hydroxy indoles in pigeon (Columba livia). *Gen. comp. Endocr.*, 6, 371-377 (1966).

235. C. L. Ralph, L. Hedlund and W. A. Murphy, Diurnal cycles of melatonin in bird pineal bodies. *Comp. Biochem. Physiol.*, 22, 591-599 (1967).

236. D. G. Klein and J. L. Weller, Indole metabolism in the pineal gland: a circadian rhythm in N-acetyltransferase. *Science*, 169, 1093-1095 (1970).

237. R. J. Wurtman and J. Axelrod, 24-Hour rhythm in the content of norepinephrine in the pineal and salivary glands of the rat. *Life Sci.*, 5, 665-669 (1966).

238. R. A. Cohen, R. J. Wurtman, J. Axelrod and S. H. Snyder, Some clinical, biochemical and physiological actions of the pineal gland. *Ann. intern. Med.*, 61, 1144-1161 (1964).

239. S. H. Snyder, M. Zweig, J. Axelrod and J. E. Fischer, Control of the circadian rhythm in serotonin content of the rat pineal gland. *Proc. nat. Acad. Sci., Wash.*, 53, 301-305 (1965).

240. W. B. Quay, Individuation and lack of pineal effect on the rat's circadian locomotor rhythm. *Physiology and Behaviour*, 3, 109-118 (1968).

241. W. B. Quay, Diagnosis of destructive lesions of the pineal; *Lancet*, ii, 42-44 (1970).

CHAPTER 3

Latitude and the Human Circadian System

H. W. Simpson and J. G. Bohlen

Chronobiology Laboratories, University of Minnesota
(Professor Franz Halberg)

ACKNOWLEDGEMENTS

We would like to acknowledge extensive help from Professor Halberg for
the stimulus, discussion and analysis of data in this chapter, and financial
support from the Medical Research Council G970/192/B.

3.1 PROPERTIES OF THE HUMAN CIRCADIAN SYSTEM WHEREBY IT MIGHT BE AFFECTED BY LATITUDE

3.1.1 The Origin of the Oscillators

The term "latitude" is here used in its geographical sense, and the term
"circadian system" implies the sum of circadian subsystems in an
organism.

In order to predict, test, and reveal any effect of latitude on the
human circadian system, it is necessary to review current knowledge of
the origin of circadian rhythms and their control systems whereby they
are externally synchronized with the environment (24 h) and internally
synchronized with each other (fixed phase relations among circadian
subsystems); also one must consider the control systems enabling them
to accommodate changes in stimuli (e.g. assumption of night work).

Whereas in the past some physiologists have tended to regard "diurnal
rhythms" as simple responses to diurnal activity and nocturnal rest,
those specializing in this field—Chronobiology [1]—have now mustered a
formidable amount of direct and indirect evidence for an origin, in part,
from endogenous oscillators. Five lines of evidence serve to illustrate this
point.

First, one might take the evidence from *in vitro* systems: thus the
pacemaker tissue in a rat's heart will continue beating (at about 70/min)
and this beat is modulated by a circadian periodicity whose phase
reflects the habits of the donor animal [2]; similarly the hamster adrenal
in organ culture may exhibit a 24 h periodicity of corticosteroid
biosynthesis for more than a week [3].

Second, in human studies on, for example, life in a deep cave without
a watch, observations of the circadian rhythms of excretion (e.g.
17-OHCS), body temperature, or the adopted voluntary sleep/awake
schedule, indicate that the "24 h" rhythms run slightly but reproducibly
slow and, moreover, there is no definite evidence that the rhythms
progressively dampen off in amplitude [4]. The suggestion here is that a
"natural" frequency of the system is being unveiled by particular
experimental manipulations. The same response is seen in blinded
mice [1] though, in them, the period is less than 24 h.

Third, there is usually a phase delay in re-phasing of the human
circadian system after long-distance flights spanning time zones (e.g. in
the acclimation of the circadian rhythm of body temperature [1].

Fourth, current evidence is against an ability of the human circadian subsystem of temperature or electrolyte excretion to acclimate to day/night cycles differing markedly from a 24 h (e.g. a 21 h) day [5].

Fifth, evidence for the inheritance of circadian oscillators has recently been obtained in *Drosophila melanogaster.* Thus Kanopka and Benzer [6] have induced mutations with ethyl methane sulfonate, isolated mutants which display no rhythmicity, a 19 h rhythm, or a 28 h rhythm (in respect of eclosion and locomotor activity), and finally have mapped the loci of the mutant genes. These findings complement earlier evidence of inheritance such as the persistence of circadian activity rhythms in generations of animals reared in conditions of altered day length or of continuous light or darkness (chickens, lizards, mice: 7).

To this evidence for a series of primary oscillators must be added the fact that rhythms in one subsystem may drive or synchronize those in another; thus, in the mammalian adrenal circadian subsytem, significant circadian rhythmicity has been observed in serum corticosterone, adrenal corticosterone, hypothalamic C.R.F., adrenal reactivity to ACTH, and pituitary ACTH [1]. These points emphasize that 24 h rhythmicity is seen at all levels of organization and, since adrenal glands *in vitro* may exhibit significant periodicity in steroidogenesis [3], there must be a hierarchical control system.

3.1.2 Modulating Influences

(i) Environmental:

While there is good evidence for a series of circadian oscillators it is also clear that mild or drastic modification of these may take place as a result of environmental effects. The heart beat is a good example since the environmental control of the rate tends to override endogenous rhythmicity, and it may take a long time series and computer analysis to reveal a circadian rhythm in pulse rate at all.

(ii) Rhythms with other periods:

Figure 3.1 illustrates the interaction between the menstrual and circadian rhythms of temperature in one healthy woman. Five hundred and thirty-seven body temperature readings were obtained at different times of day and night for seven months by voiding urine on to a quick-reading mercury thermometer (Ovulindex). The time series was analysed by fitting cosine waves (least squares) of menstrual frequencies (5d–40d x 1d) to obtain the best fit by the minimal residual error criterion. The 26d period thus derived was used to combine the menstrual periods and obtain one idealized cycle. Each graphed point represents a three day mean smoothed by a three point moving average. Note that the circadian amplitude is higher when the level is higher; note also that the peaks are later in the second half of the cycle but since this

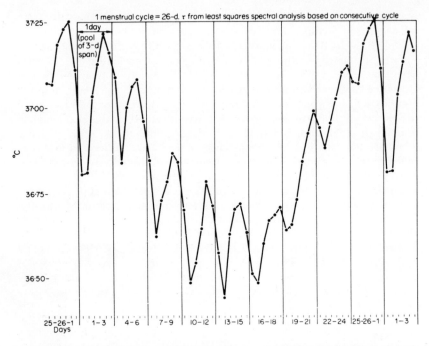

Figure 3.1. Interaction of the circadian and menstrual rhythms of body temperature in a healthy woman.

is balanced by an earlier trough the phase has not materially altered. One would expect a similar pattern for circadian/circannual interactions (e.g. for 17-OHCS excretion) since rhythms with both these periods have been separately described [1]. Another example of positive correlation between amplitude and level is seen in Figure 3.6 (potassium excretion).

3.1.3 Synchronization

(i) *Introduction*

Circadian rhythms, of course, have an environmental counterpart, but rhythms of other periods usually have no such obvious geophysical correlates. Figure 3.1 demonstrates two body temperature rhythmicities, a 26 d menstrual period and a 24 h circadian. The majority of women do not have synchronous menses but they do have synchronous 24 h temperature rhythms with the peak in the latter half of the activity span. This leads us to ask what stimulus brings about the synchronization of the circadian system or, indeed, are there different effectors for different subsystems?

(ii) *Light*

Light is currently regarded as a principal synchronizer in mammals. For example, when men sleep in the dark, and remain blindfold during the first 3 h while up and about, there is a phase delay in the circadian rhythm of water, sodium, and potassium excretion and also leucocyte count [8, 9]. In the rat, light appears to effect ACTH release [10]. If one assumes, therefore, that photic stimuli are mediated through the pituitary, it is not surprising that the sella turcica is smaller in blind humans [11].

Such effects of light on the circadian system might be mediated through the optic tracts and neuroendocrine connections to the pineal, the primitive "third eye" of reptiles: first, because the circadian rhythmicities in the pineal (e.g. serotonin content) are at least partly dependent on light, in primates as well as other laboratory animals [12, 13, 14]; second, the serotonin level may also vary with the oestrous phase [15]; and third there may be changes in gonadal and excretory function in humans with tumours of this area [16]. Also evidence that the pineal may be an integrator or an origin of endogenous oscillations comes from work on sparrows. Pinealectomized birds retain a gross circadian rhythm of activity when subjected to alternations of light and dark but not in a constant environment (though the rhythms do not disappear immediately) [17]. Information about the circadian system in blind human subjects is rather inconsistent and there is a need for comprehensive longitudinal and transverse studies with full time series analyses. At present it seems that at least some circadian rhythms are present in blind people but with variable alteration in amplitude and phase [18, 19]. Reliable estimates of period do not appear to be available and are important in view of Halberg's studies on experimentally blinded mice. These animals displayed a free-running circadian rhythm of body temperature whose period averaged 23.5 h [1]. It is interesting that man's habitually adopted routine begins with reveille about dawn whereas, if artificial light is available, bedtime is usually well after sunset. Rarely does he get up before sunrise and go to bed at sundown. The human failing of rising late and in a hurry, with retiral often delayed, fits the explanation that the "inherited" human circadian frequency (to judge, for example, from cave studies) has a natural period slightly longer than 24 h [4] but is constrained to 24 h by synchronization from routine diurnal activities and nocturnal rest. It also emphasizes the relative importance of events at dawn or reveille as the most potent synchronizing influences.

(iii) *Social synchronizers*

Another question is whether human interactions or "fixed" daily events have a synchronizing effect, consciously or unconsciously. In one cave study involving two groups of women in two tents, members of each group were synchronized with one another at a free-running period characteristic of their own particular tent [4]. However, this is not so remarkable as at first sounds since if one woman had a shorter (nearer 24 h) free-running circadian periodicity of sleep/wakefulness then she might disturb the others on awakening and tend to advance their phases towards hers.

(iv) *Other Synchronizers*

Secondary synchronizers such as 24 h ambient temperature variation have been demonstrated in plants and insects [20, 21] but they can hardly be the important factor in blind people who, as mentioned, may exhibit synchronized 24 h rhythms in the highly controlled house or hospital environments of today.

In animal studies, feeding times, noise or smell may have a synchronizing effect under certain conditions, but they have not yet been proved to be important in man [22].

(v) *Phase Dependence of Synchronizers*

The effectiveness of a synchronizer in altering the phase of a circadian subsystem depends upon the rhythm phase when the synchronizer impinges upon the system. Thus in human 21 h "day" studies, it was noted that when reveille corresponded to the early part of the normal sleep span, no phase adjustment occurred whereas if the reveille corresponded to evening, late sleep, or "day" times, then the adrenal responded with a 21 h rhythm [23]. Similarly phase shifts in the Gonyaulax luminescence rhythm, induced by light "pulses", have been shown to be dependent on circadian phase for direction and amount [24].

3.2 THEORETICAL WAYS IN WHICH LATITUDE MIGHT AFFECT THE CIRCADIAN SYSTEM

3.2.1 Introduction

Various influences which might be associated to some extent with latitude (e.g. cosmic radiation, gravity or magnetic field) will not be considered here since the evidence associating them with the human circadian system is tenuous. Similarly 24 h temperature variation is more a characteristic of climate than of latitude; very large fluctuations have been recorded in Siberia or Alaska while small variations are recorded on the polar pack-ice or in rain forests. Consequently this discussion will centre on the variation in lighting. The amplitude of the 24 h periodicity of light is maximal in equatorial and minimal in polar regions; in contrast the amplitude of the 365 d periodicity of light is maximal in polar and minimal in the equatorial regions. Consequently, in the equatorial regions there is an almost constant circadian photoperiod whereas in others there is a seasonal variation in hour of sunrise and sunset; thus these presumably significant events impinge on different phases of the circadian system over the seasons.

These facts suggest that latitude might affect the circadian system as follows.

3.2.2 Low Latitudes

In low equatorial latitudes, due to the regularity of a mean LD 12 : 12 photoperiod, one might expect relatively high-amplitude circadian rhythms phased in a precise relation to it (or to the habitual routine which will normally be based on it).

3.2.3 Middle Latitudes

In middle latitudes, the smaller 24 h amplitude of light intensity might result in a later phase in the circadian system if brightness level in the morning were important; however social factors (including the effect of putting on the electric light when the alarm clock rings) might counteract this. Moreover the change of photoperiod length with season could well result in effects on the circadian system. Dawn might be a more effective synchronizer at some phases of the circadian system than others; the length of the photoperiod could also be important as it is in plants. If these factors were important one would expect seasonal variations in phase and amplitude.

3.2.4 High Latitudes

In high latitudes, particularly near the poles where there is no 24 h periodicity in light, one might expect increasing lateness of phases with respect, say, to sleep, as the photo-synchronizer decreased in amplitude and the circadian system tended to resume the period longer than 24 h which is characteristic of cave isolation studies. Such effects might appear at the solstices at fairly high latitudes and extend to all seasons at very high latitudes. It is also conceivable that the various circadian sub-systems might be differentially labile to such weak photo-synchronization and hence phase relationships within the circadian system might be altered with respect to those observed at the equator, leading to internal desynchronization..Of course these effects might be "corrected" by a strict social schedule, constant awareness of time, and artificial light, as experienced by a polar meteorologist working during the day; but they could well affect the polar explorer who would only know the time from his watch (or the position of the sun relative to his direction of travel). The explorer sleeping in a tent is, to some extent, continuously influenced by daylight and this is very different from sleeping in a hut which is easily darkened. It is interesting to note here that the British Antarctic Survey made a tent of special black material to cut out light because of sleep disturbances reported by field personnel. Finally, at some critical high latitude (which might vary greatly between subjects and conditions), one would expect the human circadian system to "free run" at a period longer than 24 h, with the possibility that each subsystem might have its own unique period, leading to internal desynchronization. If so, could it, on the basis of an internal desynchronization of, say, C.N.S. amine rhythms, produce specific chrono-pathologies and perhaps precipitate the transient hysterical illness called pibloctoq by the Polar Eskimos?

Another possibility is an amplitude and/or phase modulation of the circadian system by an increased annual rhythm component in high latitudes in a manner similar to the amplitude interaction of the circadian and menstrual temperature cycles seen in Figure 3.1. There is described, for example, both a circadian and circannual rhythm in human 17-hydroxycorticosteroid excretion [25, 26]. Also of great biological interest is the description by Dr. Cook of an annual sex cycle in Eskimos [27]. During the polar night he records that the sexes are indifferent to one another but "there is a grand annual outburst of sexual rage soon after the return of the sun . . . this culminates during the first summer days with what might be called an epidemic of venery, when wives and husbands are frequently exchanged with becoming grace and good intentions. . . . The first discharge (*vaginal*) usually begins during the annual period of sexual excitement. It is rather profuse in quantity and highly coloured with blood. At the same time there are other haemorrhages from the mucous membranes of the mouth or nose. It continues for about five days. Every succeeding monthly flow becomes paler and less profuse, until the advancing night checks the discharges almost entirely. During the very cold and dark periods of the year there is rarely more than an unusual moistening of the vagina for a few days every month". Llewellyn [28], commenting on Cook's observation, noted also that others had verified this fact. He also suggests that the phenomenon is a result of the polar photoperiod and is mediated by the pituitary—"an annual menopause (c.f. hibernants such as bears)". Clearly such a seasonal rhythm could modulate some circadian cyclicities. Though this annual sex behaviour in Eskimos has not been commented on by contemporary observers, it is consistent with a recent statistical study which reveals an annual "surge" of births whose timing is a function of latitude. Moreover the birth rate peak predicted by this study for the appropriate latitude (78°) would be about midwinter. This, of course, fits with an increased number of conceptions in the spring [29].

In the context of evolution one might have expected the indigenous circumpolar peoples to have circadian oscillations of lower amplitude and annual rhythms of correspondingly higher amplitude. The opposite might be expected in equatorial peoples. Evidence that modification of the period of circadian rhythms may occur during evolution is forthcoming from corals; thus the occurrence of annual and daily growth ridges in certain corals suggests a year of 400 "days" (c. 22 h each) since there are 400 ridges/annual ridge. This fits with independent geological evidence of a 400 d year in middle Devonian times (i.e. 350-400 million years ago) [30].

3.3 EXAMINATION OF RHYTHM PARAMETER DATA FROM VARIOUS PARTS OF THE GLOBE

3.3.1 Previous General Observations from Low and Middle Latitudes

An article entitled "Agreement in end-points from circadian rhythmometry on healthy human beings living on different continents" [31] deals with 17-OHCS, potassium excretion, and oral temperature. The emphasis is on circadian rhythm parameters derived by cosinor analysis from different continents, hemispheres and ethnic groups. The data were not considered specifically from the aspect of latitude and so, for our purposes, they have been retabulated to highlight any correlation of:

Latitude and rhythm timing (acrophase) (Table 3.1)
Latitude and rhythm amplitude (Table 3.2)

No clear effect of latitude on acrophase emerges from this wealth of thoroughly analysed data. In contrast, the amplitude for oral temperature appears to increase with latitude. If we include the amplitude of 0.28°C derived from the Wainwright Eskimos (71°N) by JGB (see 3.3.2) the correlation coefficient of amplitude and latitude is 0.92, $P < 0.001$, and the regression equation of amplitude upon latitude is

$$\text{Amplitude,} \, °C = \text{latitude}° \times 0.0019 + 0.1438$$

This gives values of 0.17°C at the equator and 0.31° at the pole. This

Table 3.1

Latitude and timing of circadian rhythms. Summary of cosinor analyses from studies in different countries to highlight any correlation between latitude and the hours elapsing between mid-sleep and acrophase of some circadian rhythms

Latitude (°N or S)	Location of Subjects (usually indigenous)	Number of hours between mid-sleep time and acrophase		
		17-OHCS	Potassium	Oral temperature
		Excretion Rate		
0-10	Surinam	11.0(20)		•
	Ceylon	9.0(7)	11.5(7)	10.5(7)
10-20	Mexico	8.0(6)	10.0(13)	15.0(6)
	Thailand	10.0(13)	—	—
20-30	No data			
30-40	U.S.A.	9.5(25)	7.8(8)	13.0(14)
	Australia		8.5(13)	14.0(13)
40-50	U.S.A.	9.5(13)	10.5(14)	13.5(24)
	Italy	9.0(31)	—	—
	Austria	12.0(50)	10.0(50)	14.5(50)
50-60	France	9.5(31)	9.5(12)	12.0(8)
	Germany	9.5(18)	10.0(18)	—
	Scotland	11.0(20)	—	—

No. of subjects in parentheses. Data from [31].

Significance of rhythms at least $P < 0.05$ and usually much less.

surprising observation may, of course, be due not to rhythm factors *per se* but rather to the fact that the cool northern climes are more conducive to exercise. It is relevant to note here that the relative contribution from endogenous vs. environmental sources has been previously assessed by a least squares spectral analysis on temperature data in humans living on a 21 h "day" [32]. In that 3 wk study, the amplitudes at around 24 h were $0.16°C$ and $0.27°C$ for the first and second 1.5 weeks and the corresponding values at a period of 21 h were $0.17°C$ and $0.24°C$. This demonstration of substantial endogenous and environmental contributions to amplitude illustrates one way in which latitude and amplitude might possibly correlate, i.e. a variable environ-

Table 3.2

Summary of cosinor analyses from many studies in different countries to highlight any correlation between latitude and amplitude of some circadian rhythms

Latitude (°N or S)	Location of Subjects (usually indigenous)	Amplitude of 24 h cosine approximating function		
		17-OHCS (μg 17-OHCS/min.)	Potassium μ mole/min	Oral temperature °C
0-10	Surinam	1.5(20)	—	—
	Ceylon	2.3(7)	84(7)	0.13(7)
10-20	Mexico	1.0(6)	60(13)	0.18(6)
	Thailand	1.0(13)	—	—
20-30	No data			
30-40	U.S.A.	2.0(25)	84(8)	0.21(14)
	Australia	—	132(13)	0.22(13)
40-50	U.S.A.	2.2(13)	102(14)	0.25(24)
	Italy	2.8(31)	—	—
	Austria	4.8(50)	90(50)	0.25(50)
50-60	France	1.1(31)	84(12)	0.22(8)
	Germany	2.8(18)	78(18)	—
	Scotland	3.2(20)	—	—

No. of subjects in parentheses.

Same source as Table 3.1. Significance of rhythms $P < 0.05$ and usually much less.

Since different analytical methods for 17-OHCS were used by different investigators, some caution is necessary in the interpretation of the results.

mental component. In any event the observation remains an interesting one for future investigators to confirm or deny—especially against the usual prediction of decreasing circadian amplitude with latitude.

The fact that no other latitude effects emerge might mean that latitude has little association with circadian rhythmicity; but that conclusion may well be premature for the following reasons:

(1) The subjects were uncontrolled for e.g. age, sex, menstrual phase.
(2) The studies were uncontrolled for e.g. season, amount of exercise, diet, etc.
(3) High latitudes are only represented by the Wainwright Eskimos.

3.3.2 Previous specific observations of circadian rhythm parameters determined in equatorial Amerindians, natives of the Galapagos, Londoners, and Wainwright Eskimos

Our own studies (H.W.S.) of the circadian rhythm in 17-hydroxy-corticosteroid excretion in equatorial Amerindians and in Scottish students revealed remarkably similar circadian rhythm timing when midsleep was used as reference [33]. This observation is in agreement with Lobban's conclusion that there was no significant difference in the phase or amplitude (relative) of potassium excretion between residents (12 children; 23 adults) indigenous to the Galapagos (on the Equator) and 22 subjects resident in London (Lat. 52°) [34]. However Lobban [35] found a diminished relative amplitude in the circadian rhythm of water, sodium, potassium and chloride excretion in 8 Eskimos at Wainwright (71°N) in July when compared with Arctic Indians in a rather similar latitude (67°N) and British controls temporarily in Spitsbergen (78°N): she suggests an ethnic rather than a latitude effect to account for the difference in circadian rhythm amplitudes. However, the sampling frequency was less in the Eskimos (c.f. her Fig. 2 in ref. 35 and Fig. 5 in ref. 36). Hence the observation of rhythm "damping", that the differences between day and night excretory rates have virtually disappeared, might have been due, at least in part, to the integrating effect of the urinary bladder which results from infrequent sampling.

3.3.3 Circadian Rhythm Parameters in two Britons who made Self Observations near the Equator one year and in the High Arctic the next year (both around the summer solstice)

(i) *Introduction*

Summer circadian rhythm parameters (Table 3.3) were obtained from H.W.S. and his wife who made eight urine collections daily for one week while on holiday in St. Lucia (lat. 14°N) in 1968 and a further series in Devon Island (lat. 76°N) in 1969 for three weeks: comparison can thus be made between data from two latitudes, derived from the same two subjects at the same time of year and nearly the same longitude, and only one year apart. Because the subjects lived in a luxury hotel in St. Lucia (the prize for a newspaper competition) with one entire wall forming a window and also a glass door opening directly on to the beach, they were well aware of the natural lighting. In Devon Island they were living in a double tent which attenuated but did not exclude natural light. The sleep/awake routine was reasonably similar in both, the midsleep time in St. Lucia being 02.40 whereas in Devon Island it was 03.00. Exercise was also similar in both, being short walks in morning and afternoon, with bathing in addition in St. Lucia. Diet was not

Table 3.3

Circadian rhythm parameters in 2 adult subjects (1♂ 37-38y + 1♀ 38-39y) who made self observations near the equator (June/July 1968 for 1 week, 7 obs/day) and in a high polar latitude (June 1969 for 3 wks, 8 obs/day)

°N Latitude	Subject	Excretory variable			
		Potassium		Sodium	
Circadian rhythm parameters		14°	76°	14°	76°
μ mole/minute {Co = Level}	HS	56	52	135	228
	MLS	48	50	102	156
{C = Amplitude}	HS	47	22	70	97
	MLS	39	23	53	65
C/Co × 100 = Rel. Amp.	HS	84%	42%	51%	43%
	MLS	81%	46%	52%	42%
Phase {Φ ref. 00°° / 24 h = 360°}	HS	−186 −160 to −206	−229 −215 to −244	−197 −170 to −222	−274 −260 to −288
	MLS	−167 −140 to −186	−226 −210 to −243	−186 −140 to −225	−269 −253 to −284
φ ref. hours from midsleep, with 95% confidence limits	HS	9.8 8.1 to 11.1	12.3 11.3 to 13.3	10.5 8.7 to 12.2	15.3 14.3 to 16.2
	MLS	8.5 6.7 to 9.8	12.1 11.0 to 13.2	9.8 6.7 to 12.4	14.9 13.9 to 15.9

controlled but Table 3.3 reveals similar daily potassium excretion rates in the two locales.

(ii) *Potassium excretion*

It is therefore of interest that a large difference in amplitude (both absolute and relative) is found between the equatorial and polar study amounting to a 50% damping in the latter (Table 3.3). Also, consistent with the predictions in 3.2.2 and 3.2.3, is the finding that the acrophases for the two subjects are, on average, 3.1 h later from midsleep in the Arctic.

(iii) *Sodium excretion*

For sodium excretion there is an unfortunate difference in level of output between the studies (due to the high salt content of the polar rations) and hence interpretation of those amplitudes is problematic. The relative amplitudes were lower in the polar study, but interpretation of these lower amplitudes depends upon acceptance of the phenomenon, which we and others have noted previously in circadian rhythm studies, that there is a positive correlation between amplitude and level in electrolyte studies when the level is within the physiological range (see Fig. 3.7). Amplitude and level have a less close connection with phase, and it is of interest that the interval between midsleep and the peak of the cosine function approximating the rhythm is much longer in the polar study (average 5 h). The standard errors indicate that this increase in interval is significant.

3.3.4 A Comprehensive Circadian Rhythm Study of Wainwright Eskimos with Observations at all four seasons

(i) *Introduction*

One of us (J.G.B.) made a study of eight adult Eskimos (5M + 3F) in Wainwright, Alaska (lat. 71°N) as a thesis project [37]. Variables observed included urinary excretion rates, oral temperature, heart rate, blood pressure, time estimation, eye-hand skill, and hand-grip strength. Some detail of the numbers of observations and the age and sex of the subjects is seen in Table 3.4. Since observations were made every 2 h by "day" and 8 h by "night" the length of the individual observation spans may be worked out. A principal objective of the study was to evaluate the relative prominence of any rhythmic circadian and/or circannual components. The subjects were studied intensively for one month at the four seasons i.e. the solstices and equinoxes because of the difficulty of continuous sampling throughout the year. (See Table 3.4.)

Table 3.4

Dates of collection and number of measurements of each variable during each season

Subject	Age	Summer 28.7.68- 2.8.68	Winter 15.12.68- 31.12.68	Spring 14.3.69- 2.4.69	Summer 13.6.69- 7.7.69	Autumn 15.9.69- 6.10.69	Totals
Males							
1	(13)		52-58	80-81	78-80	83-85	293-304
2	(14)	20	58-65	80	85-87		243-252
3	(43)	20	30-32	59-64	73-74	33-34	215-224
4	(61)	18	47	73-75	70-78	75-76	283-294
5	(63)	20	50-51	77-78	79-80	76-77	302-306
Females							
6	(13)		44-46	75-76	79-80		198-202
7	(57)	19	49-50	79	77-80	74-77	298-305
8	(57)	20	58-63	78-79	80	77-78	313-320

Rhythm parameters in these data were analysed by the mean cosinor method and the results are depicted in Fig. 3.2.

(ii) *Oral temperature*

The analyses reveal, first, that the Eskimos have a circadian rhythm of temperature with a period at or close to 24 h and with phase and frequency synchronized between individuals; second, they show that the amplitudes are similar at all seasons and that they are greater than those reported in studies at lower latitudes (Table 3.2). Certainly there is no "damping" of the circadian amplitude of temperature even though the data were collected well north of the Arctic Circle. Further support for this finding comes from observations on ourselves (4-5 subjects), while encamped on Devon Island (lat. $76°$N) a few degrees further North than Wainwright ($71°$N). These, analysed by weekly cosinor, revealed amplitudes of 0.33, 0.33, 0.43 and $0.36°$C. All of these values are substantially greater than those listed in Table 3.2 from more southerly latitudes. Third, the circadian cosinor-derived temperature acrophases of the Wainwright Eskimos are as follows (hours after midsleep): Winter 14.7 h, Spring 12.4 h, Summer 14.4 h, Autumn 12.3 h. These values fall within the range of acrophases for inhabitants of more southerly latitudes (10.5 h to 15.0 h, see Table 3.1) and therefore we have no clear evidence here that the phasing of the Eskimos' circadian temperature is different though the solstice figures are at the upper limit. Fourth, there is a possible seasonal variation in the phasing since the acrophases are later in midsummer and midwinter than in spring and autumn. The

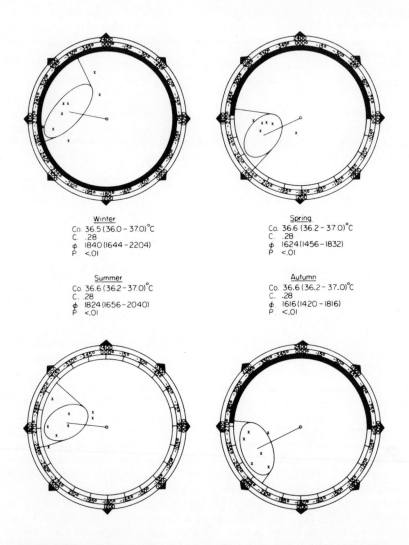

Winter
Co. 36.5 (36.0 – 37.0)°C
C. .28
φ   1840 (1644 – 2204)
P   <.01

Spring
Co. 36.6 (36.2 – 37.0)°C
C. .28
φ   1624 (1456 – 1832)
P   <.01

Summer
Co. 36.6 (36.2 – 37.0)°C
C. .28
φ   1824 (1656 – 2040)
P   <.01

Autumn
Co. 36.6 (36.2 – 37.0)°C
C. .28
φ   1616 (1420 – 1816)
P   <.01

Figure 3.2 Cosinor summaries of oral temperature rhythm in Wainwright Eskimos. Crosses represent individual subjects. Ellipses and parentheses indicate 95% confidence limits for the group, and P the probability that there is no phase and frequency synchronized rhythm for the group. φ is acrophase in clock hours; to use midsleep as reference subtract 4 h. Black portion of inner circle indicates hours of darkness.

reality of this shift is strengthened by a similar shift in the potassium acrophases and it is tempting to suggest that this could be associated with the more noticeable equinoctial sunrise and sunset, providing a more effective synchronizer.

(iii) *Potassium excretion*

Cosinor results for potassium excretion from the 2237 urine samples collected from 8 Eskimos (Fig. 3.3) show first that a rhythm was demonstrable at all four seasons with phase and frequency synchronized between individuals and with relative amplitudes (amplitude/level) of 40%, 45%, 40%, 52%. These are similar to the potassium excretion data from ourselves (Caucasians) a few degrees further north in Devon Island (lat. 76°N cf. Wainwright 71°N): 42%, 46% for the subjects referred to in Table 3.3 and 38% for a third adult subject; and they are about half the amplitude obtained for the equatorial data (Table 3.3) from the same subjects. Second, cosinor analysis shows the potassium acrophases for the Wainwright Eskimo group (71°N) to be as follows: winter 11.8 h, spring 10.1 h, summer 12.0 h, and autumn 10.1 h after midsleep. The summer figure of 12 h also bears a close similarity to our own data from Devon Island (76°N) collected at the same season i.e. 12 h, 12 h, and 14 h estimated by least squares in three individuals. The midsleep-acrophase intervals at the solstices are larger than those in Table 3.1 from many different latitudes, 7.8 h-11.5 h, for which the season was not stated, whilst the intervals obtained at the equinoxes fall within the range of Table 3.1.

(iv) *Sodium excretion*

The circadian cosinors of sodium excretion by the Wainwright Eskimos show a similar trend to those for potassium. The acrophases are as follows: winter 12.7 h, spring 11.8 h, autumn 12.2 h. Though the summer cosinor is not significant at the 5% level it is clear from the visual plot of Fig. 3.4 that the acrophases of most Eskimo subjects are later than at the other seasons. The average vector indicates an acrophase 15 h after midsleep and this compares very closely again with our Devon Island (76°N) data for 3 Britons collected at a similar season (i.e. 15.1 h, 14.9 h and 18.0 h); two of the same adults, as mentioned above (3.3.3) had substantially earlier acrophases near the equator (10.5 h and 9.8 h).

The relative amplitudes are somewhat less at all seasons than those recorded in ourselves in summer (Eskimos 27%, 25%, 17%, 26% at lat. 71°; ourselves 43%, 42%, 33% at lat. 76°, but 51%, 52% at lat. 14°. It would be possibly premature to interpret a real difference here between Eskimos and ourselves, since the Eskimo figures are based on group

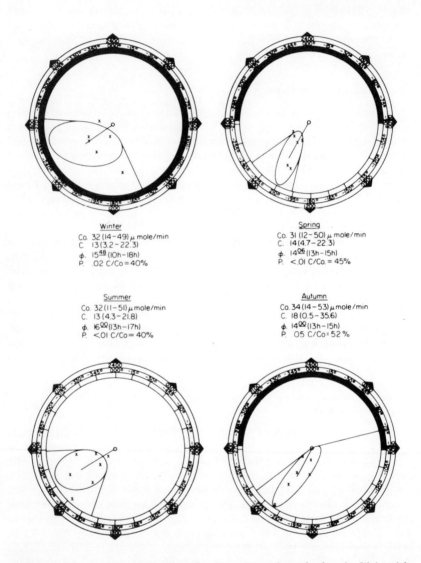

Winter
Co. 32 (14 – 49) μ mole/min
C.   13 (3.2 – 22.3)
φ.   15$\frac{48}{}$ (10h – 18h)
P.   .02 C/Co = 40%

Spring
Co. 31 (12 – 50) μ mole/min
C.   14 (4.7 – 22.3)
φ.   14$\frac{06}{}$ (13h – 15h)
P.   < .01 C/Co. = 45%

Summer
Co. 32 (11 – 51) μ mole/min
C.   13 (4.3 – 21.8)
φ.   16$\frac{00}{}$ (13h – 17h)
P.   < .01 C/Co = 40%

Autumn
Co. 34 (14 – 53) μ mole/min
C.   18 (0.5 – 35.6)
φ.   14$\frac{00}{}$ (13h – 15h)
P.   05 C/Co = 52 %

Figure 3.3. Cosinor summaries of urinary potassium rhythm in Wainwright Eskimos. Conventions as in Fig. 3.2.

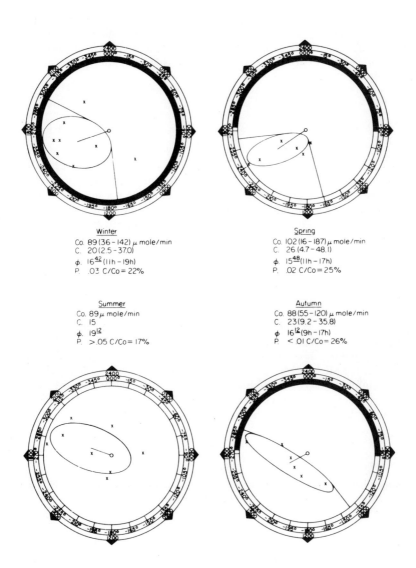

Winter
Co. 89 (36 - 142) μ mole/min
C. 20 (2.5 - 37.0)
φ. 16⁴² (11h - 19h)
P. .03 C/Co = 22%

Spring
Co. 102 (16 - 187) μ mole/min
C. 26 (4.7 - 48.1)
φ. 15⁴⁸ (11h - 17h)
P. .02 C/Co = 25%

Summer
Co. 89 μ mole/min
C. 15
φ. 19¹²
P. > .05 C/Co = 17%

Autumn
Co. 88 (55 - 120) μ mole/min
C. 23 (9.2 - 35.8)
φ. 16¹² (9h - 17h)
P. < .01 C/Co = 26%

Figure 3.4. Cosinor summaries of urinary sodium rhythm in Wainwright Eskimos. Conventions as in Fig. 3.2.

cosinor estimates at the four seasons while ours are individual least squares estimates. For example a between-Eskimo difference in phase will result in a lower amplitude estimate. One provisionally concludes that the arctic environment may be associated with later phases, relative to midsleep, and with lower amplitudes.

3.3.5 A Longitudinal Winter Study of an Adult Male Eskimo at Resolute Bay (75°N)

We asked the school janitor at Resolute Bay, 75°N, if he would collect regular urine samples for us. He was 3/4 Eskimo, 1/4 whaling stock. His adherence to the regular school routine and his literacy made him a good subject for observation. The study was supervised during two visits by one of us and also by the village nurse and administrator. The project began in late September and apart from two gaps (one of nearly a month in October; the other of two weeks around Christmas), it was complete until early February. The data have been statistically analysed in two ways. First, a least squares spectral analysis was carried out in which cosine waves in the circadian range (20-28 h) were fitted to the whole time series of each variable and, second, a serial section analysis was also performed (24 h cosine wave with 20 d interval, 24 h increment). The results of the former analysis are summarized in Table 3.5. This least squares spectral analysis revealed sharply defined highly significant 24 h rhythms for all variables examined, urinary potassium, sodium, chloride, and volume, except for calcium (see 3.3.6). It is highly relevant that the acrophases of these variables, except Ca, are remarkably consistent through the winter season and there is no evidence of phase drift.

Table 3.5

Winter circadian rhythm parameters in one adult male Resolute Bay (75°N) Eskimo who made collections of his urine from September 1968 to February 1969. Least squares spectral analysis of urine variables

	Level = Co	Amplitude = C	C/Co%	Φ*	P=	N=
Sodium	88 μmole/min	41	46	13.7	<0.01	290
Chloride	62 μmole/min	18	29	11.9	<0.01	324
Volume	1.06 ml/min	0.35	32	13.7	<0.01	368
Potassium	20 μmole/min	12	59	9.6	<0.01	323

* Hours elapsed between midsleep and acrophase.

Fig. 3.5 is representative. The potassium acrophase (10 h after midsleep) is earlier than the group estimates for the Wainwright Eskimos for winter (nearly 12 h after midsleep) and his relative amplitude (59%) much greater (40%). He is, however, well within their 95% confidence interval. It seems likely that any effects of latitude were obscured by his regular school routine.

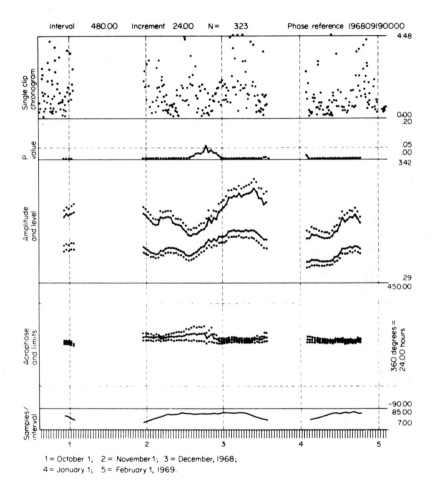

1 = October 1;   2 = November 1;   3 = December, 1968;
4 = January 1;   5 = February 1, 1969.

Figure 3.5. Serial section of urinary potassium in one healthy Canadian Eskimo. For further explanation see Ch. 1, pp. 15-19.

3.3.6 Observations of the Circadian and Circannual Subsystems of Calcium Excretion in the Wainwright Eskimos (lat. 71°N) and in a Resolute Bay Eskimo (lat. 75°N)

JGB studied calcium excretion rates in the same Eskimos as in 3.3.4 (Wainwright 71°N) at the solstices and equinoxes [37]. Cosinors with a period of 24 h failed to show a phase and frequency synchronized rhythm at the 5% level at any season, but the fairly close grouping of individual acrophase estimates between 08.00 and 14.00 suggested that with a larger number of subjects this significance might have been obtained. The fit of a 24 h cosine for each individual to the short time series representing each season was significant at the 5% level on 19/40 occasions (and at the 1% level on 14 of these), which is more than might have been expected by chance if no rhythm really existed. One concludes that a circadian rhythm in calcium excretion is masked by considerable "noise", but may be statistically validated for at least some of the Eskimos, and that individual variation of parameters probably occurs. There may also be an annual rhythm, for when a cosine wave with a period of 365 d was fitted to each of the 10 individual data spans

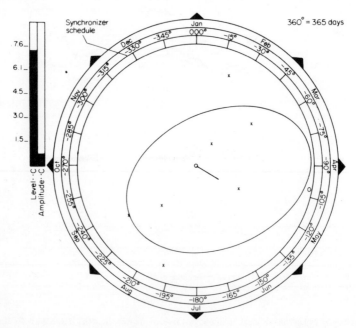

Figure 3.6. Circannual cosinor of urinary calcium excretory rate in eight Wainwright Eskimos. Crosses represent individuals, ellipse indicates 95% confidence limits for the group.

from the four seasons, in 8/10 of the analyses there was a significant
rhythm (P < 0.01); however, the group cosinor analysis of these was not
significant because the acrophase in different individuals varied from the
middle of January to nearly September (Fig. 3.6). Further interest in
calcium is raised by our study of the school janitor at Resolute Bay
(75° N) (3.3.5). Analysis of his calcium excretion in the urine (Fig. 3.7)
revealed an apparent rhythm with a period of 22.9 h. In the least squares
spectral analysis this component is significant at P < 0.009, and there is
no component at 24 h or indeed any other significant component in the

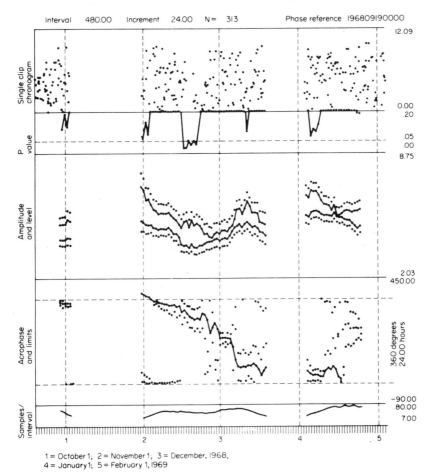

1 = October 1; 2 = November 1; 3 = December, 1968,
4 = January 1; 5 = February 1, 1969

Figure 3.7. Serial section of urinary calcium in one healthy Canadian Eskimo. For
further explanation see Ch. 1, pp. 15-19.

spectral range 20-28 h. Particularly important is the fact that urine volume, potassium and sodium all show precise 24 h rhythms of high amplitude in the same samples so failure to find that period in calcium is not a sampling artefact. The demonstration of a "free running" circadian periodicity with a period shorter than 24 h in man on a normal routine is almost without precedent and obviously it will have to be confirmed. This unusual period for a man raises the possibility of circadian desynchronization and the attendant possibility of "beat phenomena" with other circadian subsystems.

Interestingly enough, a "calcium deficiency" hypothesis has been suggested by various arctic authorities as a basis for the hysterical disease "Pibloctoq", but unfortunately based only on the supposed similarity between the attacks and "tetany" and with no evidence of abnormal blood calcium [38]. Moreover osteomalacia and rickets are rare in Eskimos. As Wallace points out, though, relative hypocalcaemia might compound with an inborn epileptogenic tendency, possibly exacerbated by inbreeding. Alternatively an epileptogenic focus might result from chronic ear infection, which is common in Eskimos or from *Trichinella spiralis* cysts in the brain from eating walrus or bear meat.

3.3.7 1969 North Pole Expedition. Circadian Rhythm Studies on the Personnel

Probably the most northerly systematic study of urinary circadian rhythms was that carried out by a North Pole Expedition (1969) organized by one of us (H.W.S.). Three adults (2M, 1F) travelled on skis from northernmost Canada (83° N) to 84° 42, before turning back after the radio generator burnt out since there was no air support for the field party [39, 40]. In two respects the study was far from ideal: the party were on alternating 12 d spans of isocaloric diets of high and low protein for another project; also, because of the extremely difficult conditions, there were only 3-5 urine voids/day with the inevitable consequence of a damping in amplitude of any urinary rhythm. On the other hand the party were living in the natural polar lighting being either out on the ice (Plate 3.1) or in a translucent tent. After polar "dawn" on March 6th, light seemed continuous since the party were so near the pole and because the surface of the pack was so reflective. The span can therefore be considered LL.

Chronograms of the excretion rates pooled for each subject for this LL span of the expedition (March 6th to April 6th 1969) are seen in Fig. 3.8. It is at once apparent that while the timing of subject H.W.S.'s circadian subsystems agrees reasonably with the routine (activity in the daytime clock hours, recumbency in the "night" hours), equally the timing of subject R.T.'s rhythms is completely out of phase with

Roger Tufft and Myrtle Simpson at 84°N sledging towards the North Pole

[To face page 108

routine. Subject M.S.'s phasing seems internally and externally desynchronized; but her results must be viewed with caution in view of the smaller number of urine collections.

The urine volume, 17-hydroxycorticosteroids, sodium, potassium and temperature have been analysed by subject and variable, using the least

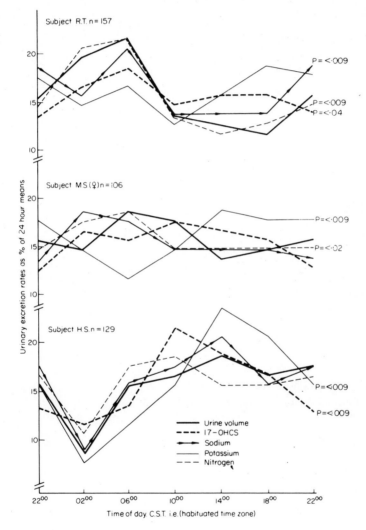

Figure 3.8. North Polar Expedition 1969. Urinary excretion rate calculated over 4 h intervals as percentage of 24 h mean.

squares spectral analysis and fitting cosine waves from 20-28 h, again over the span March 6th to April 6th. This analysis reveals that the best fitting periodicities in the spectral range 20-28 h are fairly evenly scattered around 24.0 h (i.e. 8 below 24 h, 2 at 24 h and 5 above 24 h) but with 9/15 in the 23.5 to 24.5 range. The distribution of scatter does not support a "free running" rhythm response with a period longer than 24 h since, in that event, one would normally expect the periodicities to be grouped above 24 h. A tabulation of the significant results is given in Table 3.6. It will be noted that the number of data points only averages

Table 3.6

North Pole Expedition 1969

Tabulation of significant circadian rhythmicities revealed by least squares spectral analysis of all urinary variables. Parameters quoted are for the 24 h cosine fit but T indicates the period of the best-fitting cosine wave in the circadian spectrum. ϕ reference is $00^{\circ\circ}$. Time series from March 6th-April 6th 1969.

	P	Co	C	C/Co	ϕ	T
Subject HS, ♂, 38y N = 129						
17-OHCS excretion	<0.009	8.4	3.5	42%	12^{31}	23.9
Potassium excretion	<0.009	36	13	36%	15^{49}	24.0
Urine temperature	<0.009	36.3	0.32	–	12^{39}	24.1
Subject RT, ♂, 38y N = 157						
17-OHCS excretion	<0.04	13.1	2.3	17%	04^{39}	24.1
Sodium excretion	<0.009	158	56	35%	03^{47}	22.9
Nitrogen excretion	<0.009	13.3	6.1	46%	03^{28}	24.0
Subject MS, ♀, 39y N = 106						
Potassium excretion	<0.009	30	7	24%	17^{42}	24.1
Nitrogen excretion	<0.02	6.9	1	14%	03^{40}	23.6

3-5/d. The analyses provide evidence that subject H.W.S. had three well-defined urinary rhythms which could be described as relatively normal. Subject R.T. also has evidence of several significant circadian rhythms and it is clear that while circadian rhythms of these variables seem to be present, their phases indicate a peak during the sleep span which is highly unusual.

The third subject has two significant circadian rhythms both of which have phase estimates rather later than usual.

In summary, then, a field study of urine variables in three polar explorers indicated continuing circadian rhythmicity at high latitude on the polar pack ice. Though the period estimates did not support the possibility of "free running" circadian rhythmicity, nevertheless in some instances the peak phases were in a completely different position relative to the sleep/awake routine than is ordinarily observed. These rather dramatic phase abnormalities may be a clue to the effect of the lighting in the very High Arctic. Any firm conclusion must, however, be tentative in view of the infrequent urine sampling schedule.

3.4 REFERENCES IN THE NORTH POLAR LITERATURE TO TEMPORAL DISORIENTATION AND MENTAL AND PHYSICAL CHANGES WHICH MIGHT BE DUE TO THE EFFECTS OF THE ARCTIC LIGHTING ON THE CIRCADIAN SYSTEM

3.4.1 Temporal Disorientation

An entry in the diary of a member of the Soviet North Pole Expedition [41] serves to emphasize the eerie feeling of timelessness experienced by travellers in high polar latitudes:

> "28th May, 1937. We lost all consciousness of time. It was light all the twenty-four hours; whatever the time, the sun was always at the same height, there was neither East nor West nor North. Everywhere, in all directions, on all sides, there is only South. Often when we woke up we would wonder; is it 4 a.m. or p.m.? Our problem was solved by Feodorov, who had a twenty-four hour chronometer in his tent. We applied to him to settle what time of day or night it was."

When travelling in high latitudes it is a matter of common personal observation that some time disorientation occurs, not to such an extreme extent as in the example quoted above but rather as a tendency to go to bed later each day. For example, when asked about the effects of the midnight sun, the manager of Resolute Bay airstrip (75°N in N.W.T. of Canada) volunteered: "I decided to do a routine inspection of our fire services and suddenly I noticed it was 4 a.m.". The point is, of course, that one doesn't usually carry out a routine task in the middle of the night. This hint of a tendency to later retiral calls to mind Quay's study on the primate *Macaca mulatta* in which he showed that if the lights were left on in the evening the pineal serotonin continued to rise; a fall is normally noticed during sleep [13]. A further hint of a possible effect of the polar lighting may be deduced from the diary of Dr. Frederick Cook [42]. After returning from his polar expedition he was forced to spend the winter of 1909 on Devon Island (76°N) in exceptionally primitive circumstances. Though he was a good diarist, it is of interest that during that winter (or possibly in the preceding months since he was out of touch for 14 months) his diary entries are three days short. We have no means of knowing whether he just forgot these days but this could be a clue that the sleep/awake routine was following the natural human circadian periodicity which is longer than 24 h. Elsewhere he makes a comment about one of his companions. "About once a month Rudolph gets a notion that we have lost a day and so we watch with eager interest the phases of the moon to keep tally on our date." Nansen [47] also writes on the effect of the continuous summer lighting: "We slept the clock round without knowing it, and I thought it was six in the morning when I turned out. When I came out of the tent, I thought there was something remarkable about the position of the sun, and pondered over it for a little while, until I came to the conclusion that it was six in the evening, and that we had slumbered for twenty-two hours". Nansen interprets this as the long sleep of tired men; but was it? Is this a hint that the continuous daylight had caused his and his companions' rhythms to run free, unsynchronized by the sun, with a period longer than 24 hours?

3.4.2 Mood Changes

Apart from time disorientation, the first-hand accounts of explorers contain many references to personality and mood changes. Indeed one is left in little doubt that the human difficulty in reaching the poles was not only due to the challenge of logistics, to the cold, and to scurvy, but also to mood changes in the men themselves. Of course it is problematical whether such changes can really be ascribed to the effect of light on the synchronization of the circadian system or to more obvious factors such as privation, cold, subnutrition, confinement or isolation. One of us, (H.W.S.) as a medical officer on a British Antarctic base for over two years, was impressed with the effect of lighting on the general mood of the station; on a clear calm sparkling day everyone responded by being cheerful, bright and enthusiastic whilst the opposite was seen on grey gloomy days. To some extent the institution of celebrations at midwinter could well be a constructive attempt to counter the gloom of the polar night. One does not have to go to polar regions to see this, but there the effect is increased by the high reflectivity of the snow and ice surface, especially in the spring when fresh snow tends to fall in many areas. Dr. Cook, after nine winters at 78° N on the N.W. Coast of Greenland, described these lighting effects to the Brooklyn Medical Society in 1896 [27]. It should be pointed out that Cook was also the first doctor to winter in the Antarctic and therefore is well qualified to speak on the effects of the polar "night".

> "The light of the Arctic summer is in itself an interesting study. The sun, always low on the sky, sends rays over snow, ice, and water at such angles as to reflect and bend the beams which fairly burn the skin even in low temperatures. It is to this mirrored reflection of polar sunbeams that we must ascribe the intense bronzing of the

exposed skin of explorers and natives. This luminous glow is also an efficient tonic to the mind and body; but before the night begins the stimulation is replaced by a progressive depression. *The darkness, cold, and isolation then drive the mental faculties on to melancholy."* (our italics.)

In a similar way Dr. Lindhard, wintering at 77° N as medical officer to a Danish expedition to Greenland, described the condition of the men in his care [43].

"I very soon found that the 'dark period' had in several ways a very distinct influence on the general condition and working capacity, although on the 'Danmark Expedition', in contrast to most other expeditions to polar regions, the darkness did not prevent us going out to a considerable extent; short journeys on the sledge were undertaken, and hunting, as also numerous walks for the sake of fresh air and exercise; indoors also, sports were cultivated so far as the limited space permitted. The appetite in general was good; only one lost in weight during the winter. On an average the bodily weight was greater in the winter than in the summer half-year.

For many sleep was difficult, broken and accompanied by dreams; one often lay for hours without being able to fall asleep. Heaviness followed in the morning, as if after a night on the watch; endeavours during the day to make up for the sleep lost resulted simply in a worse night afterwards. Whereas, under ordinary conditions, *a fall of the rectal temperature occurs during sleep, one found here not rarely a rise in the winter,* (our italics), in agreement with the above conditions; in one case, where the sleep was much broken, this was indeed the rule. The contrast from the deep heavy sleep during the 'off-watches' at sea was very striking.

Both the desire and capacity for work were reduced, especially the latter and especially for brain-work. As mentioned, one wakened up heavy and indisposed, with the feeling which is not unknown as 'wooden'; one of the most energetic workers of the expedition used this expression in a conversation with me, after I had written it down as characteristic of my own condition. The wish was present to begin work seriously; the pipe was lit and the pen filled, but one got no further. The pen became dry and was again dipped in the ink; the pipe was smoked out, refilled and started again; always in the belief that something really was to be done or being done. It was only when the bell rang for dinner some hours later, with nothing done, that the inevitable conclusion was accepted, that one was unable to work. The 'mental' work that suited us best was playing at cards; the object here to some extent corresponded with our powers. Home sickness in the true sense was not felt; but with some there was a tendency to have 'dark forebodings' with regard to their homes. This however was only a form taken by the mental depression and discomfort. Feelings of depression were experienced by almost everyone, but only in a single case did these deepen to slight melancholia. One felt oneself under the influence of the weather; depressed when the weather was bad, lighter when it again cleared up; and some investigations indicated that the meteorological factors really had an appreciable importance on the state of health.

One felt irritable and not disposed to be friendly, and the same qualities magnified were read into others. Conversation at table had a tendency to become fragmentary or even to assume a pointed form.

It may be admitted that certain other conditions, which are not necessarily connected with the polar night and which were also altogether unknown during the light period, may have had something to do with the state mentioned. The lack of cleanliness, for example; thus, the common feeding and living room was likewise used as the working room, as also for various other purposes, often not quite aesthetic; manners at table were as a rule 'conspicuous by their absence' and the service was extremely primitive, at times unappetizing. There can scarcely be any doubt, that when all forms are dispensed within the common living room and on meeting, it will be difficult for anyone, whose experience and training have not taught him complete self-control, always to show a presentable side to the surroundings. *There can be no doubt, however, that the polar night, the cold and the dark, especially the dark were indeed the true cause.* (Our italics.) The same discomforts were all present in the summer, though in less degree, and yet one's whole existence both outside and inside was quite different during the light period. We did not feel so overcome and crushed by the natural forces as in the dark period; even with an open eye for all that was grand and imposing in nature one felt an impulse to exertion and to play our part. There was a desire to be active and we could work. We went unwillingly to bed, preferring to sleep just as the occasion offered; the sleep was short yet refreshing. One had the feeling of never resting and never tiring, which contrasted strongly with the physical helplessness of the dark period. We were disposed to be friendly towards one another, very willing to listen and likewise to talk."

This, in addition to mentioning mental changes, also hints at specific rhythm abnormalities during the polar night namely:

1. Altered sleep rhythm: it is well established that sleep placed on a different stretch of the circadian system is of altered character from the point of view of the proportion of stages I-IV and R.E.M. [44].

Similar winter and also summer sleep alterations were noted by Dr. Lewis on the British North Greenland Expedition [45]; also Shurley and Pierce have evidence of a period slightly longer than 24 h in sleep/awake habit in a South Pole meteorologist whose routine was self-chosen [46].

2. Nocturia suggesting an alteration in the phase of peak urine excretion.
3. A high rectal temperature on awakening strongly suggests a re-phasing of the body temperature rhythm.

These points must be viewed against Lindhard's and Cook's observation that body weight was increased, on average, in winter indicating that subnutrition was not a factor; it is interesting, too, that in the Danmark Expedition an effort was made to counter the tendency to winter inanition. Short sledge journeys were undertaken and indoor sports encouraged.

Yet more observations on the effect of polar lighting are found in Nansen's *Farthest North*. Of his epic journey in November 1896 he writes [47]:

"I had hoped to get so much done this winter, work up my observations and notes, and write some of the account of our journey; but very little was done. It was not only the poor, flickering light of the oil-lamp, which hindered me, nor yet the uncomfortable position—either lying on one's back or sitting up and fidgeting about on hard stones, while the part of the body thus exposed to pressure ached; but altogether these surroundings did not predispose one to work. The brain worked dully, and I never felt inclined to write anything."

This certainly was not Nansen's normal mood—the mood of a man later to be so dynamic in the League of Nations. Could it be that his circadian subsystems were "free running" at their own particular periodicities with no alteration of light to "latch" on to and, hence, out of phase with one another? Or was it merely due to the dejection and intellectual starvation of being confined to the stone hut on Franz Joseph land for nine months along with one companion?

3.4.3 Gross Mental Changes

These reports of mood changes lead us on to the question of whether more overt manifestations of mental breakdown occur. To return to the writing of Dr. Cook [27] we read:

"The earliest effects become manifest in the mental realm. The physical changes become evident slowly, often not at all until near the end or at sunrise of the next year. Long-continued storms, with intervening periods of dark cloudy skies precede the passing of the sun in its final stages of descent for the long doom of blackness. At this time the native women become explosive in temperament and hysterical in cycles of weird activity. The men, though excitable, do not reach a climax of despair until the first weeks of night have passed. We did not feel the disaster of mental storm until midnight (midwinter H.W.S.). Being in a desperate environment, we were largely hiding our peculiarities from one another because we were trying to prevent discouragement. Each of us, however, had very grave worries which were prolonged and intensified into hallucinations."

Cook had an advantage over many other polar writers in that he was often describing anonymous Eskimos. One has the impression that many other explorer-authors have had to be more restrained!

Apart from these observations, the sagas of polar exploration contain accounts of rows and possible murders (e.g. Captain Hall [48] or Peary's Yale University scientist Professor Marvin [49] and while these could be accounted for by more obvious privations, personality clashes, etc., one cannot rule out a contribution from the peculiar polar lighting and an effect on the circadian system, such as C.N.S. amine rhythms (see e.g. [50]).

Rather more definite are the accounts of the transient hysterical mental illness called pibloctoq. Vivid accounts of this condition were communicated to the New York Neurological Society in 1913 by Dr. Brill, a chief of the psychiatry clinic at Columbia University [51]. His information was gleaned from Peary's writing, and personal discussions with Professor MacMillan of Peary's polar expeditions. Peary speaking of Thule district of Greenland (78°N) recalls that:

". . . someone among the adult Eskimos would have an attack every day or two, and that one day there were five cases."

Also, of the twenty women on board the Roosevelt (Peary's expedition ship), eight were affected with pibloctoq. He goes on to describe the attack as follows:

"The patient, usually a woman, begins to scream and to tear off and destroy her clothing. If on the ship, she will walk up and down the deck, screaming and gesticulating, and generally in a state of nudity, though the thermometer may be in the minus forties. As the intensity of the attack increases she will sometimes leap over the rail upon the ice, running perhaps half a mile. The attack may last a few minutes, an hour, or maybe even more, and some sufferers become so wild that they would continue running about on the ice perfectly naked until they froze to death if they were not forcibly brought back."

The attack ends in an hour or so with the patient sobbing or falling asleep. Brill considered epilepsy, and certainly the susceptibility of Eskimos to upper respiratory and therefore chronic ear infections would make this a possibility; however the absence of any definite "grand mal" made him conclude that pibloctoq was basically a hysterical condition. The polar literature abounds with accounts of the condition or similar conditions with other names. Thus Professor Rodahl writes of "transitional madness" [52], Dr. Vallee of quajimaillituq (similar to the term used to describe rabid dogs) [53], and Dr. Folk of "spring fever" [54]. This last name suggests a seasonal incidence—important if one suggests that the arctic lighting might be an aetiological factor. Dr. Vallee, in agreement with Brill, also regarded it as a recurrent illness but with a slightly earlier onset in late winter. Cook, as mentioned, noticed that depressive mood changes, hysterical outbursts and hallucinations were worst in late winter or at sunrise. While the evidence linking mental phenomena in explorers and in indigenous Eskimos with the polar lighting remains non-proven, much of the evidence is consistent with such an effect. Theoretically the lack of effective synchronization could result in an external and/or more serious internal desynchronization of the circadian system. Particularly relevant might be an alteration in the phasing of the C.N.S. amine rhythms.

3.5 CONCLUSIONS

Data of circadian rhythm parameters obtained at different latitudes available up to the end of 1972 permit the following conclusions:

1. There appears to be an increase in the amplitude of body temperature rhythm with increasing latitude. However the studies on which these results are based did not closely control voluntary exercise.
2. No detectable change is noted in rhythm parameters when data for low and middle latitudes are compared. However the studies on which this statement is based did not control factors such as age, sex, diet, menstrual phase, season, etc. so any changes may have been obscured.
3. At high latitudes, particularly when subjects were living in the open (e.g. in tents), there is abundant evidence of acrophase delays especially at the solstices (using midsleep as reference) and/or of amplitude damping. A South Pole meteorologist on a voluntary schedule was found to have a sleep/awake periodicity desynchronized from 24 h. Moreover the polar literature contains numerous references to temporal disorientation, sleep disturbances, high body

temperature on awakening, nocturia, mood changes, and gross mental breakdown.

4. The possible chronopathological basis of the "hysterical" disease "Pibloctoq" is discussed.

5. The finding of a desynchronized circadian rhythm of 22.9 h in calcium excretion of one Eskimo is discussed.

REFERENCES

1. F. Halberg, Chronobiology. *Ann. Rev. Physiol.*, 31, 675-725 (1969).
2. D. B. Jackson, Circadian variations in endogenous neurotransmitter levels in the isolated sino-atrial node. Ph.D. Thesis Univ. North Dakota (1967).
3. R. V. Andrews and G. E. Folk, Circadian metabolic patterns in cultured hamster adrenal glands. *Comp. Biochem. Physiol.*, 11, 393-409 (1964).
4. A. Reinberg, Methodologic considerations for human chronobiology. *J. Interdiscipl. Cycle Res.*, 2, 1-15 (1971).
5. H. W. Simpson, M. C. Lobban and F. Halberg, Arctic Chronobiology. *Arctic Anthropol.*, 7, 144-164 (1969).
6. R. J. Konopka and S. Benzer, Clock mutants of drosophila melanogaster. *Proc. natn. Acad. Sci., U.S.A.*, 68, 9, 2112-2116 (1971).
7. J. Aschoff, Exogenous and endogenous components in circadian rhythms. *Cold Spring Harb. Symp. quant. Biol.*, 25, 11-18 (1960).
8. G. W. G. Sharp, The effect of light on diurnal leucocyte variations. *J. Endocr.*, 21, 213-218 (1960).
9. G. W. G. Sharp, The effect of light on the morning increase of urine flow. *J. Endocr.*, 21, 219-223 (1960).
10. C. Fortier, Dual control of adrenocorticotrophin release. *Endocrinology*, 49, 782-788 (1951).
11. F. Hollwich, The influence of light via the eyes on animals and man. *Ann. N.Y. Acad. Sci.*, 117, 105-131 (1964).
12. W. B. Quay, 24-hour rhythms in pineal 5-hydroxytryptamine and hydroxy-indole-o-methyl transferase activity in the macaque. *Proc. Soc. exp. Biol. Med. (N.Y.)*, 121, 946-948 (1966).
13. W. B. Quay, Significance of darkness and monoamine oxidase in the nocturnal changes in 5-hydroxytryptamine and hydroxyindole-o-methyltransferase activity of the macaque's epiphysis cerebri. *Brain Res.* 3, 277-286 (1966).
14. R. J. Wurtman, J. Axelrod, G. Sedvall and R. Y. Moore, Photic and neural control of the 24 h epinephrine rhythm in the rat pineal gland. *J. Pharmac. exp. Ther.*, 157, 487-492 (1967).
15. W. B. Quay, Circadian and estrus rhythms in pineal and brain serotonin. In *Prog. in brain res.*, 8. Himwich and Himwich (Ed.). Elsevier, New York, 61-63 (1964).
16. D. S. Russell, L. J. Rubinstein and C. E. Lumsden, *The Pathology of Tumours of the Nervous System*. Arnold, London. 318 pp (1959).
17. S. Gasten and M. Menaker, Pineal function: the biological clock in the sparrow. *Science*, 160, 1125-1127 (1968).
18. M. C. Lobban and B. E. Tredre, Perception of light and the maintenance of human renal diurnal rhythms. *J. Physiol., Lond.*, 189, 32-33P (1967).

19. E. D. Weitzman and L. Hellman, Temporal organization of the 24 h pattern of the hypothalmic pituitary axis. Conf. on biorhythms and human reproduction in Tuxedo. R. Vande Wiele (Ed.). Wiley, New York. In Press (1973).
20. F. W. Went, Photo and thermo periodic effects in plant growth. *Cold Spring Harb. Symp. quant. Biol.*, 25, 221-230 (1960).
21. C. Pittendrigh, V. Bruce and P. Kaus, On the significance of transients in daily rhythms. *Proc. natn. Acad. Sci. U.S.A.*, 44, 965-973 (1958).
22. F. Halberg, E. Halberg, C. P. Barnum and J. Bittner, Physiological 24 h periodicity in human beings and mice, the lighting regimen and daily routine. In *Photoperiodism and related phenomena in plants and animals*. R. B. Withrow (Ed.), *Am. Ass. advd. Sci. Washington D.C.*, 55, 803-877 (1959).
23. H. W. Simpson, Studies on the daily rhythm of the adrenal cortex. Ph.D. Thesis, Glasgow University. 2 volumes (1965).
24. R. Christianson and B. M. Sweeney, The dependence of the phase response curve for the luminescence rhythm in Gonyaulax and the light intensity in constant conditions. *Chronobiology*, In press (1973).
25. M. Okamoto, K. Kohzuma and Y. Horiuchi, Seasonal variation of cortisol metabolites in normal man. *J. Clin. Endocr. Metab.*, 24, 470-471 (1964).
26. G. Watanabe, Seasonal variation of adrenal cortex activity. *Arch. Envir. Hlth*, 9, 192-200 (1964).
27. F. A. Cook, Some physical effects of Arctic cold, darkness and light. *Med. Record*, 51, 833-836 (1897).
28. J. Llewellyn, Light and sexual periodicity. *Nature*, 129, 868 (1932).
29. E. Batschelet, D. Hillman, M. Smolensky and F. Halberg, Angular linear correlation co-efficient for rhythmometry and circannual changes of human births at different latitudes. *Chronobiology*. In press (1973).
30. C. J. Scrutton, Fossil Corals—Calendars for the ancient earth. *Spectrum*, July 6-7 (1966).
31. F. Halberg, J. Reinhardt, F. C. Bartter, C. Delea, R. Gordon, A. Reinberg, J. Ghata, M. Halhuber, H. Hoffmann, R. Gunther, E. Knapp, J. C. Pena and M. Garcia-Sainz, Agreement in endpoints from circadian rhythmometry on healthy human beings living on different continents. *Experientia*, 25, 106-112 (1969).
32. H. W. Simpson, Cycles in Social Behaviour. *Nature*, 231, 463 (1971).
33. F. Halberg and H. W. Simpson, Circadian acrophases of human 17-hydroxy-corticosteroid excretion referred to midsleep rather than midnight. *Hum. Biol.*, 39, 405-413 (1967).
34. M. C. Lobban, Human renal diurnal rhythms at the equator. *J. Physiol., Lond.*, 204, 133-134P (1969).
35. M. C. Lobban, Daily rhythms of renal excretion in arctic-dwelling Indians and Eskimos. *Quart. J. exp. Physiol.*, 52, 401-410 (1967).
36. M. C. Lobban, The entrainment of circadian rhythms in man. *Cold Spring Harb. Symp. quant. Biol.* 25, 325-332 (1960).
37. J. G. Bohlen, Circumpolar chronobiology. Ph.D. Thesis, University of Wisconsin (1971).
38. A. Wallace, Mental Illness, Biology and Culture. Chapter 9. In *Psychological Anthropology; Approaches to Culture, and Personality*, Francis L. K. Hsu (Ed.). Dorsey Press, Homewood, Illinois. 255-293 (1961).
39. H. W. Simpson, The Exploring Scientist. *Br. J. Sports Med.* In press (1973).
40. M. Simpson, *Due North*. Gollancz, London (1970).
41. L. Brontman, *The Soviet Expedition to the North Pole, 1937-1938*. Covici Friede, New York (1938).
42. F. A. Cook, *Return from the Pole*. F. H. Pohl (Ed.). Burke, London (1953).

43. J. Lindhard, Contribution to the physiology of respiration under the Arctic climate. *Meddelesler om Grønland*, 41, 78-81 (1913).
44. J. I. Evans, The effect on sleep of travel across time-zones. *Clin. Trials J.*, 7, 64-75 (1970).
45. H. E. Lewis, Physiology. In *Venture to the Arctic*. 145-166. R. A. Hamilton (Ed.). Pelican Books Ltd., Harmondsw orth, Middlesex (1958).
46. J. Shurley and C. Pierce, Ref. to unpublished work in *Biological rhythms in Psychiatry and Medicine*. G. Luce (Ed.), for National Institute of Mental Health. *Pub. U.S. Department of Health* (1970).
47. F. Nansen, *Farthest North*. Constable, London. 2 volumes. 1220 pp (1897).
48. F. K. Paddock, C. C. Loomis and A. K. Perkons, An inquest on the death of Charles Francis Hall. *New Engl. J. Med.*, 282, 784-786 (1970).
49. P. Freuchen, *Ice Floes and Flaming Waters*. Travel Book Club, trans. Hambro. Gollancz, London (1930).
50. A. H. Friedman and C. A. Walker, Circadian rhythms in rat mid brain and caudate nucleus biogenic amine levels. *J. Physiol., Lond.*, 197, 77-85 (1968).
51. A. A. Brill, Piblokto or Hysteria among Peary's Eskimos. *J. nerv. ment. Dis.* 40, 514-520 (1913).
52. K. Rodahl, *North, 1953*. Harper, New York, 237 pp (1953).
53. F. G. Vallee, Eskimo theories of mental illness in the Hudson Bay region. *Anthropologica*, 8, 53-83 (1966).
54. G. E. Folk, *Introduction to Environmental Physiology*. Lea Febiger, Pha (1966).

CHAPTER 4

Chronopharmacology

Alain Reinberg*

Laboratoire de Physiologie
Fondation Adolphe de Rothschild
29, rue Manin-75-Paris-19, France

4.1 CHRONOBIOLOGY: A SCIENTIFIC BACKGROUND
FOR CHRONOPHARMACOLOGY

Biological responses to various agents, including chemical substances such as drugs, are not constant over time but vary rhythmically; in human beings, as well as in experimental animals circadian rhythms of susceptibility have been demonstrated [1-11] and the number of functions showing such rhythms is rapidly increasing as research progresses.

As a first example let us consider the effects of ACTH on adrenal secretion in the mouse [12]. For at least one week before the study male and female mice of the Bagg albino (C) stock were maintained on a regimen standardized for periodicity analysis [13] with: 12 h of light (from 0600 to 1800) alternating with 12 h of darkness (from 1800 to 0600). The animals were kept at a temperature of $24 \pm 0.5°C$, in

---

* Maître de Recherches au Centre National de la Recherche Scientifique

individual cages, with food and water *ad libitum*. In each experiment sampling was done with different groups of animals that were comparable for genetic background, sex, and past history. Sampling took place at 4-hourly intervals, from 0800 on one day to 0800 on the next day. Each group was divided into three subgroups. Mice of the first subgroup were killed immediately. Mice of the second subgroup received an injection of saline (0.2 ml/20 gm body weight, i.p.). Those of the third group received an injection of ACTH (0.4 i.u./20 gm b.w.t., i.p.). Animals from these subgroups were killed 15 min after injection. Determinations of corticosterone concentration were made using pooled sera (5 mice) and adrenal glands from individual animals.

Untreated animals showed a circadian rhythm of serum and adrenal tissue corticosterone concentrations with a peak at about 1600 and a trough at about 0400. The animals injected with saline or ACTH—both in a constant dose and at fixed times—also showed a circadian rhythm. However, the response varied as a function of the timing of the treatment—that is the phase of the adrenal cycle in which the treatment was introduced. The serum and adrenals showed the greatest response to saline injection four to eight hours before the circadian corticosterone peak, while the greatest response to ACTH occurred at about 2400.

In further experiments Ungar and Halberg [14] demonstrated a circadian rhythm in the in vitro response of the mouse adrenal to ACTH. Corticosterone secretion from adrenals incubated with a constant dose of ACTH in the medium depends upon the time (clock hour) at which the glands were removed. When mice are standardized on a regimen with L 0600-1800 D 1800-0600, the circadian peak of adrenal susceptibility to ACTH occurs at about 0400 while the circadian trough occurs at about 1600.

These facts may seem paradoxical at first glance, but they are explained by recent findings in chronobiology [8, 9, 15-18]. No longer can one postulate that an animal organism is constant within the limits of a day, a month, or a year. The present data emphatically contradict the notion of constancy.

Observed rhythms may be fitted, [inter alia, by approximating functions of the form:

$$f(t) = C_0 + C \cos (\omega t + \phi)$$

where ω = the angular frequency and t = time as sketched in Fig. 4.1], and objective estimates of the parameters, with their confidence limits, may be obtained [43-46] (see also Chapter 1).

Data taken from *in vitro* adrenal reactivity to ACTH, mentioned above, were re-analysed [8] according to the cosinor method [45, 46, 47]. Period fitted = 24 h = 360°; 1 h = 15°. The midpoint of the daily light span—a point on the synchronizing environmental

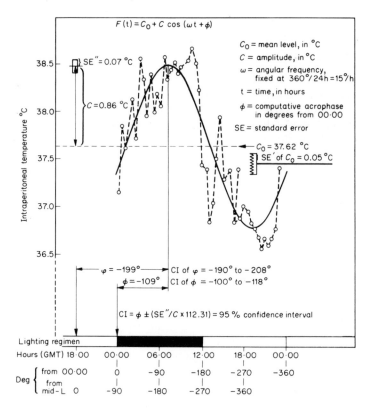

$F(t) = C_0 + C \cos(\omega t + \phi)$

C_0 = mean level, in °C

C = amplitude, in °C

ω = angular frequency, fixed at 360°/24h = 15°/h

t = time, in hours

ϕ = computative acrophase in degrees from 00·00

SE = standard error

SE″ = 0.07 °C

C = 0.86 °C

C_0 = 37.62 °C

SE′ of C_0 = 0.05 °C

φ = −199°   CI of φ = −190° to −208°

ϕ = −109°   CI of ϕ = −100° to −118°

CI = $\phi \pm (SE''/C \times 112.31)$ = 95 % confidence interval

Lighting regimen

Hours (GMT) 18·00   00·00   06·00   12·00   18·00   00·00

Deg { from 00·00   0   −90   −180   −270   −360

from mid−L   0   −90   −180   −270   −360

Figure 4.1. Least squares fit of 24-h cosine function (continuous line) of intraperitoneal temperatures (o) of an adult female MSD rat. Telemetry at ~30′ intervals for 24 h. (From Halberg [8], Fig. 1).

cycle—is taken as phase reference for these night-active rodents. Mice were synchronized for L 0600-1800 D 1800-0600. Mid-L was 1200 (noon-midday) or −180°. In these defined experimental conditions φ (statistical approximation of the circadian peak) of the *in vitro* adrenal reactivity to ACTH is −235° (−210 to −258), 95% confidence limits, while serum corticosterone, φ, is −66° (−42 to −89) and adrenal corticosterone φ, is −79° (−47 to −112).

Most of the described cyclical changes in susceptibility to toxic and/or pharmacological agents have been circadian rhythms, which are the focus of this paper. It should be added that low frequency rhythms in pharmacological effects have been reported by Aron *et al* [49], Haus and Halberg [26], Beauvallet *et al*. [27], Petrovic and Kayser [39], Miller *et al*. [41, 118].

Within certain limits, the period, amplitude and phase of circadian rhythms can be influenced by the cyclic variations of certain environmental factors. There are the alternation of day and night, heat and cold, noise and silence [1, 13, 19-25, 29-38]'. For man, the hours of work and repose, related to the duties of social life [1, 6, 8, 31, 45, 46, 50, 51], comprise important synchronizers [52].

From a methodological point of view, any study on biological rhythms (including circadian susceptibility rhythms) has to be realized under conditions standardized for periodicity [13, 126]. The following environmental conditions must be controlled: (1) lighting regimen for experimental animals, sleep-wakefulness schedule for human subjects; (2) duration and stability of the synchronization prior to the study. Without these criteria the analysis of data may be meaningless and it is possible to misinterpret the results.

Experiments in chronopharmacology and chronotoxicology have been selected for summary in this review on the following basis: (1) appropriate methodology for the control of synchronizers and data gathering; (2) accurate technical procedures for determinations, measures, etc., on each biologic variable selected as an index of periodicity; and, most important, (3) statistical analyses of time series data, by microscopic rather than macroscopic methods of parameter estimation. At the present stage of knowledge in chronobiology, a quantitative description of rhythmic phenomena should take priority over other problems such as the interpretation of observed rhythms.

4.2 THE HOURS OF CHANGING RESPONSIVENESS OR SUSCEPTIBILITY

The phenomenon of rhythmically changing responsiveness has been experimentally demonstrated in animals [1-3, 10, 11, 141] and in man [5-7]. The discovery of the hours of changing responsiveness has led to studies of circadian chronotoxicity and circadian chronopharmacology. As a matter of fact, regular and predictable circadian changes in biologic susceptibility now can be viewed as a very common phenomenon. For this reason a review of temporal toxicology is the best introduction to the temporal effects of drugs.

F. Halberg et al. [10, 11, 55-62] carried out experiments with mice of the inbred D_8 (Dilute Brown, subline 8) and C (Bagg albino) stocks. During a standardization span of at least one week before and during the study, mice were kept at $24 \pm 0.5°$C, with L 0600-1800 D 1800-0600. They were housed individually and given food and water ad libitum. Different groups of mice, comparable for body weight, age, and sex were injected (or submitted to stimulation) at four-hourly intervals, starting at 1800 on one day for the first group and ending 24 to 48 h later for the last one.

Several types of agent were tested: *E. coli* endotoxin [55, 56]. C Mice in seven groups, each composed of 15 to 20 individuals, were given intraperitoneal injections of *E. coli* lipopolysaccharide (Difco) in doses of 100 mg in 0.2 ml per 20 mg of body weight. In each group mortality was recorded at four-hourly intervals for two days after injection.

Ouabain [57, 58]

Injections were intraperitoneal: 0.5 to 0.15 mg of ouabain (Lilly) in 0.2 ml of saline was given per 20 g of body weight to D_8 and C mice respectively. The number of deaths was recorded 10 min after the injection (when most deaths occurred) and at several later time points until one week post injection.

White Noise [10, 11]

At four-hourly intervals D_8 mice were transferred individually from their cages to a stimulator (the stimulus, yielded by a white-noise generator, was about 104 db above 0.0002 dynes/square centimetre r.m.s. pressure). Each stimulation lasted 60 sec. The rate of audiogenic convulsions and of deaths were recorded in each group.

Chlordiazepoxide (Librium, Roche) [59]

Intraperitoneal injections of this neuroleptic drug were given every four hours to different groups of D_8 mice (5.4 mg/20 g body wt). Survival time (measured under standard conditions) was recorded.

Ethanol [60]

C mice were injected intraperitoneally with 25% v/v ethanol solution (0.8 ml/20 g body wt). Rate of mortality was recorded.

Metyrapone (SU4885, Ciba) [61]

Groups of C mice were given this compound (6.5-8 mg/20 g body wt) which is a specific inhibitor of 11-β-hydroxylase. The percent of deaths in each group has been recorded.

Dimethylbenzanthracene [62]

This was used to produce breast cancer in groups of D_8 mice (0.05 ml of 0.5% solution). The percentage of tumors induced was recorded under standardized experimental conditions.

Circadian rhythms of susceptibility to these potentially noxious agents were observed. The results drew attention to two points: (a) agents as different as a bacterial endotoxin (*E. coli*), a plant alkaloid (ouabain), a neuroleptic (chlordiazepoxide), an enzyme inhibitor (metyrapone), or a noise (acoustic stimulation) can be used to reveal hours of changing responsiveness; (b) the timing of crests and troughs in susceptibility vary from one agent to another.

The same temporal series were re-analysed by the cosinor method and the results are summarized in Fig. 4.2. These analyses reinforce the conclusion that different agents can used to reveal rhythms of responsiveness and that the timing of the crests and troughs in susceptibility vary, depending upon the agent. Moreover, as the cosinor analyses indicate, the circadian rhythms detected and studied have been demonstrated to be statistically significant, and C (amplitude) and φ (acrophase) are given with 95% confidence limits.

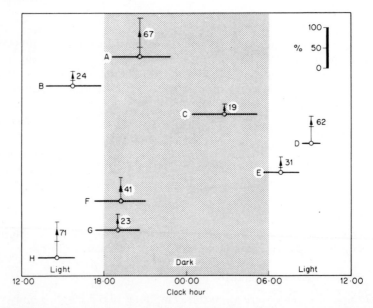

Figure 4.2. Acrophase and amplitude of circadian rhythms in susceptibility to potentially harmful agents. Horizontal lines indicate acrophase and its 95% confidence limits, by reference to scale below. Arrows on vertical lines indicate amplitude as % of mean also indicated by a numeral, and see inset scale. Horizontal bars indicate 95% confidence limits of acrophase. Position on ordinate is irrelevant. A-D, D_8 mice. E-H, C mice. Agents tested were: (A) White noise, ∼104 db; (B) Dimethyl-benzanthracene, 0.05 ml 0.5%; (C) Chlordiazepoxide, 5.4 mg/20g b.wt; (D) Ouabain, 0.5 mg/20g b.wt; (E) Ouabain, 0.5 mg/20g b.wt; (F) Ethanol, 0.8 ml 25% v/v; (G) SU 4885, 6.5-8 mg/20g b.wt; (H) *E. coli* endotoxin, 100 μg/20g b.wt.

Such evidence suggests that circadian rhythms of susceptibility may exist for other agents and for other animal species.

Experiments have been conducted (from macroscopic and/or microscopic point of view) on plants, several species of insects, and other strains of mice and rats, using a wide variety of chemical agents as shown in Table 4.1. In all these various experimental circumstances circadian rhythms of chronotoxicity were reported.

Circadian susceptibility rhythms to several potentially harmful physical agents have also been published: leaf resistance of a plant (*Kalanchoe*) to heat [78], recovery rate of *Drosophila* from heat [42], mortality rate of *Drosophila* [79] and mice from X-rays [1, 80, 81, 82], mortality rate of rats and of CBA mice from gamma irradiation [83] —102 body weight loss from partial body X-rays [140].

This coherent body of knowledge leads one to question the validity of evaluating acute, subacute, and the chronic toxicity of a given drug by the so-called *lethal dose 50* (LD_{50}), corresponding to the quantity of drug which has to be administered to each individual of an homogenous group of animals in order to obtain 50% mortality [2, 3, 6, 7]. The LD_{50} test must be defined in terms of its timing, as well as other conventional references such as species, sex, and age of test animals. Thus, one could refer to a maximal LD_{50} for a specific drug at the circadian acrophase of toxicity and to a minimal LD_{50} located 12 h earlier or later.

4.3 CIRCADIAN CHRONOPHARMACOLOGIC EFFECTS CAN BE DEMONSTRATED AT ALL LEVELS OF ORGANIZATION AND FOR AN ALREADY WIDE VARIETY OF CHEMICAL AGENTS AND ANIMAL SPECIES, INCLUDING MAN

By reference to Tables 4.2 and 4.3, the considerations summarized in the above subtitle can be drawn from animal experiments with mice, rats, cats, and monkeys as well as from studies on human beings (healthy and sick, children and adults) [91-103].

In some experiments the biologic variable investigated is the susceptibility of the body as a whole. Thus Scheving, Vedral, and Pauly [74, 86] demonstrated that the duration of sleep induced by pentobarbital sodium injection (35 mg/kg) in the white adult rat is a function of the hour of administration. The circadian phase of administration also determines the length of time required for the onset of constant tremor in all members of experimental groups of white rats injected at fixed times with fixed doses (64 mg/kg) of tremorine (1,3-pyrrolidino-2-butyne), according to Scheving and Vedral [73, 86]. With L 0600-1800 D 1800-0600, it took 150 min before tremors occurred at about 1400, whereas it took only 35 min at about 2200.

The susceptibility of organ systems has been explored *in vivo* from a

Table 4.1

Cicadian susceptibility rhythms to chemical agents given in fixed dose at different hours, assessed by percentage of tumors [*] or of death, or by survival time [**], or knockdown as well as mortality used as index [***]

Species	Chemical agent	Synchronizer schedule	Mean	Amplitude C	Acrophase (clock hour)	References
Hamster*	Dimethylbenzanthracene	L 0600-1800 D 1800-0600	100%	[25%]	1200	Halberg [48]
White rat	D-amphetamine sulphate Nicotine	L 0600-1800 D 1800-0600	100%	21.5 (15.7-27.3)	2408 (2208-0212)	Scheving, Vedral [75]; Scheving [76]
White rat	Pentobarbital sodium	L 0600-1800 D 1800-0600	48	[25]	2200	Pauly, Scheving [73]
White rat	Tremorine (1,3-pyrrolidino-2-butyne)	L 0600-1800 D 1800-0600	58	[30]	400	Scheving, Vedral [74]
White rat	Strychnine	L 0600-1800 D 1800-0600	100%	[25%]	2100	Tien Ho Tsai et al. [77]
C mice	Brucella somatic antigen	-	-	-	-	Halberg, Spink et al. [68]
C mice	E. coli lipopolysaccharide	L 0600-1800 D 1800-0600	100%	71% (47-95)	1430 (1312-1348)	Halberg, Stephens [55]; Halberg et al. [56]
C mice	Ethanol	L 0600-1800 D 1800-0600	39	16 (9-23)	1908 (1740-2056)	Haus, Halberg [60]
C mice	Metytapone (SU-4885)	L 0600-1800 D 1800-0600	100%	23% (11-35)	1900 (1730-2030)	Ertel, Halberg, Ungar et al. [61]
B₆ mice	Acetylcholine	L 0600-1800 D 1800-0600	~76	[14]	2000	Jones et al. [69]

Subject	Drug	Synchronizer	Mean	Circadian Amplitude	Acrophase	Authors
B₁ and C mice	Fluothane	L 0600-1800 D 1800-0600	~17	[13]	2400	Matthews et al. [70]
D₈ mice	Ouabain	L 0600-1800 D 1800-0600	100%	62% (57-67)	0900 (0820-0936)	Halberg, Stephens, Haus [57-58]
D₈ mice*	Dimethyl benzanthracene	L 0600-1800 D 1800-0600	~36	[9]	1600	Haus, Halberg [62]
D₈ mice**	Chlordiazepoxide	L 0600-1800 D 1800-0600	100%	19% (6-24)	0240 (0016-0532)	Marte, Halberg [59]
Swiss mice** ♀	Pentobarbital sodium	L 0730-1930 D 1930-0730	~108	–	1300	Lindsay et al. [128]
Mice	Aurothioglucose	–	–	–	–	Wiepkema [71]
CF₁ mice ♀	Lidocaine hydrochloride	L 0605-1805 D 1805-0605	16 to 20	[38%]	2105	Lutsch, Morris [72]
House fly***	Pyrethrum	L 0500-2000 D 2000-0500	100%	18% (15-21)	1510 (1346-1634)	Sullivan et al. [67]
Madeira cockroach***	Pyrethrum	L 0500-2000 D 2000-0500	100%	19% (15-23)	1734 (1606-1854)	Sullivan et al. [67]
Two-spotted spider mite	DDVP (Insecticide)	–	–	–	–	Pollick et al. [66]
Boll-weevils (insect)	Methylparathion (insecticide)	L 1000-2400 D 0000-1000	~20	[40%]	Crest of resistance at 1000	Cole, Akisson [64, 65]

SYNCHRONIZER: lighting regimen. MEAN: mean over all sampling times and subjects; "100%" indicates data were reported at different points only as percentage deviations from the overall mean value. CIRCADIAN AMPLITUDE: difference between the highest [or lowest] value and mean value in a sinusoidal oscillation, determined by harmonic analysis. Values (in parentheses) are 95% confidence limits. Values in [brackets] give one-half the range of group means over the circadian period [included as an approximation of circadian amplitude, when only group means at different clock hours were available]. ACROPHASE: cited by different authors indicates maximum of susceptibility or of resistance. Values (in parentheses) are approximate 95% confidence limits.

Table 4.2

Effects of chemical agents, given in fixed dose at different hours as a function of circadian system phase

Species	Chemical agent	Conventions as in Table 4.1					
		Variable investigated	Synchronizer schedule	Mean	Amplitude	Acrophase (clock hour)	References
Cotton (plant)	Herbicides (dicryl, etc.)	Inhibition of growth of 12-day-old cotton seedlings	L 0600-1800 D 1800-0600	—	—	0600	Gosselink, Standifer [63]
House cricket (insect)	Ether, chloroform, carbon tetrachloride	Recovery time of 50% of subjects	L 0800-2000 D 2000-0800	~40	[3]	2300	Nowosielski et al. [136]
Lower vertebrates (Fish, Frog, Lizard). White-throated sparrow (bird)	Prolactin	Fattening response	LD 16 : 8	—	—	Middle of light period	Meier et al. [132-134]
C mice	ACTH	Corticosterone response; standardized groups of animals	L 0600-1800 D 1800-0600	~210	[110]	2400	Haus, Halberg [12]
C mice	ACTH	Incubated adrenal removed from group of animals at fixed hours. Corticosterone response	L 0600-1800 D 1800-0600	100%	[75%]	0340 (0200-0512)[14]	Ungar, Halberg [14]
Mice (immature ♀)	Gonadotropin	Uterine and ovarian weights	L 0600-1800 D 1800-0600	—	—	1700	Lamond, Braden [84]
C mice	Methylprednisolone (sodium succinate)	Temperature, serum corticosterone, liver glycogen, Body weight, etc.	L 0600-1800 D 1800-0600	—	—	When timed to the predictable circadian peak of the serum corticosterone, steroid injection causes no or minimal disturbance in the circadian rhythms investigated	Smolensky [85]

Species	Drug	Parameter measured	Lighting conditions			Clock hour	Reference
C_{57} B_1 mice	Pentobarbital sodium	Duration of anaesthesia	L 0800-2000 D 2000-0800	100%	[30%]	1400	Davis [129]
Swiss-Webster mice ♂ Holtzman rats	Aminopyrine, 4-dimethyl-amino-azobenzene, p-nitroanisole, hexobarbital	Changes in circadian reactivity rhythm of hepatic drug-metabolizing enzymes	L 0630-2000 D 2000-0630	—	—	0200	Radzialowski, Bousquet [107]
Sprague-Dawley rats ♂	Hexobarbital (p-nitroanisol)	Changes in circadian reactivity rhythm of hepatic drug-metabolizing enzymes	L 0600-1800 D 1800-0600	—	—	2200	Nair, Casper [135]
Sprague-Dawley rats ♂	Hexobarbital	Duration of sleep as a function of the hour of drug administration	L 0600-1800 D 1800-0600	~54 min	[10 min]	1400	Nair, Casper [135]
White rats	Pentobarbital sodium	Duration of sleep as function of the hour of drug administration	L 0600-1800 D 1800-0600	~67 min	[24 min]	1900	Scheving, Vedral, Pauly [73, 74, 86]
White rats	Tremorine (1,3-pyrrolidino-2-butyne)	Time required for the onset of constant tremor in all members of experimental group	L 0600-1800 D 1800-0600	~90 min	[60 min]	1400	Scheving, Vedral, Pauly [73, 74, 86]
White rats	Colchicine	Rhythms of cell division in the cornea	—	—	—	—	Scheving, Pauly [87]
White rats	Acetylcholine	Right atrium beat frequency in vitro	L 0600-1800 D 1800-0600	—	—	1100	Spoor, Jackson Jackson [88-89]
Peromyscus	Pentobarbital sodium	Rate of recovery	LD : natural	—	Recovery is more rapid during the active (in D) span of the cycle	—	Emlen, Kern [127]
Cats	Atropine: Short acting barbiturate	Inhibition of circadian rise in plasma 17—OHCS level	LD : natural	—	Injection just prior the rise of 17-OHCS in plasma is more effective than at any other tested clock hour	—	Krieger & Krieger [92]

Table 4.3

Chemical agent effects upon circadian rhythms of experimental animals

Species	Chemical agent	Variable investigated	References
Canary	Barbiturates, reserpine, monoamine-oxidase inhibitors, etc.	Behaviour (circadian rhythm of self-selected rest and activity)	Wahlström [130]
Mice	Morphine chlorhydrate	Alteration in blood glucose and in liver glycogen circadian rhythms	Sable *et al.* [131]
Holtzman rats	Corticosterone, phenobarbital	Alteration of circadian rhythm of hepatic drug-metabolizing enzyme activity	Radzialowski, Bousquet [107]
Rats	Parachlorophenylalanine, Disulfiram, Monoamine-oxidase inhibitors	Alteration in sleep circadian rhythms (24 h EEG)	Mouret [91]
Rats	Chloramphenicol	Changes in circadian rhythms of histamine excretion	Wilson [104]
Rats	Cytosine arabinoside (antimetabolite)	Inhibition of the mitotic acrophase in corneal epithelium	Cardoso *et al.* [139]
Monkeys	Reserpine	Changes in circadian rhythms of circulating eosinophils and of urinary 5-hydroxyindol acetic acid	Anderson [93]
Monkeys	Anti-depressant (an analog of thioridazine)	Experimentally induced neurosis and psychosis with alteration of physiological rhythms. Rapidity of the animal recovery	Stroebel [90]

chronopharmacologic point of view more often than has whole body responsivity. For example, in animals circadian variations of uterine and ovarian weights have been determined after timed injections of gonadotropin in immature female mice [84]; the rate of inhibition of the circadian rise in plasma 17-OHCS level has been determined following a single timed injection of atropine or short acting barbiturates, etc., in cats [92].

At this level chronopharmacological effects have also been demonstrated in man [94-97]. Adults—apparently healthy subjects as well as allergic patients—were placed on a standardized routine of diurnal activity and nocturnal rest (from 2300 to 0700) for at least one week before testing. Subjects' profiles were obtained at four-hourly intervals during 24 h on several functions including the evaluation of erythema area measured 15 min after intradermal standardized injections (flexor surfaces of the forearms) of histamine ($10\mu g$ in 0.1 ml of saline solution) for healthy subjects or of house dust extract (0.1 ml of a freshly prepared 1/50,000 dilution) for allergic subjects. Standardized scratch tests were done with sodium benzylpenicillin (0.4 ml of a freshly prepared solution of 25,000 units/ml) on patients sensitized to this antibiotic. Some such results are seen in Tables 4.4 and 4.5.

A circadian rhythm characterizes the erythematous response of healthy subjects to a fixed dose of histamine as shown in Fig. 4.3 (*P* that response is non-rhythmic < 0.001). Under the conditions of the study this rhythm amplitude is 24% (between 19 and 29%). The acrophase is found at 2308 with 95% confidence limits from 2144 to 0048, when

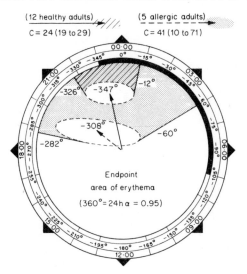

Key: C = amplitude in % of series mean
        Histamine : intradermal injection
        Penicillin : scratch test

Figure 4.3. Cosinor analysis of circadian rhythm in skin reactions to histamine and to penicillin of mature human beings. Note remarkable agreement in timing of acrophases for susceptibility rhythms to histamine and penicillin, agreement apparent from the closeness of the arrows and the overlap of areas delineated as 95% confidence intervals by tangents drawn to the error ellipses. (From A. Reinberg, Z. Zagulla-Mally, J. Ghata, F. Halberg [95], Fig. 2.)

Table 4.4

Effects of chemical agents, as a function of circadian system phase in healthy or allergic (*) human beings. Conventions as in Table 4.1

No of subjects	No of days [Sampling frequency, h]	Chemical agent	Variable investigate	Synchronizer schedule	Amplitude as per cent of 24 h mean	Acrophase	References
12	1[4]	Histamine	Skin reactions (erythema and wheal) to intradermal injection	L 0730-2330 D 2330-0730	24% (19-29)	2308 (2144-0048)	Reinberg et al. [94, 95]
6	1[4]	48/80 (histamine liberator)	Skin reactions (erythema and wheal) to intradermal injection	L 0730-2330 D 2330-0730	[38%]	2300	Reinberg et al. [94]
5*	1[4]	Penicillin	Immediate skin reaction (erythema) to the antigen	L 0700-2300 D 2300-0700	41% (10-71)	2032 (1848-0400)	Reinberg et al. [95]
6*	1[4]	House dust extract	Immediate skin reaction (erythema) to the antigen	L 0700-2300 D 2300-0700	38% (5-72)	2152 (1632-0108)	Reinberg et al. [95]
6	4[6]	Sodium salicylate	Total duration of salicylate excretion in urines (oral administration)	L 0700-2300 D 2300-0700	12.3% (7.5-16.8)	0641 (0145-1052)	Reinberg et al. [97]
7	1[6]	Acetylcholine	Bronchial reaction	L 0700-2300 D 2300-0700	29% (4-53)	1427 (1101-1753)	Reinberg et al. [142]

Table 4.5

Chemical agent effects upon circadian rhythms of human beings and as a function of circadian system phase

Group	Chemical agent (fixed dose, multiple test times)	Variable investigated	References
Mentally deficient adult	Reserpine	Alteration of body temperature circadian rhythm	Halberg [3]
Healthy adults	Cyproheptadine (per os) (antihistaminic drug)	Circadian changes in the duration of the inhibitory effect on skin reaction to histamine	Reinberg, Sidi [96]
Healthy adults	Dexamethasone (per os)	Alteration of circadian rhythms for plasma and urine 17-hydroxycortico-steroid	Nichols et al. [98] D'Agata et al. [99], Bricaire et al. [100]
Asthmatic children	Prednisone (per os)	Phase shift in acrophase of circadian rhythms for peak expiratory flow rate and urinary K and Cl (no change when drug is given at the acrophase of the plasma 17-OHCS circadian rhythm)	Reindl et al. [102], Halberg [15]
Healthy adults	ACTH (I.V. infusion)	Phase shift in circadian rhythm of 17-OHCS excretion (no change when ACTH infusion corresponds to the time of the expected peak of 17-OHCS excretion)	Martin et al. [105, 106]
Adrenal insufficiency	Cortisol (per os)	Phase shift in acrophase of circadian rhythms for heart rate, grip strength, and K, Na, 17-OHCS, 17-KS urinary excretions	Reinberg, Ghata, Halberg, Gervais, Apfelbaum et al. [143]
Healthy adults	Ethanol (per os)	Circadian changes in blood ethanol concentration. Alteration in psychophysiologic test circadian rhythms	Rutenfranz et al. [101]
Diabetic adults	Regular insulin	Alteration of cortisol circadian rhythm in plasma	Serio, Della Corte et al. [137]

local midnight is taken as phase reference, and at $-292°$ (-271 to -317) when mid-sleep (0300 or $-45°$) is taken as phase reference. Figure 3 also shows a circadian rhythm in allergic cutaneous responses to penicillin ($P < 0.025$), with amplitude of 41% (10 to 71%) and acrophase at 2032 (1848 to 0400). A circadian rhythm is also demonstrated for the erythematous response of allergic patients to fixed doses of house dust extract (Fig. 4.4).

The duration of urinary salicylate excretion after the oral administration of a fixed dose (1 gm of sodium salicylate) of the drug was studied in six human subjects standardized on a routine of nightly sleep from 2300 to 0700 [97]. Urine was collected from each subject at 4-hourly intervals for 48 h. Salicylate in each sample was determined by two closely related methods after extraction with dichloroethane. The average duration of drug excretion depended on the time of its administration. A circadian rhythm in this pharmacological phenomenon was detected ($P < 0.002$) and analysed by the cosinor method. Results thus obtained were compared with impressions gained from a conventional time plot of the data (Fig. 4.5) and found to be in agreement. The longest duration of salicylate excretion followed drug administration at 0641. The 95% confidence arc for this circadian acrophase extends clockwise from 0244 to 1004.

Let us now consider examples of circadian variation in pharmacological effects at the level of organs or tissue. It has been mentioned already that circadian changes have been demonstrated in the response of the isolated adrenal to a constant dose of ACTH. Jackson and Spoor [88, 89] studied circadian changes in the sensitivity of isolated rat atria to acetylcholine (ACh). Animals were standardized with L 0600-1800 D 1800-0600. At fixed times the heart was removed from animals of different groups and the *in vitro* effect of ACh was tested. The per cent decrease in rate at 1 and 10 μg/ml in the medium was greater if the atria were isolated at 1100 than if isolated at 2300. The 50% effective concentration, determined graphically, for the atria isolated at 2300 is 1.6 times greater than for those isolated at 1100 (5.7 μg/ml and 3.6 μg/ml respectively).

Circadian changes also have been reported at cellular and subcellular levels. For example the rhythm of cell division in the white rat cornea can be influenced by timed injections of colchicine [87]. Smolensky [85] demonstrated in the mouse—L 0600-1800 D 1800-0600—that methylprednisolone treatment can shift the phase of the circadian liver glycogen rhythm. Analysis of data showed that a single injection of this steroid at 1600 induced a phase advance, i.e., an earlier acrophase of liver glycogen when compared with the acrophase of controls. Similar treatment at 0400 resulted in a phase delay, a later timing of the acrophase by comparison with controls.

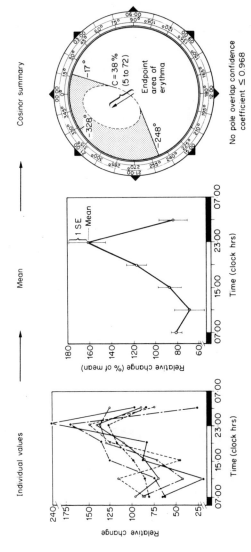

Figure 4.4. Circadian susceptibility rhythm to house dust of six mature human beings (two women, four men), and cosinor analysis. Ordinate: area of erythema as % of mean. The direction of the arrow in the cosinor and the adjacent shaded area indicates the span of greater susceptibility; the length of the same arrow indicated the extent of the predictable periodic change, i.e., the rhythm's amplitude. That a rhythm does indeed occur is detected by the cosinor method when the error ellipse, the white space within the shaded area, does not cover the middle of the plot, the so-called pole. (From A. Reinberg, Z. Zagulla-Mally, J. Ghata, F. Halberg [95], Fig. 1.)

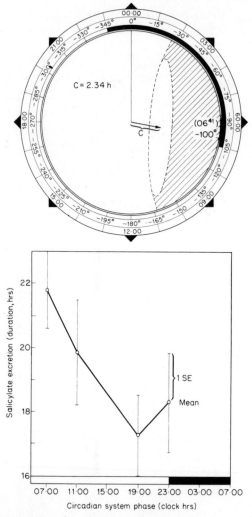

Figure 4.5. *Below:* Circadian rhythm in duration of excretion of salicylate after administration to human subjects at different hours. *Above:* Cosinor analysis of this rhythm, see Chapter 1 [97].

Radzialowski and Bousquet [107] have studied circadian rhythm activities of hepatic drug-metabolizing enzymes such as: aminopyrine N-demethylase, 4-dimethyl-aminoazobenzene reductase, p-nitroanisole O-demethylase, in male and female Holtzman rats and male Swiss-Webster mice, L 0630-2000 D 2000-0630. Groups of livers were removed at two-h intervals and microsomal or 9000 x g supernatant fractions were prepared and assayed. Under the condition of this study the activity of each of the enzyme systems studied showed a circadian rhythm with a peak at about 0200 and a trough at about 1400. Adrenalectomy, or the maintenance of elevated plasma corticosterone levels by administration of this steroid, resulted in the alteration of the circadian rhythm in the drug metabolism. Such alterations were also observed in rats pretreated (4 days) with phenobarbital, although the circadian rhythm of the plasma corticosterone level was not affected.

Cardoso [138, 139] suggested the use of circadian mitotic rhythms as a guide for the administration of antimetabolites. Cytosine arabinoside (Aracy) administered to rats (kept in L 0600-1800 D 1800-0600) at 2300 h profoundly inhibits the mitotic acrophase in corneal epithelium, whereas no inhibition of this acrophase is seen with the same dose of Aracy administered at 1400 h of the same day. Increased tolerance of leukemic mice to Aracy with schedule adjusted to circadian system has been shown by Haus et al. [145]. Mice (BDF_1) inoculated with L 1210 leukemia survive for a statistically significantly longer span when four courses of Aracy are administered at 4-day intervals—not in courses consisting of eight equal doses at 3-h intervals, but in sinusoidally increasing and decreasing 24-h courses, the largest amount being given at previously mapped circadian and circannual times of peak host resistance to the drug.

Insofar as circadian rhythms exist at all levels of organization, appearing in subcellular metabolic activity such as that of RNA, DNA, phospholipids, enzymes, etc. [1, 41, 80, 108-112], it is not surprising that circadian rhythms can be detected and studied in pharmacological effects. Nevertheless other considerations and tentative explanations should await further quantitative studies in chronopharmacology.

4.4 CHARACTERISTICS OF SUSCEPTIBILITY RHYTHMS AND CHANGES IN ENVIRONMENTAL FACTORS

Environmental changes, particularly the manipulation of a synchronizer, can influence circadian susceptibility rhythms just as such manipulation can modify the parameters characterizing other physiological and biological circadian rhythms. Experimental demonstrations have been published, and selected examples follow.

Suppression of known synchronizers

Known synchronizers have been suppressed in experimental animals by exposure to constant darkness (DD), constant light (LL), and blinding [1-4, 19-24, 31-38, 52-54]. Such suppression of synchronizers has been accomplished in man in isolation underground or in isolation from time cues, etc. [1, 2, 4, 6, 28, 29, 40, 50, 113-117]. These studies have led to the following observations: (1) that circadian rhythms persist despite the absence of synchronizers; and (2) their period changes and differs statistically from 24 h.

In their experiment on the susceptibility rhythm of C mice to ethanol, Haus and Halberg [60] studied the effect of constant darkness. Groups of mice were kept in DD for 5 days before four-hourly injections were begun. A statistically significant susceptibility rhythm was detected under the new experimental conditions of continuous darkness.

Scheving and Vedral [75, 76] studied the circadian susceptibility rhythm of white rats to D-amphetamine sulphate under different conditions: (a) L 0600-1800 D 1800-0600, (b) in blinded animals (bilateral enucleation, four weeks before experiments), and (c) in constant light (LL) (Fig. 4.6, Table 4.6). A circadian susceptibility rhythm can be objectively detected in LD (12 : 12), as well as LL and DD (blinded). The acrophase of each group is statistically different from the others. In another experiment on rats, Scheving *et al.* [86] observed that even after blinding animals show circadian changes in duration of sleep induced by pentobarbital.

The point to be emphasized is that the suppression of a synchronizer does not obliterate the circadian susceptibility rhythms; they persist, although they eventually show changes in the estimated values of period, amplitude and acrophase.

The phase shift of a synchronizer is followed by a phase shift of

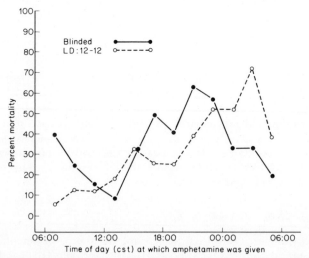

Figure 4.6. Percentage mortality over 24 h after administration of amphetamine (26 mg/kg) at different clock hours. Comparison of circadian rhythm between animals maintained under an artificial LD cycle, L 0600-1800 D 1800-0600, with blinded rats [75] Fig. 2.

Table 4.6

Cosinor analysis of circadian variation of mortality of rats after fixed dose of
d-amphetamine sulphate at different hours [75, 76]. Cf. Figure 4.6

Condition of experiment	LD 12 : 12	DD (blinded)
No. of rats	16	12
Mean % mortality and S.E.	21.5 ± 5.8	17.4 ± 4.7
Acrophase and 95% confidence limits	0008 (2208-0212)	2132 (1932-2324)

rhythms with a greater or lesser delay, depending upon the studied
function, the animal species, etc. [1, 4, 6, 8, 19-21, 31, 33, 38, 40, 144].

Peak susceptibility of D_8 mice to ouabain [57] at about 0800 occurs
only when the lighting conditions are as follows: L 0600-1800
D 1800-0600. This peak may be phased to any desired clock hour by the
appropriate shift in the lighting schedule and by allowing for a sufficient
shift time. For instance, 17 days after light inversion (a phase shift of
12 h) L 1800-0600 D 0600-1800 peak susceptibility to ouabain in D_8
mice came at 2000 while the trough appeared at 0800, the within-day
changes showing statistical significance.

Several observations of theoretical as well as practical interest should
be drawn from these facts.

(1) We must insist once again that it is imperative to exert optimum
control over the environmental rhythms in any biorhythm study,
including studies of susceptibility rhythms.

(2) With "microscopic" analysis and its power to resolve temporal
structure it is now possible to use a characteristic point of the
environmental rhythm as a phase reference (i.e., mid-L for
experimental animal, mid-sleep for men) or a characteristic point
on the organism's circadian temporal structure such as the
acrophase of body temperature or 17-OHCS excretion acrophase [8].

Such biological quantification may be especially useful if one knows
that the acrophase of susceptibility varies from agent to agent in a given
species (Fig. 4.2) and from species to species with a given agent.

However, one must not interpret all statistically significant differences
in the timing of susceptibility rhythms to a given agent as being the
result of strain or species differences. Before the latter can be accepted,
one must rule out the effects of age, of circannual rhythms [26] and
other factors.

Qualifications also apply to the reports of a circadian peak in duration
of pentobarbital-induced anaesthesia occurs at about 7 h from mid-L in
white rats [86] while it occurs at about 1 h from mid-L in C_{57} B mice

[129], both rats and mice being synchronized to LD 12 : 12. In this connection it should be emphasized that susceptibility rhythms are system phase dependent. In particular, the circadian activity of the adrenal can be considered one of the possible component rhythms that contributes to the rhythm of susceptibility—at least for several agents. This seems to be the case in the animal experiments of Smolensky [85], Krieger [92], Radzialowski [107], and in some of our studies of human beings [94, 95, 119].

For instance, the acrophase of skin susceptibility to histamine, to 48/80 (a histamine liberator: Burroughs-Wellcome & Co.) and the acrophase of allergic patients to antigens (house dust extracts, penicillin) differ roughly by 180° (12 h) from the acrophase of adrenal secretion. In other words blood corticosteroid levels were, on the average, lowest when the cutaneous response to histamine and other agents was highest.

We have seen that certain manipulations of synchronizers and certain drug administrations both can influence physiologic circadian rhythms. For example: a circadian rhythm in the temporal series of peak expiratory flow rate (P.E.F.R.) is detectable in healthy persons as well as in patients (children) suffering from asthma. P.E.F.R. acrophase can be shifted [46] by changing the healthy subjects' time schedule, as after a transmeridianal flight from Paris to Minneapolis and vice-versa (Halberg, Reinberg unpublished data) and (2) by giving the asthmatic children a timed administration of prednisone [102, 15] (Fig. 4.7).

Such a pharmacologically induced change in physiologic rhythms may be desirable or undesirable. A timed administration of drug might be desired: (1) to treat a circadian desynchronization occurring in certain diseases as in adrenal insufficiency (Reinberg, Ghata, Halberg, Apfelbaum, Gervais [143]); (2) to treat the circadian desynchronization that follows transmeridianal flight [4, 8, 120-122], and to shorten the resynchronization time.

On the other hand a drug induced alteration of physiological rhythms may be regarded as undesirable, and a considerable amount of work remains to be done in chronopharmacology on undesired rhythm changes brought about by drugs.

Under certain circumstances the timed administration of drug can be considered as a powerful synchronizer in influencing physiologic rhythms.

4.5 PERSPECTIVES IN CHRONOPHARMACOLOGY

One of the aims of chronopharmacology could be a better control and quantification of drug effects. If circadian rhythms of susceptibility are not taken into consideration—and usually they are not considered in

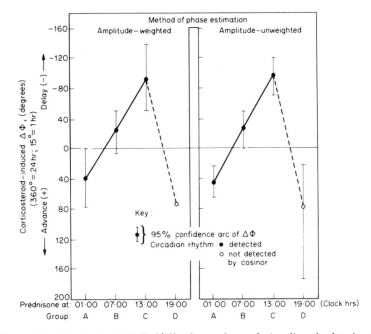

Figure 4.7. Drug-induced shift ($\Delta\phi$) of aerophase of circadian rhythm in peak expiratory flow, after giving prednisone at different hours [15] Fig. 8 and [102] Fig. 6.

conventional testing procedures— predictable changes will be ignored and will be perceived as noise in analysed values. Obviously, this will enlarge the confidence limits of measured effects. The timing of experiments and synchronization of animals can be arranged systematically to improve the accuracy of usual methods in pharmacology.

Another aim of chronopharmacology is to investigate the circadian variations of intensity or duration of pharmacologic effects. Examples of circadian changes in intensity already have been presented in *in vitro* adrenal response to ACTH [14] and in acetylcholine effects on the spontaneous activity of isolated heart right atria [88, 89]. Let us now consider another example of circadian changes in the duration of a pharmacologic effect. Antihistaminic effects of drugs can be evaluated in man by studying the extent to which they inhibit local skin reactions to intradermally injected histamine. This test was applied temporally by Reinberg and Sidi [96]. Six healthy subjects were standardized on a routine of diurnal activity and nocturnal (2300-0700) rest. For each subject a control curve was first obtained showing the circadian variation of skin reactions to histamine in the absence of any antihistaminic drug administration. A second and third group of profiles were obtained after a single (4 mg) dose of oral cyproheptadine (Periactine), Merck, Sharp and Dohm), an antihistaminic drug. Cyproheptadine was taken at 0700 for one of the last two profiles and at 1900 for the other one. For each subject and each time, the changes in response after cyproheptadine administration were expressed as percentage deviation from the control response.

The antihistaminic affect of the drug taken at 0700 lasted for 15-17 h. This inhibitory effect lasted only six to eight hours when Periactine was administered at 1900. The difference in the duration of this effect is statistically significant.

Usually, active synthetic drugs have several effects. Some of them can be interesting while others are considered vexing or incapacitating side effects. If circadian changes of desired and undesired effects exhibit phase angle differences one can try, by timing drug administration, to maximize the useful action and/or to minimize undesired effects. This possibility was studied by Smolensky [85] for methylprednisolone. Treatment was given at two times corresponding to expected times of high (1600) and low (0400) concentration of serum corticosterone in C mice standardized to L 0600-1800 D 1800-0600. At each of the injection times separate groups of immature males were injected with either 4 mg methylprednisolone hemisuccinate (solu-medrol, Upjohn) daily or 8 mg on alternate days, for 7 days. Among the variables investigated (see Table 4.2), changes in body weight appear to be a good gauge in evaluating time and graded treatment schedules. Daily single-pulsed injections of 4 mg at 1600 resulted in least body weight reduction. The same dose of methylprednisolone at 0400 daily or 8 mg on alternate days caused greater reduction of body weight than when it was injected 12 h earlier.

Daily injections, when timed to the expected and predictable peak of the serum corticosterone rhythm, caused minimal or no disturbance in the frequency, level, amplitude, and acrophase of the circadian systems investigated. In view of the aforementioned findings, steroid treatment timed close to the peak of serum corticosteroid concentration seems warranted to reduce side effects resulting from subchronic steroid therapy.

Clearly, therapy is the ultimate aim of circadian system-phased study of pharmacological effects. Chronotherapeutics is no longer a utopian concept but will have to be based on further research. The possibility of timed therapy was foreseen by Kisch [123], Volhard [120], Menzel [124], Anderson [125], among others. The development of chrono-pharmacology will help to bring their pioneering ideas to fruition.

ACKNOWLEDGEMENTS

I would like to express my deep gratitude to Professor Franz Halberg, University of Minnesota, for his participation, encouragement and advice in the preparation of this manuscript. My sincere thanks are due to Mrs. Gay Luce, N.I.M.H., for her critical comments and her untiring efforts in correcting the English text.

REFERENCES

1. F. Halberg, Temporal coordination of physiologic function, in: *Cold Spring Harbor Symp. Quant. Biol.*, **25**, 289-310 (1960).
2. F. Halberg, Physiologic 24-hour rhythms. A determinant of response to environmental agents, in: *Man's Dependence on the Earthly Atmosphere*, pp. 48-89. K. E. Schaefer (ed.). New York: MacMillan (1962).
3. F. Halberg, Circadian (about 24-hour) rhythms in experimental medicine. *Proc. R. Soc. Med.*, **56**, 253-256 (1963).
4. F. Halberg, Organisms as circadian systems: temporal analysis of their physiologic and pathologic responses, including injury and death, in: *Walter Reed Army Inst. of Res. Symp. on Medical Aspects of Stress in the Military Climate*, pp. 1-36. Washington D.C. (1965).
5. A. Reinberg, Hours of changing responsiveness in relation to allergy and the circadian adrenal cycle, in: *Circadian Clocks*, pp. 214-218. J. Aschoff (ed.), Amsterdam: North-Holland (1965).
6. F. Halberg and A. Reinberg, Rythmes circadiens et rythmes de basses fréquences en physiologie humaine. *J. Physiol. (Paris)*, **59**, 117-200 (1967).
7. A. Reinberg, The hours of changing responsiveness or susceptibility. *Perspectives Biol. Med.*, **11**, 111-128 (1967).
8. F. Halberg, Chronobiology. *Ann. Rev. Physiol.* **31**, 675-725 (1969).
9. A. Reinberg, Perspectives en chronopharmacologie, chronotoxicologie et chronothérapeutique. *Europ. J. Toxicol.*, **6**, 319-320 (1969).
10. F. Halberg, J. J. Bittner and R. J. Gully, Twenty-four hour periodic susceptibility to audiogenic convulsion in several stocks of mice. *Fed. Proc.*, **14**, 67-68 (1955).
11. F. Halberg, J. J. Bittner, R. J. Gully, P. G. Albrecht and E. L. Brackney, 24-hour periodicity and audiogenic convulsions in mice of various ages. *Proc. Soc. exp. Biol. Med.*, **88**, 169-173 (1955).
12. E. Haus and F. Halberg, Stage of adrenal cycle determining different corticosterone responses of C mice to unspecific stimulation and ACTH. 1st Cong. Endocrinol. Copenhagen Comm. 219 (1960).
13. F. Halberg, Physiologic 24-hour periodicity: general and procedural considerations with reference to the adrenal cycle. *Z. Vitam. Horm. Ferment-forsch.* **10**, 225-296 [1959].
14. F. Ungar and F. Halberg, Circadian rhythms in the *in vitro* responses of mouse adrenal to adrenocorticotropic hormone. *Science*, **137**, 1058-1060 (1962).
15. F. Halberg, Ritmos y corteza suprarenal. IV. Simposio panamericano de farmacologia y terapeutica Mexico 1967. *Excerpta med. Int. Cong. Ser.*, **85**, 7-39 (1967).

16. F. Halberg, F. C. Bartter, W. Nelson, R. Doe and A. Reinberg, Chronobiologie. *J. Europ. Toxicol.*, 6, 311-318 (1969).
17. A. Reinberg, La chronobiologie: une étape nouvelle de l'étude des rhythmes biologiques. *Sciences*, 1, 181-197 (1970).
18. A. Reinberg, Biorythms et chronobiologie. *Presse méd.* 77, 877-8 (1969).
19. A. Reinberg and J. Ghata, *Biological Rhythms*, 138 pp. New York: Walker (1964).
20. E. Bünning, *The Physiological Clock*, 145 pp. Berlin: Springer-Verlag (1964).
21. W. Menzel, *Menschliche Tag-Nacht-Rhythmik und Schichtarbeit*, 189 pp. Basel/Stuttgart: Benno Schwabe (1962).
22. A. Sollberger, *Biological Rhythm Research*, 461 pp. New York: Elsevier (1965).
23. J. L. Cloudsley-Thompson, *Rhythmic Activity in Animal Physiology and Behavior*, 236 pp. New York: Academic Press (1961).
24. J. E. Harker, *The Physiology of Diurnal Rhythms*, 114 pp. London: Cambridge Univ. Press (1964).
25. B. C. Goodwin, *Temporal Organization in Cells*, 163 pp. New York: Academic Press (1963).
26. E. Haus and F. Halberg, Circannual rhythm in level and timing of serum corticosterone in standardized inbred mature C-mice. *Env. Research*, 3, 81-106 (1970).
27. M. Beauvallet, J. Fugazza and M. Solier, Nouvelles recherches sur les variations saisonnieres de la noradrenaline cerebrale. *J. Physiol. (Paris)*, 54, 289-290 (1968).
28. N. Kleitman, *Sleep and Wakefulness*, 552 pp. Chicago: Univ. Chicago Press (1965)..
29. J. N. Mills, Human circadian rhythms. *Physiol. Rev.*, 46, 128-171 (1966).
30. C. Kayser and A. A. Heusner, Le rythme nycthéméral de la dépense d'énergie. Etude de physiologie comparée. *J. Physiol. (Paris)*, 59, 3-117 (1967).
31. *Circadian Clocks*, 479 pp, J. Aschoff (ed), Proc. Feldafing Summer School. Amsterdam: North-Holland.
32. *Biological Clocks*, 524 pp. Cold Spring Harbor Symp. Quant. Biol. (1960).
33. *Circadian Systems*, 93 pp. S. F. Fomon (ed), Rept 39th Ross Conf. Pediat. Res. Ross Labs, Columbus, Ohio (1961).
34. *Rhythmic Functions in the Living System*, 1326 pp. W. Wolf (ed.), Ann. N.Y. Acad. Sci. (1962).
35. *Photo-Neuro-Endocrine Effects in Circadian Systems, with Particular Reference to the Eye*, 645 pp., E. B. Hague (ed), *Ann. N.Y. Acad. Sci.* (1964).
36. *Symposium on Rhythms*, pp. 886-994, 1116-1117, in Verhandl deut. Ges. inn. Med. 33rd Kongr. München: Bergmann (1967).
37. *Proc. 1st Intern. Symp. Biorhythms in Exptl. Clin. Endocrinol.* Florence May 30-31, *Rass. Neur. Veg.* (1966).
38. *La photorégulation de la reproduction chez les oiseaux et les Mammifères*, 588 pp., J. Benoît and I. Assenmacher (eds.), Colloq. Inter. C.N.R.S. Montpellier 1967. C.N.R.S. Paris. 1970 (No. 172).
39. A. Petrovic and C. Kayser, L'activité gonadotrope de la préhypophyse du Hamster (*Circetus cricetus*) au cours de l'année. *C.r. Séanc. Soc. Biol.*, 151, 996-998 (1957).
40. J. Aschoff, Comparative physiology; diurnal rhythms. *A. Rev. Physiol.*, 25, 581-600 (1963).
41. R. D. Green and J. W. Miller, Catecholamine concentrations: changes in plasma of rats during estrous cycle and pregnancy. *Science*, 151, 825-826 (1966).

42. C. S. Pittendrigh, On temporal organization in living systems. *Harvey Lectures*, 56, 93-125 (1961).
43. F. Halberg and H. Panofsky, Thermo-variance spectra, method and clinical illustration. *Expl. Med. Surg.*, 19, 284-309 (1961).
44. F. Halberg, H. Panofsky, M. Diffley, M. Stein and G. Adkins, Computer techniques in the study of biologic rhythms. *Ann. N.Y. Acad. Sci.*, 115, 695-720 (1964).
45. F. Halberg, M. Engeli, C. Hamburger and D. Hillman, Spectral resolution of low frequency, small-amplitude rhythms in excreted 17-ketosteroids. *Acta endocr. Copenh.* Suppl. 103, 5-54 (1965).
46. F. Halberg, Y. L. Tong and E. A. Johnson, Circadian system phase: an aspect of temporal morphology: procedure and illustrative examples, in: *The Cellular Aspects of Biorhythms*, pp. 20-48. Symp. on rhythmic Research, 8th Int. Cong. Anat. Berlin: Springer-Verlag (1967).
47. F. Halberg, Resolving power of electronic computers in chronopathology—an analogy to microscopy. *Scientia*, 101, 412-419 (1966).
48. F. Halberg, Grundlagenforschung zur Ätiologie des Karsinoms. *Mkurseärz H Fortbild.*, 14, 67-77 (1964).
49. Cl Aron, J. Roos and G. Asch, Données nouvelles sur le rôle joué par la ponte provoquée dans les phénomènes de reproduction chez la ratte. *Ann. Endocrinol.*, 28, 19-30 (1967).
50. M. Apfelbaum, A. Reinberg, P. Nillus and F. Halberg, Rythmes circadiens de l'alternance veille-sommeil pendant l'isolement souterrain de sept jeunes femmes. *Presse méd.*, 77, 879-882 (1969).
51. A. Reinberg, F. Halberg, J. Ghata and M. Siffre, Spectre thermique (rythmes de la température rectale) d'une femme adulte saine avant, pendant et après son isolement souterrain de trois mois. *C.r. hebd. Séanc. Acad. Sci., Paris*, 262, 782-785 (1966).
52. F. Halberg, M. B. Visscher and J. J. Bittner, Relation of visual factors to eosinophil rhythm in mice. *Am. J. Physiol.*, 179, 229-235 (1954).
53. J. Aschoff, Zeitgeber der tierischen Tagesperiodik. *Naturwissenschaften*, 41, 49-56 (1954).
54. C. S. Pittendrigh and V. G. Bruce, An oscillator model for biological clocks, in: *Rhythmic and Synthetic Processes in Growth*, pp. 75-109. Princeton N.J.: Princeton Univ. Press (1957).
55. F. Halberg and A. N. Stephens, 24-hour periodicity in mortality of C mice from *E. coli* lipopolysaccharide. *Fed. Proc.*, 17, 339 (1968).
56. F. Halberg, E. A. Johnson, B. W. Brown and J. J. Bittner, Susceptibility rhythms to *E. coli* endotoxin and bioassay. *Proc. Soc. exp. Biol. Med.*, 103, 142-144 (1960).
57. F. Halberg and A. N. Stephens, Susceptibility to ouabain and physiologic circadian periodicity. *Proc. Minn. Acad. Sci.*, 27, 139-143 (1959).
58. F. Halberg, E. Haus and A. N. Stephens, Susceptibility to ouabain and physiologic 24-hour periodicity. *Fed. Proc.*, 18, 63 (1959).
59. E. Marte and F. Halberg, Circadian susceptibility rhythm of mice to librium. *Fed. Proc.*, 20, 305 (1961).
60. E. Haus and F. Halberg, 24-hour rhythm in susceptibility of C mice to toxic dose of ethanol. *J. appl. Physiol.*, 14, 878-880 (1959).
61. R. J. Ertel, F. Halberg and F. Ungar, Circadian system phase dependant toxicity and other effects of methopyrapone (SU 4885) in the mouse. *J. Pharmac. exp. Ther.*, 146, 385-399 (1964).
62. E. Haus and F. Halberg, Interactions of a chemical carcinogen with neuroendocrine factors in mouse breast cancer. *Experientia*, 18, 340-341 (1962).

63. J. G. Gosselink and L. C. Standifer, Diurnal rhythms of sensitivity of cotton seedling to herbicides. *Science*, **158** (No. 3797), 120-121 (1967).
64. C. L. Cole and P. L. Akisson, Daily rhythm in the susceptibility of an insect to a toxic agent. *Science*, **144**, 1148-1149 (1964).
65. C. L. Cole and P. L. Akisson, A circadian rhythm in the susceptibility of an insect to an insecticide. In: *Circadian Clocks*, J. Aschoff (ed.) pp. 309-313. Amsterdam: North-Holland (1965).
66. B. Pollick, J. W. Nowosielski and J. A. Naegle, Daily sensitivity rhythm of the two spotted spider mite, *Tetranychus urticae* to D.D.V.P. *Science*, **145**, 405 (1964).
67. W. N. Sullivan, B. Cawley, D. K. Hayes, J. Rosenthal and F. Halberg, Circadian rhythm in susceptibility of house flies and madeira cockroaches to pyrethrum. *J. econ. Ent.*, **63**, 159-163 (1970).
68. F. Halberg, W. W. Spink, P. G. Albrecht and R. J. Gully, Resistance of mice to brucella somatic antigen, 24-hour periodicity and the adrenals. *J. clin. Endocr. Metab.*, **15**, 887 (1955).
69. F. Jones, E. Haus and F. Halberg, Murine circadian susceptibility resistance cycle to acetylcholine. *Proc. Minn. Acad. Sci.*, **31**, 61-62 (1963).
70. J. H. Matthews, E. Marte and F. Halberg, A circadian susceptibility—resistance cycle of male B_1 mice to fluothane. *Can. Anaesth. Soc. J.*, **11**, 280-290 (1964).
71. P. R. Wiepkema, Aurothioglucose sensitivity of CBA mice injected at two different times of day. *Nature*, **209**, 937 (1966).
72. E. F. Lutsch and R. W. Morris, Circadian periodicity in susceptibility to lidocaine hydrochloride. *Science*, **156**, 100-102 (1967).
73. J. E. Pauly and L. E. Scheving, Temporal variations in the susceptibility of white rats to pentobarbital sodium and tremoring. *Int. J. Neuropharmacol.*, **3**, 651-658 (1964).
74. L. E. Scheving, D. F. Vedral and J. E. Pauly, A circadian susceptibility rhythm in rats to pentobarbital sodium. *Anat. Rec.*, **160**, 741-750 (1968).
75. L. E. Scheving and D. F. Vedral, Daily circadian rhythm in rats to D-amphetamine sulphate: effect of blinding and continuous illumination on the rhythm. *Nature*, **219**, 621-622 (1968).
76. L. E. Scheving, Circadian variation in susceptibility of the rat to D-amphetamine sulfate. *Anat. Rec.*, **160**, 422 (1969).
77. Tien Ho Tsai, L. E. Scheving and J. E. Pauly, Circadian rhythms in plasma inorganic phosphorus and sulfur of the rat: also in susceptibility to strychnine. *Jap. J. Physiol.*, **20**, 12-29 (1970).
78. B. Schwemmle and O. L. Lange, Endogen-tagesperiodische Schankungen der Hirtzresistenz bei *Kalanchoë blossfeldiana*. *Planta (Berlin)*, **53**, 134-144 (1959).
79. L. Rensing, Ein Circadianer Rhythmus der Empfindlichkeit gegen Röntgenstrahlen bei Drosophila. *Z. vergl. Physiol.*, **62**, 214-220 (1969).
80. F. Halberg, The 24-hour scale: a time dimension of adaptative functional organization. *Perspective Biol. Med.*, **3**, 491-527 (1960).
81. D. J. Pizzarello, D. Isaak, K. E. Chua and A. L. Rhyne, Circadian rhythmicity in the intensity of two strains of mice to whole body radiation. *Science*, **145**, 286-291 (1964).
82. R. F. Nelson, Variation of radiosensitivity of mice with time of day. *Acta radiol.*, **4**, 91 (1966).
83. Y. G. Grigoryev, N. G. Darenskya, Y. P. Druzhinin, S. S. Kusnetsova and V. M. Seraya, Diurnal rhythms and ionizing radiation effects. Abstracts: Cospar XIIth Plenary Meeting, Prague. 163-164 (1969).

84. D. R. Lamond and W. H. Braden, Diurnal variation in response to gonadotropin in the mouse. *Endocrinology*, 64, 921-936 (1959).
85. M. Smolensky, Reduction of side effects and induction of phase shifts by circadian timing of daily or alternate-day single-pulsed solumedrol ® injections. Dissertation Report. Univ. Minn. (1969).
86. L. E. Scheving and D. Vedral, Circadian variation in susceptibility of the rat to several different pharmacological agents. *Anat. Rec.*, 154, 417 (1966).
87. L. E. Scheving and J. E. Pauly, The effect of colchicine on the circadian rhythm of cellular division in the cornea of rats. (In press.)
88. R. P. Spoor and D. B. Jackson, Circadian rhythms: variation in sensitivity of isolated rat atria to acetylcholine. *Science*, 154, 782-784 (1966).
89. D. B. Jackson, 24-hour variation in sensitivity of isolated rat atria to autonomic drugs: a pharmacologic study. Ph.D. thesis University of South Dakota (1967).
90. C. F. Stroebel, Biologic correlates of disturbed behaviour in the rhesus monkey, in: *Circadian Rhythms in Non Human Primates*, pp. 91-105, F. H. Rohles (ed.). S. Karger, Basel (1969).
91. J. Mouret, Rythme circadien et sommeil chez l'animal. *Europ. J. Toxicol.*, 2, 333-335 (1969).
92. D. T. Krieger and H. P. Krieger, Circadian pattern of plasma 17-hydroxycorticosteroid: alteration by anticholinergic agents. *Science*, 155, 1421-1422 (1967).
93. J. A. Anderson, Discussion: *Circadian Systems*, pp. 54-55, S. J. Fomon (ed.). 39th Ross Conf. on Pediatric Res. Ross Lab. Columbus, Ohio (1961).
94. A. Reinberg, E. Sidi and J. Ghata, Circadian reactivity rhythms of human skin to histamine or allergen and the adrenal cycle. *J. Allergy*, 36, 273-283 (1965).
95. A. Reinberg, Z. Zagulla-Mally, J. Ghata and F. Halberg, Circadian reactivity rhythm of human skin to house dust, penicillin and histamine. *J. Allergy*, 44, 292-306 (1969).
96. A. Reinberg and E. Sidi, Circadian changes in the inhibitory effects of an antihistaminic drug in man. *J. invest. Derm.*, 46, 415-419 (1966).
97. A. Reinberg, Z. Zagulla-Mally, J. Ghata and F. Halberg, Circadian rhythm in duration of salicylate excretion referred to phase of excretory rhythms and routine. *Proc. Soc. exp. Biol. Med.*, 124, 826-832 (1967).
98. T. Nichols, C. A. Nugent and F. H. Tyler, Diurnal variations in suppression of adrenal function by glucocorticoids. *J. clin. Endocr. Metab.*, 25, 343-349 (1965).
99. R. D'agata, C. Di Stephano, C. Furno and L. Mughini, Sulle variazioni del ritmo circadiano surrenalico dopo somministrazione orale di glicocorticoidi. *Riv. crit. clin. Med.*, 68, 652-657 (1968).
100. H. Bricaire, J. Leprat and J. P. Luton, Etat actuel des explorations dynamiques de la corticosurrénale. *Presse méd.*, 76, 2157-2160 (1968).
101. J. Rutenfranz and R. Singer, Untersuchungen zur Frage einer Abhägigkeit der Alkoholwirkung von der Tageszeit. *Int. Z angew. Physiol.*, 24, 1-17 (1967).
102. K. Reindl, C. Falliers, F. Halberg, H. Chai, D. Hillman and W. Nelson, Circadian acrophases in peak expiratory flow rate and urinary electrolyte excretion of asthmatic children: phase shifting of rhythms by prednisone given in different circadian system phases. *Rass. Neur. Veg.*, 23, 5-26 (1969).
103. A. Reinberg and C. J. Falliers, Circadian variations in the responses of asthmatic children to methylprednisolone. VII. Int. Congr. Allergology. Florence Oct. 1970. Excerpta med. Int. Congr. Series No. 211, 85-86.

104. C. W. M. Wilson, The occurrence of circadian histamine rhythms in the rat. *Int. Arch. Allergy*, **28**, 32-34.
105. M. M. Martin and Helman, Temporal variation in SU 4885 responsiveness in man: evidence in support of circadian variation in ACTH secretion. *J. clin. Endocr. Metab.*, **24**, 253-260 (1964).
106. M. M. Martin and D. H. Mintz, Effect of altered thyroid function upon adrenocorticotical ACTH and methopyrapone (SU 4885) responsiveness in man. *J. clin. Endocr. Metab.*, **25**, 20-27 (1965).
107. F. M. Radzialowski and W. F. Bousquet, Daily rhythmic variation in hepatic drug metabolism in the rat and mouse. *J. Pharm. exp. Therap.*, **163**, 229-238 (1968).
108. T. Vanden Driessche, The nuclear control of the chloroplasts' circadian rhythms. *Sci. Prog. London*, **55**, 293-303 (1967).
109. C. P. Barnum, C. D. Jardetsky and F. Halberg, Time relations among metabolic and morphologic 24-hour changes in mouse liver. *Am. J. Physiol.*, **195**, 301-310 (1958).
110. J. W. Hastings and Keynan, Molecular aspects of circadian systems, in: *Circadian Clocks*, pp. 167-182, J. Aschoff (ed.). Amsterdam: North-Holland (1965).
111. C. D. Jardetzky, C. P. Barnum and F. Halberg, Physiologic 24-hour periodicity in nucleic acid metabolism and mitosis of immature growing liver. *Am. J. Physiol.*, **187**, 608-616 (1956).
112. R. J. Wurtman and J. Axelrod, Daily rhythmic changes in tyrosine transaminase activity of the rat liver. *Proc. natn. Acad. Sci. U.S.A.*, **57**, 1594 (1967).
113. J. N. Mills, Circadian rhythms during and after three months in solitude underground. *J. Physiol. (Lond.)*, **174**, 217-231 (1964).
114. K. E. Schaffer, M. J. Jacey, C. R. Carey and W. F. Mazzone, Saturation excursion diving: biochemical cycle functions in lactic deshydrogenase, lactate and pyruvate responses. *Aerospace Med.*, **39**, 343-350 (1968).
115. A. Reinberg, L'homme et les rythmes circadiens, Cahiers Sandoz, Paris (1966).
116. J. Ghata, F. Halberg, A. Reinberg and M. Siffre, Rythmes circadiens désynchronisés du cycle social (17-hydroxycorticosteroides, température rectale, veille-sommeil) chez deux adultes sains. *Annls. Endocr. (Paris)*, **30**, 245-260 (1969).
117. R. Wever, Mathematical models of circadian rhythms and their applicability to men, in: *Cycles biologiques et psychiatrie*, pp. 61-72, J. de Ajuriaguerra (ed.). Symposium Bel-Air III, Paris: Masson (1968).
118. G. R. Spratto and J. W. Miller, The effects of various estrogens on the weight, catecholamine content and rate of contractions of rat uteri. *J. Pharmac. exp. Ther.*, **161**, 1-6 (1968).
119. A. Reinberg, J. Ghata and E. Sidi, Nocturnal asthma attacks: their relationship to the circadian adrenal cycle. *J. Allergy*, **34**, 323-330 (1963).
120. E. Haus, F. Halberg, W. Nelson and D. Hillman, Shifts and drifts in phase of human circadian system following intercontinental flights and in isolation. *Fed. Proc.*, **27**, 224 (1968).
121. K. E. Klein, H. M. Wegmann and H. Brünner, Circadian rhythms in indices of human performance physical fitness and stress resistance. *Aerospace Med.*, **39**, 512-518 (1968).
122. A. Reinberg, Evaluation of circadian dyschronism during transmeridian flights: life Sciences and Space research VIII, pp. 172-174, North-Holland (1970).
123. F. Kisch, Über die 24-stunden Rhythmik von Wachen—Schlafen und die kurative Bedemtung des Schlafes bei Herzkranken. *Wien. klin. Wschr.*, **9**, 270-275 (1938).

124. F. Volhard, Aussprache zum Thema Kreislauf und Atmung. *Verh. dt. Ges. Kreislaufforsch.*, 13, 127 (1940).
125. W. Menzel, Der 24-Stunden Rhythmus des menschlichen Blut Kreislaufes. *Ergebn inn. Med. Kinderhk.*, 61, 1 (1942).
126. A. Reinberg, Methodologic considerations for human chronobiology. *J. interdiscipl. Cycle Res.*, 2, 1-15 (1971).
127. T. Emlen and W. Kem, Activity rhythm in Peromyscus its influence on rates of recovery from nembutal. *Science*, 142, 1682-1683 (1963).
128. H. A. Lindsay and V. S. Kullman, Pentobarbital sodium: variation in toxicity. *Science*, 151, 576-577 (1966).
129. W. M. Davis, Day-night periodicity in Pentobarbital response of mice and the influence of socio-psychological conditions. *Experientia*, 18, 235-237 (1962).
130. G. Wahlström, The circadian rhythm of self-selected rest and activity in the canary and the effects of barbiturates reserpine, monoamineoxydase inhibitors and enforced dark periods. *Acta physiol. Scand.*, 65, Suppl. 250, 1-67 (1965).
131. R. Sable, R. Agid and D. Abadie, Effets de la morphine sur le métabolisme glucidique en fonction du rythme et du jeune chez la souris. *J. Physiol., Paris*, 62 (Suppl. 1), 214 (1970).
132. R. W. Lee and A. H. Meier, Diurnal variations of the fattening response to prolactin in the golden top minnow *Fundulus chrysotus*. *J. exp. Zool.*, 166, 307-316 (1967).
133. A. H. Meier, Diurnal variations of metabolic responses to prolactin in lower vertebrates. *Gen. comp. Endocrinol.* (Suppl. 2), 55-62 (1969).
134. A. H. Meier, Diurnal variation of the fattening response to prolactin in the white-throated sparrow *Zonotrichia albicollis*. *Gen. comp. Endocr.*, 8, 110-114 (1967).
135. V. Nair and R. Casper, The influence of light on daily rhythm in hepatic drug metabolizing enzymes in rat. *Life Sci.*, 8 (part 1), 1291-1298 (1969).
136. J. W. Nowosielski, R. L. Patton and J. A. Naegele, Daily rhythms of narcotic sensitivity in the house cricket, *Gryllus domesticus* L. and the two-spotted spider mite, *Tetranychus urticae* Koch. *J. Cell. comp. Physiol.*, 63, 393-398 (1964).
137. M. Serio, M. Della Corte, Piolanti S. Romano, L. Giglioli and G. Giusti, Transverse circadian rhythmometry of plasma cortisol in diabetic subjects in relation to therapy. *Annls. Endocr.*, 32, 403-408 (1971).
138. S. S. Cardoso, A. L. Ferreira, A. C. M. Camargo and G. Bohn, *Experientia*, 24, 569-570 (1968).
139. S. S. Cardoso and J. R. Carter, *Proc. Soc. exp. Biol. Med.*, 131, 1403-1406 (1969).
140. M. Garcia-Sainz, F. Halberg and V. Moore. *Rev. Mex. Radiol.*, 22, 131-146 (1968).
141. A. Reinberg and F. Halberg, Circadian chronopharmacology. *Ann. Rev. Pharmacol.*, 11, 455-492 (1971).
142. A. Reinberg, P. Gervais, M. Morin and C. Abulker, Rythme circadien humain du seuil de la réponse bronchique à l'acétylcholine. *C.r. Séanc. Acad. Sci., Paris*, 272, 1879-1881 (1971).
143. A. Reinberg, J. Ghata, F. Halberg, M. Apfelbaum, P. Gervais, P. Boudon, C. Abulker and J. Dupont, Distribution temporelle du traitement de l'insuffisance corticosurrénalienne. Essai de chronothérapeutique. *Annls. Endocr.*, 32, 566-573 (1971).

144. J. Aschoff, Eigenschaften der menschlichen Tagesperiodik. *Schriftenreihe Arbeitsmedizine Sozial medizine Arbeitshygiene,* 38, 21-43 (1970).
145. E. Haus, F. Halberg, L. E. Scheving, J. E. Pauly, S. Cardoso *et al.* Increased tolerance of leukemic mice to arabinosyl cytosine with schedule adjusted to circadian system. *Science,* 177, 80-82 (1972).

CHAPTER 5

Circadian Rhythms of Parasites

F. Hawking

Medical Research Council
National Institute for Medical Research
Mill Hill, London NW7

The two- and three-day rhythms of malaria have been known for over
two millenia since the age of Hippocrates, and the periodicity of
microfilariae has been known for almost a century, but for most of this
time they were accepted as medical curiosities and ceased to attract
attention. It is only during recent years that they have been studied as

biological rhythms and that their purpose and mechanisms have been elucidated. Other rhythms in parasites have come to light more recently.

With all circadian rhythms of parasites, two questions are to be asked. The first is: Why does the parasite do it? What benefit does it derive from it? In practically all cases the answer is that the rhythm somehow facilitates transmission to a new host. The second is: How does the parasite do it? How is the rhythm produced and how does the parasite know when to time it? This question is much more difficult to answer since it varies from parasite to parasite. As will be seen below, it is only in a few instances that an answer (even a partial answer) can be given. The rhythm of the parasite must take its clue from some component of the circadian rhythms of the host, but different parasites take different components. The temperature cycle of the host is often the most promising one to investigate.

The circadian rhythms of parasites may be divided into various classes:

(1) Rhythms in which the same individuals migrate backwards and forwards in the body of the host according to a circadian pattern, e.g. microfilariae, trypanosomes of frogs.

(2) Rhythms depending on the synchronous cell division of the parasites according to a 24-, 48-, or 72-h pattern, e.g. malaria parasites.

(3) Synchronous ejection of infective forms from the body of the host at some particular time of the day or night, e.g. oocysts of sparrows, ova of *Schistosoma haematobium*

(4) Migration of worms up and down the intestine on a 24-h pattern, e.g. *Hymenolepis diminuta*.

5.1 RHYTHMS DEPENDING ON MIGRATIONS OF THE SAME INDIVIDUALS

5.1.1 Microfilariae

Microfilariae are the larvae of filarial worms, which are long, thin nematodes embedded in the tissues of man, other mammals, birds and cold blooded vertebrates. The microfilariae, which are usually about $250 \times 8 \ \mu m$ (Plate 5.1), circulate in the blood stream until some of them are picked up by a blood-sucking arthropod and conveyed after appropriate development to a new host. Many species of microfilariae show "periodicity" as was discovered by Patrick Manson in 1878, i.e. the microfilariae are numerous in the blood at some part of the twenty-four hours but scanty at other times. (A sceptical wit asked Manson if he imagined that the microfilariae carried watches in their pockets so that they could tell the time; but this was long before the days of the

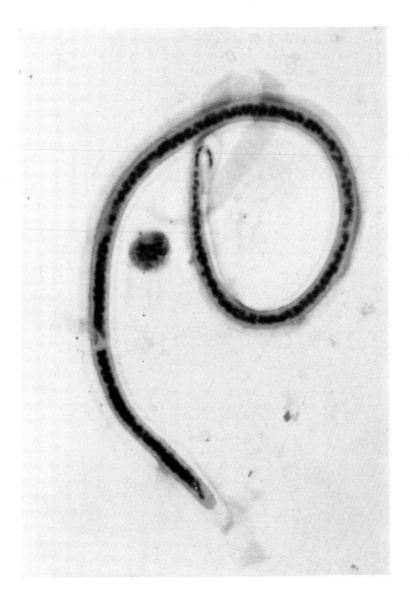

Plate 5.1. Microfilaria of *Wuchereria bancrofti*. Magnification ×1260.

[*To face page 154*

"biological clock".) This periodicity obviously constitutes a circadian rhythm, and among well-known species it may take one of four types as illustrated in Fig. 5.1:

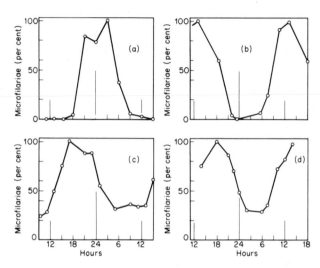

Figure 5.1. Different patterns of the microfilaria cycle. The numbers of microfilariae are expressed as percentages of the maximum number observed. (a) *W. bancrofti*, (b) *L. loa*, (c) *D. immitis*, (d) *W. bancrofti* Pacific.

(1) Microfilariae numerous in the blood at night, but rare or absent by day, e.g. *Wuchereria bancrofti,* the main human filaria. This is transmitted by mosquitoes which bite at night.

(2) Microfilariae present throughout the 24 h but more numerous in the afternoon, e.g. the Pacific type of *W. bancrofti* which is transmitted mostly by day-biting mosquitoes.

(3) Microfilariae numerous in the blood by day but rare or absent by night, e.g. *Loa loa* of man in West Africa which is transmitted by the day-biting *Chrysops*.

(4) Microfilariae which are more numerous by evening or night and less numerous in the early morning, e.g. *Dirofilaria immitis* of dogs, which is transmitted by mosquitoes which bite in the evening or night.

In all these examples the microfilariae are most numerous about the time when the insect vector sucks blood, and this part of the cycle is clearly an adaptation to insect transmission. (There are also so-called subperiodic strains of *W. bancrofti* in Malaya in which the cycle is less marked, considerable numbers of microfilariae persisting in the peripheral blood during the day. Furthermore, there are non-periodic species of microfilariae which show no circadian rhythm, and such species are excluded from the subsequent discussion.)

During the time (usually day) when microfilariae disappear from the peripheral blood, they are found in great numbers in the small vessels of the lungs, as was demonstrated by Hawking and Thurston [1]. This accumulation of microfilariae in the lungs is probably a dynamic one rather than a static one. That is, any particular microfilaria may remain in the lungs for only a relatively short time at a stretch (not continuously for 12 h); but during the period of accumulation, the number of microfilariae entering the lungs is greater than the number leaving it, and the time spent by each microfilaria in the lungs is greater than the time spent passing round the general circulation; and during the period of liberation, conversely. Apparently the lungs form a large reservoir resembling the submerged bulk of an iceberg, and the number of microfilariae seen in the peripheral blood corresponds only to the part of the iceberg above the surface. Even when the number of microfilariae in the peripheral blood seems high, the absolute number in the lungs may be greater or several times greater. (If an infected animal or man is examined after death, great numbers of microfilariae are found in the left side of the heart and in the aorta as was reported by Manson, Rodenwaldt, and other early workers; but this is due to release from the lungs during the final death agony, and it does not represent the conditions which occur during life.)

The cycle may be considered to consist of a passive phase (during the night) when the microfilariae are evenly distributed throughout the blood and so they seem to be numerous in the peripheral blood; and of an active phase in which the microfilariae that are brought to the lungs by the circulating blood somehow hold themselves there, and thus all the microfilariae in the circulation soon accumulate in the lungs. The microfilariae of *W. bancrofti* must clearly be in the peripheral blood during the night, so as to have the chance of being sucked up by the vector mosquitoes. In order however to explain their accumulation in the lungs by day, we must postulate that owing to some special physiological condition (explained below) accumulation in the lungs is favourable to the *survival* of the microfilariae even if unfavourable to their *transmission*. Accordingly the periodicity of microfilariae is a compromise which enables them to make the best of both worlds—they spend the day-time in the lungs enjoying the good conditions there—they come out in the evening hoping to meet a mosquito—and in the morning, if they have been disappointed, they go back to the lungs to rest and recuperate until the next night.

Causes of Accumulation in the Lung

This accumulation of microfilariae in the lungs during the active phase of the cycle cannot be due to any opening or shutting of capillaries by the host, since the microfilariae of *W. bancrofti* and of *L. loa* are both of the same size and shape yet the former accumulates in the human lung by day and the latter by night. The accumulation must be due to action by the microfilariae themselves. The behaviour of the microfilariae of *Dipetalonema witei* in the-capillaries of the lung of gerbils with active circulation has been studied cinematographically by Hawking and Clark [2]. The microfilariae have no hooks or suckers or other mechanical appliances for attaching themselves to the walls of the capillaries. They are never seen lying motionless in a blocked capillary. They are always wriggling. This wriggling really consists of a wave of contraction/ relaxation on opposite sides of the body, which starts at the head and passes backwards, so that sinuous waves pass backwards down the body. In an open space or wide vessel these waves lead only to futile lashing; but in a narrow vessel, where there is lateral friction from the walls, the waves carry the microfilaria quickly forward. Sometimes however the excitation centre at the head end seems to be inhibited (probably by a higher oxygen tension, see below) and then waves of contraction start at the tail end and travel forward. In a narrow vessel such waves carry the microfilariae backwards through the vessels. It is believed that the accumulation of microfilariae in the lungs during the day time is based mechanically on this backward migration in narrow vessels (as will be explained in more detail below).

Consideration of the anatomy and physiology of the pulmonary vessels indicates the point at which such accumulation is most likely to occur. Thus anatomically, the vessels consist of wide arteries tapering down through narrower and narrower arterioles until the tube-like capillaries are reached; after that the vessels broaden out again into wide venules and still wider veins. Accumulation of foreign objects is most likely to occur in the final narrowing of the arterioles; once the cylindrical capillaries are reached there is no mechanical obstacle to prevent the object being swept onwards into the widening venules. Physiologically when the blood reaches the capillaries, there is an abrupt change (increased oxygen tension, decreased carbon dioxide); but once the blood has passed half way down the capillaries, there is no reason why a microfilaria should not continue on its way with the fully oxygenated blood. Hence on anatomical and physiological grounds it seems likely that accumulation of microfilariae must be located mainly in the pre-capillary arterioles, and certainly not in the post-capillary venules.

As regards the physiological changes which take place as blood passes through the pulmonary capillaries, there are two main kinds (1) the great increase in oxygen tension (2) the moderate decrease in carbon dioxide tension, with secondary changes in pH and electrolyte distribution. During many years we have studied the effect on the microfilaria count of changing the oxygen tension or the carbon dioxide tension in the air breathed by the host. With many different kinds of microfilariae we have observed only small effects produced by alteration of carbon dioxide tension or by alkaline infusions (and these might be ascribed to the secondary changes of oxygen tension produced). On the other hand, we have usually found great changes produced by alteration of the oxygen tension, although the direction of such changes may be different with different species of microfilariae. Consequently we believe that the accumulation of microfilariae in the lungs is somehow related to the great increase of oxygen tension which occurs there. In the human lung, the increase of oxygen tension begins on the arteriolar side in vessels as wide as 2 mm in diameter [3]. This fact confirms the belief that accumulation of microfilariae (when it occurs) is located on the arteriolar side of the capillaries, and certainly not on the venule side.

Regulation of Rhythm

Having thus reviewed the general background for the accumulation of microfilariae in the lungs, we may now consider how such accumulation is regulated to form a circadian rhythm. In such consideration I will endeavour to make distinct which parts of the description are well-established experimental observations, and which parts are theoretical speculations devised to explain the observed facts [4].

Periodic microfilariae may be divided into three groups, according to their reactions to various physiological stimuli, these reactions being related to the mechanisms which control their accumulation in, and their liberation from, the lungs.

Group 1

The first group consists of *W. bancrofti, Brugia malayi,* and probably other Brugias. These are mostly parasites of man and are transmitted by night-biting mosquitoes. If an infected man (or monkey) is made to breathe oxygen (above a concentration of 30-40%) at night, when there are many microfilariae in the peripheral blood, the microfilaria count falls in a few minutes down to about 20% of its original value, i.e. 80% of the circulating microfilariae accumulate in the lungs. When the breathing of oxygen ceases, the microfilaria count regains its previous level in a few minutes (Fig. 5.2). Similarly, if the man takes violent exercise at night, the number of microfilariae in the peripheral blood falls rapidly during the exercise and rises again quickly afterwards. Certain other stimuli, such as hyperpnoea either voluntary by conscious effort, or as a result of breathing carbon dioxide, produce the same effect to a less extent; presumably this is a secondary reaction to the raising of the oxygen tension in the alveoli by hyperpnoea. Many other stimuli have been investigated, e.g. intravenous infusion of sodium bicarbonate, injection of adrenaline, pituitrin, insulin, glucose, heparin, alteration of body temperature, etc. but none of these have significant effects on the distribution of the microfilariae of this group.

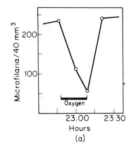 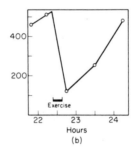

Figure 5.2. The change in the microfilaria count of *W. bancrofti* produced by (a) breathing oxygen, (b) vigorous exercise.

These observations may be explained by the following hypothesis. The microfilariae are unwilling to enter a region (such as the pulmonary arterioles and capillaries) where the oxygen tension increases sharply. When they are carried by the circulating blood down the pulmonary arterioles into such a region, the increased oxygen tension at their anterior ends inhibits the normal waves of contraction; and then contractions start from the tail and travel up the microfilariae in reverse of the normal direction. These reverse contractions carry the microfilariae backwards up the narrow vessels, so that they are held in this arteriolar zone, being pushed forwards by the circulating blood and backwards by their own reversed contractions and they therefore accumulate in this part of the lungs. Thus the increase of oxygen tension (steep oxygen gradient) acts as a metaphorical "oxygen barrier" to the microfilariae. The microfilariae are sensitively adjusted to the height of this oxygen barrier (i.e. to the difference between venous oxygen tension and arterial oxygen tension) which in man during the day time is approximately 95 mm Hg less 40 mm Hg, i.e. 55 mm Hg. By night however the oxygen tension in alveolar air and in arterial blood is a little less (about 87 mm Hg) and the mean tension in venous blood is higher (45 mm Hg) so the "oxygen barrier" (arterio-venous difference) is lower (about 42 mm Hg) and the microfilariae pass through the barrier into the general circulation (Fig. 5.3). During the night, if the man is made to breathe oxygen the arterial oxygen tension is raised to about 600 mm Hg and the microfilariae do not pass this; accordingly they accumulate in the lungs and the peripheral microfilaria count falls as described above. If the man takes strong exercise, the arterial oxygen tension remains the same or falls slightly (e.g. 95 mm Hg) but the venous oxygen tension falls greatly (e.g. to 20 mm Hg) so that the arteriovenous difference becomes 75 mm Hg and the microfilariae are again held back.

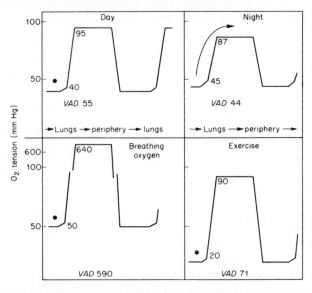

Figure 5.3. Diagrammatic illustration of the oxygen tension in lungs and peripheral circulation in various circumstances and its relation to the passage of microfilariae of *W. bancrofti* through the lungs. When the oxygen tension difference between venous and arterial blood *(VAD)* is 55 mmHg or more, microfilariae do not pass through this "barrier" and they accumulate in the pulmonary arterioles (black spot); at night, the difference is only 44 mm (approx) and the microfilariae pass through the lungs (arrow).

According to this hypothesis, the factor causing microfilariae to accumulate in the lungs is their inborn resistance to entering a zone of raised oxygen tension. Such resistance is quite comprehensible on general biological grounds. Human beings breathe 20% oxygen in the air and benefit from it, but even they are harmed by prolonged high oxygen tensions. Lower organisms are harmed by atmospheric oxygen and may readily be killed, perhaps because it oxidises vital enzymes. The pulmonary arterioles (where microfilariae accumulate) are the site in the circulation where the mean oxygen tension falls to its lowest point, just before it rises again in the pulmonary capillaries. Accordingly the favourable physiological conditions, which the microfilariae seek in the lungs as postulated above, prove to consist of the lowness of the oxygen tension.

Microfilariae have two elaborate chemoreceptor sense organs at their head end (together with eight simpler tactile or thermal receptors); and they also have two simpler chemoreceptor sense organs near the tail end [5]. It is not yet known whether these sense organs are sensitive to changes of oxygen tension, but they certainly supply an anatomical framework by which such sensitivity could be obtained. The sense organs of the microfilariae correspond to the amphids and phasmids of the adult worms.

(Note. In the Pacific type of *W. bancrofti* the microfilariae in the peripheral blood become somewhat more numerous during the afternoon, thus showing an opposite pattern to *W. bancrofti* elsewhere (see Fig. 5.1). The mechanisms controlling this cycle have not yet been elucidated. Increasing the oxygen tension in inspired air causes a rise in the peripheral microfilaria count i.e. it causes microfilariae to emerge from the lungs—so does hypoxia. Exercise causes the count to fall.)

Group 2

The second group consists of *L. loa* of man (with microfilariae numerous in the blood by day), of *Edesonfilaria malayensis* in *Macaca irus,* and of *Dipetalonema setariosum* in East African mongooses. Both of the latter show microfilariae in the blood by night. To simplify discussion the group will be considered first in terms of *E. malayensis.* The microfilariae of this group are not greatly affected by changes of oxygen tension in the inspired air but they are very sensitive to changes of body temperature. Thus with *E. malayensis* and *D. setariosum,* if the day-time body temperature (which is normally high) is lowered by 2-3°C, the microfilariae are liberated from the lungs into the circulation and the microfilaria count rises. If the night-time body temperature (which is normally low) is raised 1-2°C the microfilariae accumulate in the lungs and the microfilaria count rapidly falls almost to zero. With *L. loa* the reactions are in the converse direction. There is a lag of 10-15 min between change of body temperature and change in the microfilaria count.

These observations may be explained by the following hypothesis. The microfilariae of this group are sensitive in principle to the "oxygen barrier" in the lungs but their reactions do not depend on the size or steepness of the "barrier" or gradient. On the other hand their sensitivity is greatly affected by the temperature. With *E. malayensis* and

D. setariosum, the sensitivity is increased by a rise in the body temperature as occurs by day or with artificial heating, and the microfilariae accumulate in the lung; on the other hand the sensitivity is lowered by a fall of body temperature (as at night or with artificial cooling) and then the microfilariae are liberated into the peripheral circulation. With *L. loa* the reactions are inverted. Perhaps there is some biochemical reaction in the sense organs of these microfilariae which reaches one equilibrium at a higher temperature, and the other at a lower temperature. (Note that the microfilariae of the other groups are *not* affected by changes of temperature in this way.)

Group 3

The third group consists of *Dirofilaria immitis* and *D. repens* of dogs, *D. aethiops (corynodes)* of East African monkeys and probably of other mammalian filariae. The microfilariae of this group are liberated from the lungs during day time by various stimuli such as anaesthetics (ether, chloroform), acetylcholine and serotonin, hypoxia, and by making the animal breathe oxygen. (They are not significantly affected by change of body temperature.) The liberation by hypoxia (which reduces the "oxygen barrier") is in agreement with the above theory but the liberation by raised oxygen pressure appears to be paradoxical. It may be explained however by the following hypothesis:

We postulate that the microfilariae are stimulated to reverse their contractions (and thus accumulate in the lungs) by small increases of oxygen tension, e.g. 50 mm Hg, the normal "oxygen barrier", but that they are paralysed by increases much greater than this. If that were the case, they would normally respond to the same mechanism of the oxygen barrier as the microfilariae of *W. bancrofti* described above, passing through the barrier when it is lower by night and being held back by the barrier (and accumulating in the lungs) when it is a little higher by day. In more detail—as the microfilariae are carried into the lungs through the pulmonary artery during the daytime, they reach the zone of increasing oxygen tension in the narrow arterioles, and they respond in the normal manner by reversing their contractions. Since they are in narrow tubes these reversed contractions carry them backwards against the blood stream so that they accumulate in this part of the lungs. During the night the oxygen barrier is lower and so the microfilariae pass through into the general circulation. When however the barrier is raised to an abnormal height by making the dog or monkey breathe oxygen, the resultant situation is illustrated in Fig. 5.4. The microfilariae are carried into the lungs by the pulmonary artery and arterioles, and they meet the beginning of the increased oxygen gradient while they are in a *wider* arteriole. They respond by reversing their contractions as usual,

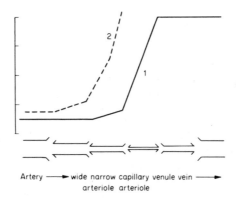

Artery ——▶ wide narrow capillary venule vein ——▶
arteriole arteriole

Figure 5.4. Diagrammatic illustration of the oxygen tension in different sizes of pulmonary vessel while the host breathed air (curve 1) or oxygen (curve 2) and its relation to the accumulation of microfilariae of *Dirofilaria* in the lungs.

but since they are in a wider vessel, they cannot get a grip on the walls and they continue to be carried along by the blood stream. Soon they reach a high tension of oxygen which paralyses or inhibits their reversed contractions, so that in spite of being now in narrow tubes suitable for reversed locomotion, they continue to be swept through the lung capillaries into the general circulation; and the microfilaria count goes up. The microfilariae of *Dirofilaria* seem easy to sweep out of the lung vessels by many different stimuli, but difficult to cause to accumulate there; those of *W. bancrofti* are difficult to sweep out, but easy to re-accumulate by heightened oxygen or muscular exercise.

Endogenous Rhythm in Microfilariae

The question may be asked whether microfilariae merely respond to host stimuli applied in a rhythmic manner, or whether they also have an endogenous circadian rhythm of their own so that in a "constant" environment they would tend spontaneously to accumulate in the lungs for 12 h and to circulate in the peripheral blood for 12 h; alternatively their sensitivity to lung conditions might endogenously increase for 12 h and decrease for 12 h. Such an intrinsic rhythm would greatly facilitate the maintenance of the migratory cycle by quite weak circadian stimuli from the host. This question is difficult to answer since the periodicity of microfilariae, i.e. their alternate accumulation in, and liberation from, the lungs cannot be studied or even manifested apart from the host which always maintains innumerable circadian rhythms. Certainly any such endogenous rhythm of the microfilariae must be relatively weak since the behaviour of microfilariae is readily and quickly altered by suitable stimuli as described above. We have not been able to find any morphological changes in microfilariae of *W. bancrofti* during day and night; but we made an attempt some years ago to detect such an endogenous rhythm experimentally [6]. A monkey infected with *E. malayensis* was placed on an inverted light/dark rhythm for some weeks until the rhythm of the monkey's activity and the cycle of its microfilariae were both inverted. Blood and microfilariae were then taken from its veins and transfused intravenously into a new monkey on a normal circadian cycle, so that the transfused microfilariae were following one cycle and their new host was following another, different by about 12 h. The behaviour of the microfilariae was watched during the next 3 days. Briefly, it was found that the transfused microfilariae followed their own cycle for approximately 6-8 h; there was a period of adaptation to the new host for about 6 h; and then (14 h after the transfusion) the microfilariae settled down to the rhythm of the recipient host. It was concluded that the microfilariae probably did have a weak endogenous rhythm of their own, but it was certainly not strong enough to sustain them in the new host for more than 6-14 h. There are many difficulties and fallacies in such experiments and the question must be considered undecided at present.

5.1.2 Trypanosomes of Frogs

Circadian rhythms have also been described by Seed and his co-workers in the trypanosomes of frogs *(Rana clamitans)* near New Orleans. Several types of trypanosomes are present, mostly not yet properly identified. In particular, there is a large one (Type I, probably *Trypanosoma rotatorium*) which is numerous in the peripheral blood by day but scanty by night. During the night the trypanosomes accumulate in the kidney [7]. These trypanosomes do not divide or multiply in the blood of the frogs; their cycle is due to migration back and forth as with microfilariae. To some extent it is explained by alternation of light and dark—the minimum period of the light phase being 4 h if a rhythm is to be maintained. On the other hand the rhythm persists even if the eyes of the frogs are destroyed, or the stirnorgan-pineal gland complex is removed; and it is not affected by any change of serotonin or melatonin [8]. Injection of adrenaline causes a rapid rise of parasitaemia but only if it is given at one particular point of the cycle. There is also another smaller trypanosome in these frogs (Type IV, perhaps *T. parvum*) which is numerous in the blood by night and scanty by day. Little is known yet as to where this trypanosome accumulates when it disappears from the blood. The first trypanosome, which is numerous by day, is probably transmitted by a leech; the frogs spend the day-time in the water to escape heat. The other may be transmitted by some night-biting insect; the frogs spend the night out of the water.

On the other hand, Bardsley and Harmsen [9] studying trypanosomes something like *T. rotatorium* in frogs *(R. catesbeiana)* in Ontario found that their trypanosomes were usually more numerous in the blood during the day-time but that this effect was due to the high temperature of the day and not to light. Their frogs seem to spend the day-time basking out of the water, so that they would be exposed to day-biting arthropods. The rhythms of trypanosomes in frogs clearly require much more study before they can be properly understood.

As far as can be seen by simple inspection, African trypanosomes such as *T. brucei* do not show any obvious circadian rhythms, although they are transmitted by day-biting tsetse flies. *T. congolense* was reported by Hornby and Bailey [10] to be more numerous in the venous blood of cows at 7 a.m. than they were at 2 p.m. But this phenomenon seems to be due to the skin being cold in the early morning and hot from the sun in the afternoon rather than to any endogenous rhythm. *T. cruzi* of S. America is transmitted by night-biting reduvid bugs, but no evidence for a circadian rhythm has yet been reported.

5.2 RHYTHMS DEPENDING ON SYNCHRONOUS CELL DIVISION—MALARIA

As mentioned at the beginning of the chapter, the typical course of malaria, with fever beginning soon after midday every second day, has been known since the time of Hippocrates. A typical temperature chart is illustrated in Fig. 5.5. After the malaria parasite or plasmodium had been discovered by Laveran in 1880 it was found that the malaria parasites in the red blood corpuscles have a 48-h cycle of development and they all come to final cell division (schizogony) about noon every second day. At this point the new young parasites (merozoites) burst out of the exhausted erythrocyte into the blood plasma and invade new erythrocytes. This bursting out and the accompanying liberation of waste products precipitate the attacks of fever observed clinically. These discoveries seemed satisfactory to explain the recurrent attacks of fever, and few people paid attention to the other remarkable features of the phenomenon—*viz.* that all the parasites should keep in step and come to cell division at the same time, that this period should be a simple multiple of 24-h, and that the cell division should always fall at the same hour (with human malaria, about midday). (Actually there are many species of malaria parasites in monkeys, other mammals and birds and most of them show synchronous cycles of 24, 48, or 72 h; on the other hand the hours at which they come to cell division vary according to the species and may happen in any part of the 24 h.) During recent years however, work on circadian rhythms in general and those of microfilariae

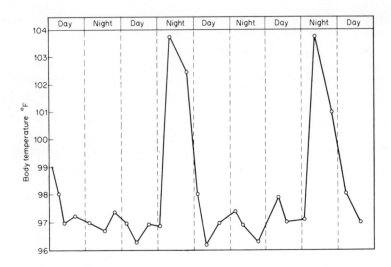

Figure 5.5. Temperature chart of a man infected with *P. vivax* malaria, showing attacks of fever on alternate nights.

in particular has directed new interest to the malarial rhythms. As we had worked on the rhythms of microfilariae (which are clearly associated with transmission by night-biting mosquitoes) and as we remembered that malaria is also transmitted by mosquitoes which bite mostly at night, we felt that the rhythms of malaria parasites ought somehow to be connected with the night-biting mosquitoes; but according to the currently accepted views of plasmodial development, this hypothesis encountered certain difficulties.

5.2.1. Different Forms of Malaria Parasites

It must be explained that there are two different kinds of plasmodia in the red blood corpuscles, as illustrated in Fig. 5.6. There are firstly the "asexual" parasites. Each of these starts as a small "ring" form in an erythrocyte and grows in volume; then the nucleus divides once, twice, three and four times to form 16 small new parasites, the completion of the process being called schizogony. The new parasites break out of the ruptured erythrocyte and invade new erythrocytes to start the cycle again. Each cycle takes 24, 48 or 72 h according to the species, all the parasites behave synchronously, and the schizogony occurs at a certain hour characteristic of the species. This is a clear, well-marked circadian rhythm; but unfortunately these asexual forms are not the ones which enter mosquitoes.

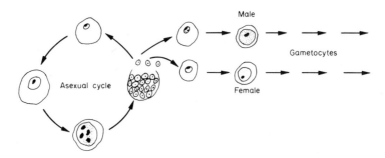

Figure 5.6. The cycles of malaria parasites in the blood; asexual, and sexual, respectively.

The other parasites are the sexual forms or gametocytes. After schizogony of the asexual forms, some of the new merozoites which invade erythrocytes develop in a different way. They grow in size so as to fill the blood corpuscle but the nucleus does not divide. Instead they differentiate to form male and female gametocytes. These are the forms which enter into mosquitoes and eventually are transmitted to new hosts. According to previous beliefs, they took 4 days to develop and then they remained unchanged in the blood for some weeks; so it seemed that the asexual forms had a cycle but did not enter mosquitoes, while the sexual forms entered mosquitoes but had no cycle. The hypothesis of association between the malaria cycle and transmission by night-biting mosquitoes has however been substantiated by recent work [11].

5.2.2 Evidence for Rhythm in Gametocytes

This work may be described in terms of *P. knowlesi* in monkeys, which has a 24-h cycle with schizogony just before midday. Batches of mosquitoes were fed on an infected monkey every 4 h for 2-3 days. The mosquitoes were kept for 6 days so that the imbibed gametocytes could develop into small cysts (oocysts) on the wall of the stomach; these oocysts can easily be counted under the low power of the microscope. The number of oocysts developing in the mosquitoes is a quantitative indication of the number of ripe or "mature" gametocytes in the blood at the time when the mosquito sucked blood. The results of a typical experiment are shown in Fig. 5.7. Briefly, in mosquitoes fed during the day-time there were few oocysts; in those fed at night there were many more; in those fed the next day there were few; in those fed the next night very many more; on the third day, relatively few. (Since the number of parasites increases 10-16 fold every noon, the scale of the graph on the second day is reduced by 10.) Clearly, the gametocytes

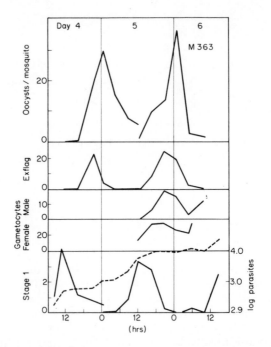

Figure 5.7. Cyclic changes of gametocytes of *P. knowlesi*, in monkey 363. Days, after inoculation of monkey. Oocysts, mean of all mosquitoes; second cycle, per 5 mosquitoes. Exflagellations per 10^6 RBC, second cycle, per 10^5 RBC. Gameto-cytes per 10^5 RBC. Stage I rings, per 10^3 RBC; second cycle, per 10^2 RBC; third cycle, per 20 RBC. Log total parasites, per 10^5 RBC (broken line).

show a 24-h cycle of infectivity with a maximum during the night when mosquito-vectors normally suck blood.

Similar evidence was also obtained by a different technique. If blood containing mature gametocytes is taken out of the body and allowed to cool to room temperature, the male gametocyte shows a remarkable development. In about 10-15 min, the nucleus divides into four and each subdivision puts out a long flagellum which lashes about furiously. The flagella quickly break away from the original mass and swim off, like spermatozoa, to find a female gametocyte to fertilize. This development takes place in nature in the mosquito's stomach; but in the laboratory it can be made to occur on a glass slide in a moist chamber. Accordingly 4-hourly slides were taken by a suitable technique from a monkey heavily infected with *P. knowlesi*, and the number of male gametocytes going into exflagellation was counted. The results are shown in Fig. 5.8. Again it is seen that the number is relatively low during the day, it

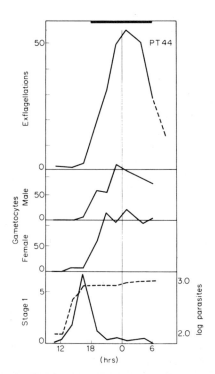

Figure 5.8. Cycle of exflagellations of *P. knowlesi*, in monkey PT 44. Exflagellations, gametocytes per 10^5 RBC. Stage I rings and log total parasites (broken line) per 10^4 RBC. The black rectangle on top represents the hours of darkness.

increases rapidly in the evening to form a plateau and then falls quickly the next morning. The distance between the points of 50% rise and 50% fall is approximately 5 h, and this is probably the duration of the mean period during which the male gametocytes are ripe for conjugation and the subsequent development in mosquitoes.

In the third place careful study of gametocytes in stained blood films taken every 4 h has shown morphological changes corresponding to this cycle. The gametocytes present at 09.00-13.00 are mostly immature, but those present between 21.00 and 05.00 are mature and fully developed. After that they degenerate and disappear.

These observations have been confirmed with several other species of malaria parasite which show a synchronous asexual cycle, *viz.* with *P. cynomolgi* of monkeys with a 48-h cycle and schizogony about midday; with *P. cathemerium* of canaries with a 24-h cycle and schizogony about 23.00; with *P. chabaudi* of rodents, with a 24-h cycle and schizogony at 04.00 and with *P. berghei* of rodents, with a poorly synchronous cycle of 24 h and schizogony about 16.00.

5.2.3 Biological Purpose of Rhythm

Apparently the mature male gametocyte is a short-lived organism. Once the nuclear apparatus has reached the development at which it will respond to cooling by breaking up into flagellated gametes or spermatozoa, it seems to be in an unstable state which cannot be maintained for much more than 6 h. If it is not taken up by a mosquito within this period, then the gametocyte breaks down and has to be replaced by another one. (It is not known whether the mature female is in a similar short-lived state of instability, or whether it can persist for some days; judging by the morphological changes in *P. knowlesi*, *P. cynomolgi* and *P. cathemerium*, the female gametocyte is also short lived.) Accordingly in the Darwinian struggle for survival, it would be advantageous if the short-lived gametocytes could concentrate their brief period of maturity into the hours when the transmitting mosquitoes usually suck blood, i.e. during the night, and this is what actually happens. In fact the biological reason for the synchronous, accurately-timed asexual cycle is to make the gametocytes match the mosquitoes.

With *P. knowlesi* it works like this: The vector mosquitoes *(Anopheles)* may be expected to bite during the night, e.g. 21.00-05.00, so this is the time at which ripe gametocytes ought to be present in the blood ready to develop in the mosquito and to be then transmitted to a new host. The gametocytes of *P. knowlesi* take about 33-36 h to develop to ripeness from the point at which they are liberated as new merozoites from schizogony of the asexual forms. Accordingly this schizogony is timed (by the synchronous asexual cycle) to take place about 12.00. (After the ripe gametocytes have been present in the blood for about 6 h, e.g. up to 05.00, they begin to degenerate but they will be replaced by a new batch on the following evening.) With *P. cynomolgi* the asexual cycle lasts 48 h and the gametocytes take about 58 h to reach maturity. In order that there will be ripe gametocytes in the blood to infect mosquitoes on Wednesday night, schizogony of the asexual forms takes place about Monday midday, with liberation of the merozoites to form the new gametocytes. With *P. cathemerium*, transmission apparently takes place by mosquitoes which bite at dawn, e.g. 06.00, and the gametocytes take about 26-29 h to develop. Accordingly schizogony is timed to occur about 23.00. With *P. chabaudi* the gametocytes are mature about 21.00-05.00 but schizogony takes place about 04.00. With each species the customary hour for schizogony is apparently fixed so that the mature gametocytes will be in the blood when the appropriate mosquito bites. Presumably all these developments have taken place by

Darwinian selection. Those strains of parasites which tended towards a matching of gametocytes to mosquitoes had an advantage in the struggle and tended to survive and to pass on their progeny. Those which did not, tended to be eliminated.

(There are some malaria parasites, e.g. *P. gallinaceum* of chickens, in which the asexual cycle is poorly synchronous and takes about 34-36 h. The male gametocytes appear to remain mature and able to exflagellate for about 18 h. A 36-h cycle would match the biting habits of crepuscular mosquitoes, which bite at dusk and at dawn (the gametocytes matching these hours alternately). This type of poorly synchronous plasmodium might be regarded as a form which is too primitive to develop a good cycle, or it might be regarded as a highly efficient form, in which the male gametocyte can be maintained for such a long period that a synchronous asexual cycle is not necessary.)

5.2.4 Rhythm in *P. falciparum*

As stated above, this principle of short-lived gametocytes, which require 9 h longer than the duration of the asexual cycle to come to maturity, has been experimentally demonstrated for all the five periodic plasmodia in which it has been sought; and it presumably applies to all plasmodia in which the asexual forms show a synchronous 24, 48, or 72 h rhythm. There is one case however which requires special consideration, and that is *P. falciparum*, the most important and most dangerous of the human malaria parasites. This differs from all other plasmodia in that the gametocytes take 12 days to develop from the asexual parasites instead of the usual 1½ or 2½ days. The asexual forms of *P. falciparum* usually show a fairly synchronous cycle of 48 h with schizogony normally about midday. With other plasmodia it is apparently easy for the gametocytes to time their development accurately to 1½ or 2½ days, but with *P. falciparum*, there might seem to be more difficulty in timing the development of gametocytes for such a long period as exactly 12½ days. An error of half a day either way would nullify the purpose of the whole arrangement.

P. falciparum is difficult to investigate, since gametocytes do not appear in laboratory infections of *Aotus* monkeys and they are usually scanty or absent in human infections after the age of about 2-5 years. In collaboration with Dr. M. E. Wilson however it has been possible to collect some data from five young children in West Africa. Appropriate blood films were taken from these children every 4 h, and the numbers of exflagellations at the different times were counted in England. The results of one such case are shown in Fig. 5.9, in which there are clearly peaks of exflagellation about midnight on the first and third days (48 h rhythm) with a minor peak indicating a minor brood on the second

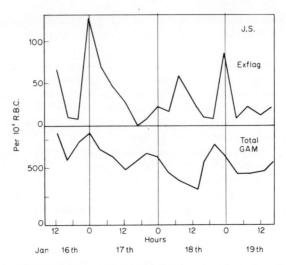

Figure 5.9. The frequency of exflagellation of the male gametocyte of *Plasmodium falciparum* in a West African child. Upper graph—numbers of exflagellations observed per 10^6 red blood corpuscles. Lower graph—numbers of total gametocytes, ditto.

midnight. In this and another case, the results indicated that a 48- (or 24-) hour rhythms was *clearly* present; in two cases they suggested it was *probably* present; and in one case they suggested that a rhythm might *possibly* be present. Taking the cases as a group it was concluded that they indicated that a 48-h rhythm is *probably* present in the gametocytes of *P. falciparum* but that further work is necessary before the question can be definitely settled. Since the period of development of the gametocytes is so long (12 days), it is possible that there might be a secondary mechanism (e.g. sensitivity to the body temperature cycle) to control the final maturation of *P. falciparum* gametocytes, in addition to the initial mechanism of starting their development at the fixed hour (midday) of schizogony of the asexual forms.

5.2.5 Mechanism for Controlling and Orientating Rhythm

Accepting then that the rhythm of the asexual parasites is biologically adapted to make the gametocytes match the mosquitoes, the next question is: How is this rhythm controlled and orientated? Clearly the rhythm of the parasites cannot be directly influenced by day and night as such but it must be entrained by some aspect of the circadian rhythms of the host, the body of the host constituting the complete environment of the parasite during these stages of its life history. The host factor which acts as Zeitgeber seems to be the cycle of body temperature [11]. If monkeys infected with malaria *(P. cynomolgi)* are anaesthetized with barbiturates from 09.00-17.00 each day (the anaesthesia being necessary to paralyse their own thermostatic mechanisms) and if the body temperature is kept high (37-38°) during this period so that the body temperature cycle is not greatly altered, the malaria asexual cycle also continues unaltered with schizogony persisting about midday. But if the monkeys are cooled to 34-35° during the day (i.e. still within extreme physiological limits) so that the body temperature cycle is inverted, then the malaria asexual cycle also becomes inverted after a lag period. In the first cycle, schizogony takes place about the usual hour (midday); in the second cycle it is postponed; by the third and fourth cycle it is practically at midnight (Fig. 5.10). If cooling is then discontinued, and the monkey is allowed to resume its normal temperature cycle, the malaria cycle gradually returns to its normal timing either by lengthening its period to about 49 h, or by shortening it to about 47 h. Schizogony thus gets back to midday after about 12 cycles. This entrainment of the malaria cycle by the body temperature cycle has been demonstrated with *P. knowlesi, P. cynomolgi,* and *P. cathemerium.* In the last instance, the infections were maintained in chick embryos in which the temperature

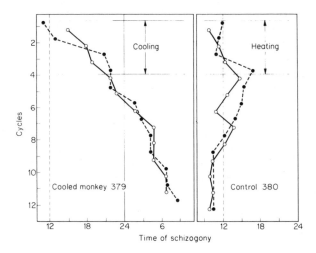

Figure 5.10. The shift of schizogony of *P. cynomolgi* during cooling in monkey 379 (and control monkey 380) with subsequent adjustment of the cycle. The spots with black lines indicate the major broods, and the circles with broken lines indicate the minor broods. Each cycle represents two days. Since schizogony at the 6th and following cycles in 379 occurs *after* midnight, this takes place a day later than the cycles suggest.

cycle can easily be manipulated by changing the eggs between hotter and cooler incubators.

Relation Between Temperature Cycle and Malaria Cycle

The most simple relationship between the temperature cycle and the asexual malaria cycle would be that some particular stage of plasmodial development (e.g. division of the nucleus) could occur only at some given temperature or following some particular temperature alteration; but this simple hypothesis is not likely to be correct. A continuous high temperature or a continuous low one does not bring the malaria cycle to a halt. Moreover the hours of schizogony in different species of plasmodia are not all related in the same way to the temperature cycle (although they always end in placing ripe gametocytes in the right relationship to the mosquito cycle). Thus in human and simian malarias, schizogony occurs about midday when the body temperature is high. In some species of avian malaria, e.g. *P. cathemerium,* schizogony occurs about 23.00 (when the temperature is low) and in others about midday when the temperature is high. This gives the impression that the asexual malaria cycle is "tuned in" to some point on the body temperature cycle which suits the requirements of that particular species of plasmodium— but the malaria cycle is not forcibly dictated by any particular temperature.

Feedbacks

Probably there can be a feedback connection between parasite infection and body temperature. In a man infected with *P. vivax,* the attack of fever which follows schizogony, alters the timing of the normal cycle of body temperature and greatly increases its range. This abnormal temperature cycle influences the time of the next schizogony and so on. Such a feedback connection would explain why the cycle of *P. vivax* malaria is often shorter than 48 h, e.g. 45 h, so that the attacks of fever occur earlier and earlier in the day.

5.2.6 Endogenous Rhythm

In the case of malaria parasites, there is clearly a strong endogenous rhythm in the parasites themselves, depending on the time taken by the asexual forms to develop from one cell division to another. Presumably the basic time is approximately 24 to 48 h according to the species of the parasite, and the actual timing is modified to exactly 24 or 48 h by the 24-h cycle of body temperature. Under conditions of abnormal temperature rhythms, the malaria rhythm can be prolonged and contracted by one or two hours, but the amount of such alteration is narrowly limited.

Furthermore this endogenous rhythm of the asexual parasites is a great help to the maintenance of a synchronous development. Once all the parasites have been synchronized (by whatever means) they will have a great tendency to maintain their synchronous development for many generations, even in the absence of any further synchronizing influences. Straggling and dispersion will develop only slowly. This tendency to maintain a synchronous infection (once it has been established) often results in a kind of "pseudosynchronization" in laboratory infection of plasmodia which naturally are not very synchronous, e.g. *P. lophurae, P. gallinaceum,* or *P. berghei.* When these infections are passed from animal to animal by syringe passage of infected blood (as is usually done) the parasites which develop in the new host and start the new infection are mostly the ripe schizonts which are liberating new merozoites to invade the red blood corpuscles of the new host. Consequently the parasites in the new host tend to begin their development in step. If the blood passage is routinely made at a certain hour of the day, and at an interval of days which is in harmony with the development of the parasite, a synchronized blood infection soon develops; but the resultant cycle is orientated, not to the circadian rhythm of the host but to the laboratory routine of the experimentalist. (With infections such as *P. knowlesi, P. cynomolgi,* or *P. chabaudi* in which the cycle is more strongly influenced by the host, the cycle is usually dominated by the host rhythm more than by the experimentalist rhythm.)

5.2.7 Other Work on Malaria Cycles

Much work on the rhythm of *P. berghei* in mice has been carried out by Arnold and his colleagues [12, 13] but their results are not always easy to harmonize with the views on malaria cycles which have been described above. This may be partly due to differences of technique, since Arnold measures the timing of the asexual cycle in a different way, especially by measuring the percentage frequency of Type III forms (large forms with no division of the nucleus) at one or two fixed hours during the day-time, while Hawking, Worms and Gammage [11] determine the hours at which the number of new small parasites (following cell division) greatly increase in the blood. Under normal laboratory conditions, *P. berghei* in mice shows little evidence of periodicity; but Arnold finds that if the mice are subjected to a 24-h cycle of alternate darkness and bright light, a distinct synchrony in growth and cell division can be observed. In male mice, he finds that the synchrony is dependent on an intact visual reception of light, on an intact autonomic nervous supply to the pineal gland, on an intact pineal gland, and on an intact testis. In female mice, synchrony of growth and division is apparent only at the low point of oestrogen secretion during the oestrus cycle. It is not clear whether this mechanism of 24-h photo-period plus intact visual and pineal gland apparatus affects the malaria parasites directly, or whether this mechanism modifies the circadian rhythms of the mouse which then have a secondary influence upon the malaria parasites. The latter explanation would seem the more probable.

5.3 RHYTHMS OF DISCHARGE OF INFECTIVE FORMS FROM THE HOST

5.3.1 Coccidia of Birds

The best known example of this is given by the coccidia of sparrows (probably *Isospora larazei*) which have been studied by Boughton [14], Levine [15] and Schwalbach [16]. These parasites are protozoa which develop in the epithelial cells of the intestine; they multiply through several generations like malaria parasites and then produce sexual forms which conjugate and change into oocysts. The oocysts are passed out in the feces. Outside the body they undergo a short period of further development and then they form resistant spores. When the spores are ingested by a new host as a food contaminant, the cycle begins anew.

The oocysts of the *Isospora* of sparrows are discharged almost exclusively in the late afternoon and early evening, e.g. from 19.00 to 03.00, during a regime of L 07.00-24.00 D 24.00-07.00 (Fig. 5.11). Sparrows disperse widely for feeding during the day and oocysts passed in the fields might have little opportunity of being ingested by another sparrow. On the other hand they congregate at evening in great numbers and this is when (and where) the oocysts are discharged. Before settling down for the night, the sparrows bathe in the dust (and coccidia spores) and then preen their feathers; during the process of preening, some of the spores presumably contaminate their beaks and are ingested.

Similar rhythms of discharge have been found in other coccidia but the timing may be different. In the common pigeon infected with *Eimeria labbeana* the oocysts are discharged mostly between 09.00 and 15.00 and hardly at all between 21.00 and 09.00; although feces are discharged all the 24 h (slightly more at night) [17]. In 7-week-old Leghorn chickens, artificially infected and studied during July by Levine [15], the oocysts of *Eimeria hagani* were discharged almost exclusively between 15.00 and 21.00. Those of *E. praecox* showed a similar but flatter curve. Those of *E. mitis* and *E. maxima* were excreted mostly between 09.00 and 21.00, with few during the night. The oocysts of *E. necatrix* were discharged mostly during the night, but on account of technical factors it is uncertain what their true hours of liberation are.

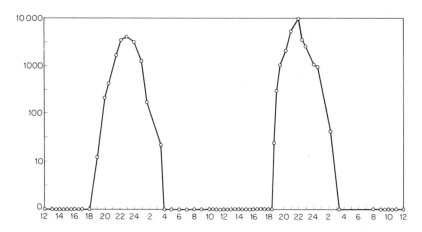

Figure 5.11. The excretion of oocysts from a sparrow from 12.00 September 17, 1959 to 12.00 September 19. The vertical (logarithmic) scale gives the number of oocyst per cm of feces. The horizontal scale gives hours of the clock. The cycle of illumination consisted of daylight shining into a room 07.00-18.30; weak lamplight 18.30-24.00; darkness 24.00-07.00.

(These oocysts are not liberated into the intestine like the others but into two caecal pouches which empty themselves independently of the general intestine.) Similar cycles have also been seen in *Isospora sylvianthina* of coaltits *(Parus major)* [16] .

Thus rhythms of discharge have been observed in many avian coccidia, but they have not yet been studied in accordance with modern conceptions. Presumably the different hours of discharge are somehow related to the different feeding and roosting habits of the different birds. Nothing is known as yet of the mechanisms by which this rhythm of discharge is produced. Perhaps when the infective spores are initially ingested by the host, the first schizogony is delayed (possibly by the body temperature cycle) until a certain hour before it can take place (D. C. Boughton—personal communication). In this way, the subsequent development of all the parasites might be synchronized and the timing of the final discharge of oocysts might be determined by the timing of the first schizogony. The whole phenomenon requires further investigation.

5.3.2 Schistosomes of Man

A somewhat similar rhythm of discharge from the host occurs with the ova of *Schistosoma haematobium* in man. The adult worms live together in the plexus of veins around the bladder and ureters. Ova are deposited by the female worms in the fine venules. These ova digest their way through the tissues until they finally emerge into the lumen of the bladder; and then they are passed out in the urine. If the urine and ova fall into fresh water which contains suitable snails, the ova hatch out into forms (miracidia) which invade the snails and develop further. The number of ova coming out in the urine is not in simple proportion to the volume of urine, but it shows a distinct peak in the morning about 10.00 a.m. (Fig. 5.12) [18]. The snail hosts, required for the further

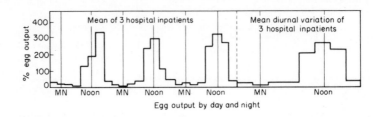

Figure 5.12. Output of eggs of *Schistosoma haematobium* in man by 3 h periods. The figures on the vertical scale are expressed as percentages of the mean value for the whole 3 days. On the right, the mean values for the 3 days are represented.

development of the parasites, are more readily available during the light warm day than they are during the cold dark night. Nothing is known about the mechanisms responsible for the rhythm in discharge of *S. haematobium* ova. It is tempting to speculate whether the adult worms of this and other schistosomes show a circadian rhythm of migration backwards and forwards in the veins which form their abode (cf. intestinal worms, 5.4). Such a rhythm might be quite independent of the rhythm of discharge of ova.

5.3.3 Pinworms

The pinworm, *Enterobius vermicularis,* is a common parasite of the lower intestine of children. The ripe female migrates out of the anus in the early hours of the night (causing considerable pruritus), lays her eggs on the surrounding skin and then dies.

This nocturnal migration has recently been confirmed by the author. Perhaps the worms take their signal from the fall in the body and rectal temperatures which occurs at night. In many ways, this behaviour resembles that of the coccidia of sparrows described above and the biological purpose would be the same, *viz.* to concentrate the infective ova or oocysts in the nocturnal resting places of the host, rather than allowing them to be dispersed over wide expanses by day.

5.4 RHYTHMIC MIGRATION OF WORMS IN THE INTESTINES

Evidence has recently been obtained that certain worms in the intestine (particularly the tape worm *Hymenolepis diminuta* of rats) migrate up the intestine during part of the 24 h and slip backwards again during the other part. The movement is one which involves the whole worm. Read and Kilejian [19] counted the mean number of scolices (i.e. heads of tape worms) in the anterior 25 cm of the intestine in groups of four rats with approximately 30 worms each. At 08.00, there were 18.2; at 12.00, 13.7; at 17.00 there were 6.7; and at 03.00 there were 16.3. (Rats are active by night and somnolent by day.) Hopkins [20] studied 14-day-old single worm infections; he cut the intestine simultaneously into 5 cm sections and he recorded the position of the scolex (head) and the tail of the worm in rats starved from 07.00 and killed at 20 min intervals from 08.00 to 20.00. Although unfortunately 24-h cycles cannot be judged satisfactorily by data restricted to 12 h, the results nevertheless indicated a posterior migration between 11.00 and 15.00 of both scolex and tail by about 10-20 cm. Apparently the tape worms migrate forward during the night which is the rat's active phase and they slip back (perhaps passively) during the day (which is the rat's passive phase). The relation of these movements to the ingestion of food by the rats is not clear. Normally the rats ingest most of their food by night. Read and Kilejian [20] found that if they fed their rats by day (instead of by night) they were able to reverse the rhythm and they conclude that these movements are associated with the feeding pattern of the host. On the other hand if food was withheld from the rats altogether, the worms still migrated up the intestine, although the start of the migration was delayed. Further work on *Hymenolepis diminuta* by Bailey [20a] has confirmed that the worm moves forward in the intestine in the early morning, the scolex being most forward from 08.00 to 10.00, and furthest back at 17.30. In addition the worm shortens itself, being shortest from 00.00 to 05.00, and longest at 14.00. These movements are correlated with the rat's ingestion of food, which is greatest from 20.00 to 08.00, and they do not occur if no food is ingested.

The whole subject requires much more study from the aspect of circadian rhythms before definite conclusions can be reached. At first sight this would appear to be the first reported circadian rhythm of parasites which is not associated with facilitation of transmission to a new host.

5.5 CIRCADIAN RHYTHMS IN OTHER PARASITES

Apart from the well-known examples discussed above, most parasites have not yet been studied from the aspect of circadian rhythms. Perhaps many parasites may not possess such rhythms, but there is no point at this place in giving lists of negatives. On the other hand, many parasites especially helminths will probably be found to show rhythms at some stage of their life history, once such rhythms are sought. Among trematode worms, the excretion of ova of *Schistosoma haematobium* into the urine particularly during the morning has already been mentioned (5.3.2). The worms in trematode ova later develop inside snails, and further infective forms (cercariae) are subsequently liberated from the snails which then invade vertebrate hosts. Under natural conditions of day and night, the cercariae of *S. mansoni* and *S. haematobium* are liberated mostly between 11.00 and 14.00 and those of *S. mattheei* between 06.00 and 08.00 (when vertebrate hosts are likely to be available), but those of *S. rodhaini* are liberated during the night [21]; perhaps this is an adaptation to the nocturnal activity pattern of its rodent host. It is not known at present whether this rhythmic liberation is a direct response to changing temperature (and light) during day/night or whether it would still be manifested to some extent even under conditions of constant temperature and illumination. The circadian migration in the intestine of the cestode worm, *Hymenolepsis diminuta*, have already been discussed (5.4); perhaps other tapeworms may show similar rhythmic migrations.

In *Onchocerca volvulus,* a human filarial worm, the microfilariae accumulate in the skin, from which they are sucked up by *Simulium,* a day-biting fly. There is some evidence that the microfilariae migrate towards the superficial layers of the skin by day and to the deeper layers by night. Duke *et al.* [22] reported a slow undulation in the number of microfilariae found in snips of the superficial skin, with lower counts between 24.00 and 06.00 and highest counts between 11.00 and 19.00; the general curve was similar to that of the dry-bulb air temperature. Lartigue [23] found that the number of microfilariae fell from 08.00 to 10.00 (when it was usually zero) and then rose steeply until 18.00. These studies however are liable to various technical difficulties and the migration should not yet be considered as proved. Such a migration might be a direct response to the changing temperature of the skin surface.

The ova of certain nematode worms such as *Ancylostoma* (hookworm) and *Nippostrongylus* are passed out in the feces. In damp soil they hatch, and larvae develop. When the larvae reach a certain infective stage they crawl to the surface of the soil and stand upon their tails waiting to make contact with a new host (man or rat). When this happens they quickly bore through the skin and start a new cycle of infection. The migration of the larvae to the surface of the soil occurs under natural conditions almost entirely by day (which is the time a new host is most likely to pass by). It has not yet been investigated whether such migration is a simple response to the warmth and light of day, or whether it would still happen with a rhythmic tendency under constant conditions of temperature and illumination.

5.6 RHYTHMS OTHER THAN CIRCADIAN ONES

Although this book is really limited to circadian rhythms, it is tempting briefly to mention rhythms of parasites which involve longer time intervals. There are some parasites which exhibit annual rhythms. One of the best known is manifested by the microfilariae of *Dirofilaria immitis* and *D. repens* in the dog. If the number of microfilariae in the blood is estimated at the same hour of the day every week or two weeks throughout the year, it is found that in temperate zones the number increases greatly during August and September and then diminishes (Fig. 5.13) [4]. This has been observed in Manchuria, in Japan, in

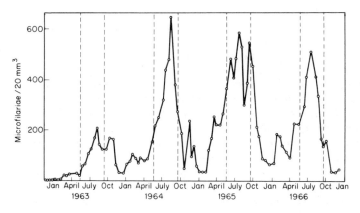

Figure 5.13. Annual variation in the microfilaria count of *Dirofilaria immitis*. The dog, No. 60, female, received two female and three male adult worms by transplantation on 8 December 1962.

England, and in the United States. It also occurs with *D. repens*. The biological advantage of this increase is clear, since August/September is the time when vector mosquitoes such as *Anopheles maculipennis* are most common. The mechanism of causation is less clear. It occurs equally in male and female dogs, so it is not due to oestral changes. Recent work by Katamine, Aoki, and Iwamoto [24] suggests that this August increase in the number of microfilariae is due to an increase of external temperature at this season and that it can be cut short by a fall of temperature; the increase seems to be due to a redistribution of microfilariae between deep tissues and peripheral blood, rather than to a production of new microfilariae.

Another rhythm of a different nature is shown by temperate zone strains of *Plasmodium vivax*. When infection occurs following mosquito bite there is usually an initial attack of fever (but this may be suppressed by drugs or by immune reactions). When this fever subsides, there is a long period of freedom; but 9-10 months after the original mosquito infection, the parasite reappears in the blood and there may be a succession of relapses at 1-2 month intervals. The biological advantage of this to the parasite is clear. Under natural conditions the initial mosquito infection might often take place in September. The malaria infection remains quiescent during the winter but it reappears about June/July when mosquitoes are also reappearing. In this case, the 9-month cycle is dependent entirely on the date of mosquito infection (which in experimental conditions may be made to occur in any month of the year), and it is not related to external climate conditions, or to physiological conditions in man. During this long period, the malaria parasites persist as special forms (exoerythrocyte forms) in the liver parenchymatous cells. But how do they measure a period of 9 months?

A different form of adaptation of ecto-parasite to periodic changes in the environment is illustrated by the rabbit fleas investigated by Miriam Rothschild [25]. (Perhaps a purist would not admit this as a biological rhythm.) These fleas live on the ears of rabbits and normally they show no sexual or reproductive activity. When a female rabbit becomes pregnant, however, the male and female fleas rapidly mate, the females lay eggs, from which young fleas emerge, and these young fleas pass over to the newborn rabbits where they settle down to repeat the cycle. The signal for the development is given by the adrenocorticotrophic hormones and progestins which appear in the blood of the pregnant rabbit and which pass over into the fleas when they suck blood.

A somewhat similar annual cycle occurs in *Leucocytozoon simondi* [26, 27]. This is a malaria-like parasite of mallard ducks and is spread by a black-fly, *Simulium*. The number of gametocytes in the blood increases rapidly in the months of March and April, i.e. at the beginning of the *Simulium* season. These gametocytes arise from large schizonts in the tissues which have been dormant all the winter. Apparently the sequence of events is as follows. In the spring the longer periods of daylight stimulate the reproductive organs of the ducks both male and female; and then something from the reproductive organs (or perhaps from the pituitary) stimulates the schizonts to produce the gametocytes. Perhaps a similar annual cycle also occurs with *P. relictum* (a common plasmodium of passerine birds); during a 2-year period of observation of acute infections produced by blood inoculation in canaries, Bishop [28] found that the number of gametocytes (in proportion to asexual forms) rose sharply in the spring, being greatest from April to July, and then dropped sharply during the autumn months. The spring relapse of *P. relictum* in sparrows living in the wild has been further studied by Applegate [29]; it appears to be related to increased secretion of corticosteroids. An annual cycle also occurs in a malaria parasite *(P. mexicanum)* of Californian lizards [30].

Lunar rhythms are also theoretically possible although I know of no actual example. If one indulges in phantasies like science fiction, it would be easy to construct plausible systems of a hypothetical tapeworm in the intestine of a wart hog; the eggs might be picked up by *Cyclops* living in a water hole and after suitable development they would be reingested by the wart hog. Supposing the *Cyclops* swarmed on moonlit nights, most of all at full moon, and that the wart hog was also more active on moonlit nights, the tapeworm might well be adapted to release eggs on a lunar rhythm. This hypothesis may be only an amusement of the imagination, but if the possibility is realized, an actual example may be uncovered and brought to light.

5.7 DIFFICULTY OF SEPARATING THE RHYTHMS OF PARASITES FROM THOSE OF ENVIRONMENT

In the classical study of circadian rhythms, it is important to determine whether the apparent rhythm is merely a response to a rhythmic environment or whether it is endogenous. This determination is made fairly easily by placing the organism in a constant environment and watching whether the rhythm of the organism ceases at once, or whether it continues as a free-running rhythm. When, however, the attempt is made to apply these concepts to parasites, difficulties are immediately encountered. The rhythms of many parasites, e.g. the migrations of microfilariae and of frog trypanosomes, cannot be manifested apart from their hosts. With other parasites, e.g. malaria, the significance lies, not in the parasite taking 24 or 48 h to develop, but in all the parasites keeping in step and developing synchronously. It is technically almost impossible to maintain most parasites under constant conditions which are yet suitable for them to manifest their rhythm. Even if parasites, such as malaria, are introduced into chick embryos maintained at a constant temperature, the process of injection will probably produce a synchronizing effect which will last until the egg hatches and another syringe-transmission must be made. It is, therefore, generally useless to spend time in argument or experimentation to decide whether a given parasite rhythm is endogenous or only a response to its changing environment. In most instances the parasite and its host form a unit as far as rhythms are concerned. All that can be accomplished is to determine which factor in the rhythms of the host is predominant in the entrainment of the parasite.

5.8 THE WHY? AND HOW? OF RHYTHMS AND THEIR MECHANISM

5.8.1 Purpose

If a biological rhythm is present in an organism, it may be expected that it confers some advantage in the struggle for survival. Presumably such rhythms have arisen as mutations which have then been picked out, propagated and intensified by Darwinian selection. Nevertheless in contemplating and discussing such rhythms it is often more convenient and dramatic to use a teleological approach and to consider the parasite as an intelligent creature which has to make the best way possible in a difficult world. Using this informal approach we may ask *why* a particular parasite has a circadian rhythm. As indicated above, the usual answer is that the rhythm facilitates transmission to a new host (this being the parasite version of "propagation of the species"). The most obvious illustrations of this occur where a parasite is transmitted by an insect vector which bites only by night or only by day, e.g. microfilariae or malaria parasites, transmitted by night-biting mosquitoes. Obviously the parasite must be available when the vector mosquitoes bite. In order for a rhythm to be advantageous however, it is necessary to have some counteracting factor which makes constant presence in the blood difficult or undesirable. Thus with microfilariae, the high oxygen tension of arterial and capillary blood is hard to tolerate and microfilariae survive better in venous blood; with malaria, the ripe gametocytes are unstable and cannot be kept in a state of readiness for all the 24 h. In such species, a rhythm which concentrates the parasites in the blood during the critical period of the 24 h is obviously advantageous. In other parasites however, e.g. African trypanosomes transmitted by day-biting tsetse flies, or leishmaniae transmitted by night-biting sand flies, there seems to be no reason why parasites should not be in the blood all the time; at any rate they apparently do not have a rhythm.

5.8.2 Mechanism

Continuing to use teleological terminology, one asks, "How does a parasite orientate its rhythm?" and the answer for all endoparasites must be that they take their clue from some factor in the rhythm of the host. In different language, the rhythm of the parasite is entrained by one of the host's rhythms. Different members of the same group of parasites, e.g. microfilariae, may take their clues from different factors as described above. Consequently each case must be worked out separately and this part of the investigation takes more labour. In such an investigation, it is often easy to invert the rhythm of a parasite by exposing the animal host to darkness by day and to light by night, and thus inverting the rhythm of the host. The relation between parasite and host is unaltered; both have been delayed 12 h and little information has been gained. It is necessary somehow to dissociate the factor being studied from the general circadian rhythm of the host, e.g. by altering the oxygen tension in the lungs or the temperature of the body, without altering the rhythm of the whole animal. Furthermore the alteration of part only of the host's rhythms must often be maintained for several cycles before concluding that no effect is being produced. There may be a lag period of two or more cycles as in malaria before anything is observed. In any unknown relationship, the temperature cycle of the host is the most hopeful one to investigate.

5.9 DIFFERENT RHYTHMS OF DIFFERENT HOSTS

In studying the entrainment of parasite rhythms by host rhythms, it is important to realise that all mammal and bird hosts do not have the same rhythms [31]. Thus man and monkeys show a well-marked rhythm of physical activity and of many other systems in which activity is high and almost continuous during most of the day but it is greatly reduced or absent by night. Many rodents show a similar but reversed cycle, i.e. they are active by night and quiet by day. Some are crepuscular in their activity, i.e. they are most active at dusk and dawn. By contrast hunting animals such as dogs and cats show quite a different behaviour. Their activity occurs in periods of half to five hours interspersed with similar rest periods, which are spread almost equally over the whole nychthemeron. Placing a dog or cat under constant illumination or constant darkness makes little difference to its behaviour, although constant darkness depresses the activity of a macaque monkey almost to zero. From an inspection of a dog's activity pattern, it is difficult to decide whether it is fundamentally a "diurnal" or a "nocturnal" animal; since the urinary excretion pattern shows more urine excreted by night, it is probably more "nocturnal" than "diurnal". Even apparently similar organisms like a chicken and a duck have different rhythms. The temperature of a chicken goes up by day and down by night in regular fashion just like that of man or monkey; a chicken feeds by day and roosts by night. The temperature of a duck is very irregular, going up and down every 15-60 min, and there is little constant difference between day and night; wild ducks often feed and fly by night. Clearly parasites in these different hosts may receive very different signals from the different rhythms of their environment.

GLOSSARY

1. Cercariae—The stages of trematode worms which emerge from snails, swim in water and infect man or animals.
2. Exflagellation—The process by which the male gametocyte of malaria (when it is cooled to room temperature) liberates four flagella-like gametes that can fertilize the female gametocyte.
3. Gametocytes (male and female)—The sexual forms of malaria parasites in the blood, which can later develop in mosquitoes.
4. Microfilariae—The first larval stages of filarial worms, which are capable of developing in suitable blood-sucking insects.
5. Miracidia—The stages of trematode worms which emerge from the ova and penetrate into fresh water snails.
6. Merozoites—The small parasites which are formed by cell division (schizogony) of asexual malaria parasites in the blood; they invade new erythrocytes and start the next cycle of infection.
7. Oocysts—the forms which develop after the conjugation of sexual stages of Sporozoa. With malaria parasites, they develop on the stomach of mosquitoes; with coccidia they are passed out in the feces.
8. Rings—The early stages of malaria parasites developing inside erythrocytes; so named because they resemble a signet ring in appearance.
9. Plasmodia—Members of the genus, Plasmodium, i.e. malaria parasites.
10. Schizogony—The final stage of cell division of asexual malaria parasites in the blood, leading to the formation of 16 merozoites; with most species of malaria, cell division of all the parasites in the blood takes place at approximately the same hour of the day.
11. Scolex (pl. scolices)—The head of a tape worm, by which it attaches itself to the wall of the intestine.

REFERENCES

1. F. Hawking and J. P. Thurston, The periodicity of microfilariae. I. The distribution of microfilariae in the body. II. The explanation of its production. *Trans. R. Soc. trop. Med. Hyg.*, 45, 307-328, 329-340 (1951).
2. F. Hawking and J. B. Clark, The periodicity of microfilariae. XIII. Movements of *Dipetalonema witei* microfilariae in the lungs. *Trans. R. Soc. trop. Med. Hyg.*, 61, 817-826 (1967).

3. A. G. Jameson, Gaseous diffusion from alveoli into pulmonary arteries. *J. appl. Physiol.*, 19, 448-456 (1964).
4. F. Hawking, The 24-hour periodicity of microfilariae: biological mechanisms responsible for its production and control. *Proc. Roy. Soc. B.*, 169, 59-76 (1967).
5. D. J. McLaren, Ciliary structures in the microfilaria of *Loa loa*. *Trans. R. Soc. trop. Med. Hyg.*, 63, 290-291 (1969).
6. F. Hawking, M. J. Worms and P. J. Walker, The periodicity of microfilariae. IX. Transfusion of microfilariae *(Edeson filaria)* into monkeys at a different phase of the circadian rhythm. *Trans. R. Soc. trop. Med. Hyg.*, 59, 26-41 (1965).
7. G. C. Southworth, G. Mason and J. R. Seed, Studies on frog trypanosomiasis. I. A 24-hour cycle in the parasitemia level of *Trypanosoma rotatorium* in *Rana clamitans* from Louisiana. *J. Parasit.*, 54, 255-258 (1968).
8. G. Mason, The diurnal rhythm of *Trypanosoma rotatorium* in *Rana clamitans*: investigation of photo receptors and physiological control. *J. Parasit.*, 56, 228 (1970).
9. J. E. Bardsley and R. Harmsen, The effects of various stimuli on the peripheral parasitaemia of the *Trypanosoma rotatorium* complex in the bullfrog *(Rana catesbeiana* Shaw) of eastern Ontario. *J. Parasit.*, 56, 20-21 (1970).
10. H. E. Hornby and H. W. Bailey, Diurnal variation in the concentration of *Trypanosoma congolense* in the blood vessels of the ox's ear. *Trans. R. Soc. trop. Med. Hyg.*, 26, 557-564 (1931).
11. F. Hawking, M. J. Worms and K. Gammage, 24- and 48-hour cycles of malaria parasites in the blood; their purpose, production and control. *Trans. R. Soc. trop. Med. Hyg.*, 62, 731-760 (1968).
12. J. D. Arnold, A. Berger and D. C. Martin, The role of the pineal in mediating photo periodic control of growth and division synchrony and capillary sequestration of *Plasmodium berghei* in mice. *J. Parasit.*, 55, 609-616 (1969).
13. J. D. Arnold, Some mechanisms of entrainment of *P. berghei* to the photoperiodic rhythm. *J. Parasit.*, 56, 10 (1970).
14. D. C. Boughton, Diurnal gametic periodicity in avian *Isospora*. *Amer. J. Hyg.*, 18, 161-184 (1933).
15. P. P. Levine, The periodicity of oocyst discharge in coccidial infection of chickens. *J. Parasit.*, 28, 346-348 (1942).
16. G. Schwalbach, Die Coccidiose der Singvogel. I. Der Ausscheidungsrhythmus der Isospora-Oocysten biem Haussperling *(Passer domesticus)*. *Zbl. f. Bakt. I. Orig.*, 178, 263-276. II. Beobachtungen an Isospora-Oocysten aus einem Weichfresser *(Parus major)* mit besonderer Berucksichtigung des Ausscheidungsrhythmus. *Ibid.*, 81, 264-279 (1960).
17. D. C. Boughton, Studies on oocyst production in avian coccidiosis. III. Periodicity in the oocyst production of Eimerian infections in the pigeon. *J. Parasit.*, 23, 291-293 (1937).
18. D. C. Dukes and L. Davidson, Some factors affecting the output of schistosome ova in the urine. *Cent. Afr. J. Med.*, 14, 115-122 (1968).
19. C. P. Read and A. Z. Kilejian, Circadian migratory behaviour of a cestode symbiote in the rat host. *J. Parasit.*, 55, 574-578 (1969).
20. C. A. Hopkins, Diurnal movement of *Hymenolepis diminuta* in the rat. *Parasitology*, 60, 255-271 (1970).
20a. G. N. A. Bailey, *Hymenolepis diminuta*: Circadian rhythm in movement and body length in the rat. *Exp. Parasitology*, 29, 285-291 (1971).
21. R. J. Pitchford, A. H. Meyling, J. Meyling and J. F. Du Toit, Cercarial shedding patterns of various schistosome species under outdoor conditions in the Transvaal. *Ann. trop. Med. Parasit.*, 63, 359-371 (1969).

22. B. O. L. Duke, P. D. Scheffel, J. Guyon and P. J. Moore, The concentration of *Onchocerca volvulus* microfilariae in skin snips taken over twenty-four hours. *Ann. trop. Med. Parasit.*, 61, 206-219 (1967).
23. J. J. Lartigue, Variations du nombre de microfilaires d'Onchocerca volvulus contenues dans des biopsies cutanées pratiquées à differentes heures de la journée. *Bull. Wld. Hlth Org.*, 36, 491-494 (1967).
24. D. Katamine, Y. Aoki and I. Iwamoto, Analysis of microfilarial rhythm. *J. Parasit.*, 56, 181 (1970).
25. M. Rothschild, Fleas. *Sci. American*, 216, No. 6, 44-52 (1965).
26. C. G. Huff, Schizogony and gametocyte development in *Leucocytozoon simondi*, and comparisons with *Plasmodium* and *Haemoproteus*. *J. infect. Dis.*, 71, 18-32 (1942).
27. E. Chernin, The relapse phenomenon in the *Leucocytozoon simondi* infection of the domestic duck. *Amer. J. Hyg.*,56, 101-118 (1952).
28. A. Bishop, Variation in gametocyte production in a strain of *Plasmodium relictum* in canaries. *Parasitology*, 35, 82-87 (1943).
29. J. E. Applegate, Population changes in latent avian malaria infections associated with season and corticosterone treatment. *J. Parasit.*, 56, 439-443 (1970).
30. S. C. Ayala, Lizard malaria in California; description of a strain of *Plasmodium mexicanum*, and biogeography of lizard malaria in western North America. *J. Parasit.*, 56, 417-425 (1970).
31. F. Hawking, M. C. Lobban, K. Gammage and M. J. Worms, Circadian rhythms (activity, temperature, urine, and microfilariae) in dog, cat, hen, duck, *Thamnomys* and *Gerbillus*. *J. Interdisciplinary Cycle Research*, 2, 455-473 (1971).

CHAPTER 6

Circadian Rhythms in Insects

Janet E. Harker

*Zoology Department, University of Cambridge,
Cambridge, UK*

Extensive studies have been made of insect circadian rhythms, and have contributed largely to our knowledge of those basic characteristics which are common to such rhythms in all animals. Laboratory studies have, in the main, been concerned with defining these characteristics and most recently the majority have been concerned with the exploration of the basic controlling system. Running alongside these experimental studies, but on the whole parallel with, rather than converging upon them, have been the even more extensive studies made in the natural environment.

Despite the advances in both of these fields each remains so complex that we are still a long way from dealing with either in terms of the other. Nevertheless it is important from time to time to consider where the evidence from each of the two fields interrelates, and where indeed it appears to be in direct conflict. Such an attempt is made in this chapter. In doing so it is neither possible to cover all the extensive literature in each field, nor to discuss in detail certain aspects which would be of major importance in a discussion confined to only one of these fields. Neither has it been possible to discuss the role of circadian rhythms in orientation phenomena or photoperiodism. In the last section one group of insects, the mosquitoes, whose natural behaviour has been studied more extensively than that of any other group, is considered separately in the light of the evidence presented in the rest of the chapter.

Circadian rhythms are apparent in the total locomotory activity of many insects; activity which may include walking, flight and various taxes. Rhythms are also apparent in the patterns of feeding, respiration, excretion, reproduction (including copulation and oviposition), and in the timing of ecdysis and the emergence of many larvae, or pupae, as adults. Associated with these rhythms are others of change in cell or nuclear volume, of laying down of materials, and of various biochemical events.

Some of these rhythms may not, in themselves, be of great biological importance to the insect; they may perhaps only be produced as a consequence of other rhythms with more adaptive significance. Some rhythmical functions which, in the field, appear to be under endogenous control, may in the laboratory show only a correlation with environmental fluctuations; on the other hand some field rhythms which are apparently influenced strongly by environmental fluctuations, may, in the final analysis, prove to have an underlying endogenous circadian component.

Perhaps the most difficult rhythms to interpret are those of physiological processes which are greatly affected by the total activity of the animal. Yet there is some evidence that the rhythmical fluctuation in the concentration of some metabolic substances, for example blood sugar, is independent of either locomotory or feeding rhythms. Such rhythms need to be considered against the background of homoeostatic control.

Despite the fact that insects are cold-blooded, and therefore have little control over the major physical factor of temperature, they have, like other animals, a high degree of homoeostatic control. Like other animals too, many of the fluctuations in the so-called stable state occur about a mean which varies with a circadian periodicity. The adaptive significance of such rhythms can be seen when homoeostatic control is regarded as a system which enables the insect to receive information from the external environment, and to react in such a way as to stabilize the internal environment. Many major environmental changes occur, however, at regular times of day, and, since biological processes take finite time, it is clearly advantageous to the insect to have a system which fluctuates appropriately and enables the control mechanism to be prepared for such changes [1].

The converse of this situation should also be considered, for if the homoeostatic control system is rigidly tied to a circadian rhythm then fluctuations in the more irregular environmental variables at the "wrong" time of day might affect the animal deleteriously. The separation of some behavioural rhythms which need to be suppressed in unfavourable environments, from metabolic rhythms, may thus be advantageous. In addition, although all circadian rhythms may be entrained by environmental variables, major changes in phase frequently show a time-lag so that an occasional variation in the environment does not, on the whole, reset a rhythm to an inappropriate time for the following day.

6.1 LOCOMOTORY RHYTHMS

By far the greater number of laboratory experiments on circadian rhythmicity in insects have been concerned with locomotory activity. Since such activity affects, or is affected by, most of the physiological systems, and virtually defines the way of life of the animal, an understanding of activity rhythms is an essential step towards the ultimate aim of biology, an understanding of living organisms in their natural environment.

6.1.1 Walking Rhythms

Although relatively few adult insects confine their movements to walking, for practical reasons it is this type of activity, rather than flight, which has gained most attention. Possibly because so many flightless insects are nocturnal rather than diurnal, there has also been a tendency to concentrate on nocturnal insects.

Detailed studies have been made, for example, of the activity rhythms of several species of cockroach [2-11], the cricket *Gryllus* [12] and the beetle *Tenebrio* [13]. In all these insects active walking begins close to the start of the dark phase in LD 12 : 12, when the temperature is held constant, although even between individuals of the same species there may be considerable variation in the phase-relationship to the light-off signal.

In natural conditions, of course, such environmental conditions are rarely, if ever, present, so that it is of considerable interest to follow the effect of different photoperiods and varying temperature on such rhythms. But before considering these environmental variables some attention should be given to another variable which occurs in laboratory experiments.

The Effect of the Actograph

One aspect of rhythmical activity in laboratory conditions which has received little attention is the effect which the container in which the animal lives, or the method of recording activity, has on the form of the rhythm displayed by the animal. It was earlier suggested [14] that animals might, by their own behaviour patterns, reinforce the effect of weak Zeitgeber, or even compensate for a lack of variability in their environment. For instance, in LL the presence of a shelter into which the animal can move and be shielded from the light might result in the animal receiving an effective LD cycle, should it actively seek darkness at one phase of its endogenous behaviour rhythm. Since this suggestion was made such a rhythm of light preference has been found in the crab *Uca* [15], and self-selected rhythms have been studied in the canary [16], but studies on insects are still badly needed.

Lipton and Sutherland [17] have tested the variability of rhythms in the cockroach *Periplaneta americana* in two different types of recorder, one being a capacitance monitor and the other a running wheel. Although the authors have not entirely clarified the point, it appears that in the capacitance monitor a rhythm entrained by LD 12 : 12 was characterized by greater activity during the first 2 or 3 h of darkness, followed by a relatively quiescent stage during the rest of the dark period, and by intermittent activity during most of the light period. Perhaps too much emphasis should not be placed on these results as only seven cockroaches tested in this way were rhythmic, and the record as illustrated might not be accepted by some authors as showing a normal activity rhythm. In the running wheel, of the 126 animals which showed an entrained rhythm, 31.8% showed a marked peak of activity at the beginning of the light period in addition to the peak associated with the onset of darkness. The great majority of insects continued to show a biphasic rhythm when placed in DD, although in most animals the secondary peak moved closer to the main peak, and some animals lost it altogether. A similar loss of the secondary peak has been reported in the cockroach [18] when its activity is measured by means of a thread attached from the pronotum to a writing lever.

It is also worth noting that Lipton and Sutherland present their results in a different form for each type of recorder; the activity in the capacitance monitor is shown by the use of histograms, whereas activity in the running wheel is shown by the direct record from an event-recorder. Because the activity occurring in any one hour is summed for the histogram presentation a great deal of information is lost, and indeed if some of the event-recorder figures are put into histogram form no rhythmical activity at all can be seen.

There can be little doubt that the now common practice of presenting results by direct event-recorder traces has led to a considerable increase in our knowledge of rhythmical activity, and the results of some earlier work needs revision in the light of the more detailed records now obtainable with modern equipment.

An initial study of the role of the recorder itself in reinforcing, or stimulating, the rhythm has been made by Brady [8] working in this laboratory. He found that when cockroaches are removed from the actograph and subjected to enforced periods of activity, the subsequent circadian activity peak is markedly increased, the peak being much more affected than is the background activity during the rest of the 24 h.

Animals placed in a running-wheel are able to run more actively than animals in a restricted box of the type used in tipping actographs, capacitance monitor devices, or in conjunction with photo-cells; furthermore any active movement in a running-wheel tends to stimulate the animal to further activity, so that a rhythm of activity becomes more marked. From the point of view of the observer this type of recorder is therefore far superior, and certainly gives much clearer evidence of both entrained and free-running rhythms; but more work is needed before any relationship can be established between the natural behaviour of the animal in the field and the type of rhythmicity recorded in the laboratory. Indeed Kavanau [19] as a result of his experiments on the behaviour of *Peromyscus* in a running-wheel of a type which could be motor-driven, motor-driven but controlled by the animal, or free-turning, concludes that records from running-wheels may be misleading.

Photoperiod and Light Intensity

Turning now to the effect of the more frequently studied environmental variables, those of photoperiod and light intensity are of particular interest.

The cockroach *Leucophaea,* when in an LD 23 : 1 regime, shows a peak of intense activity about 7-8 h after the usual peak associated with the onset of darkness: this secondary peak of activity lasts, under these conditions, for about 4-5 h. Similar secondary peaks occur whenever the length of the dark period is less than between 7-8 h, that is whenever light falls on the animal 7-8 h after the initiation of the dark-active peak. This secondary peak must represent a phase of extreme sensitivity to light, as is clearly shown when the insect is kept in LD 1 : 1 (Fig. 6.1). In this latter regime the LD cycle does not entrain the activity rhythm and the cockroach free-runs, but when the light comes on during the light-sensitive phase intense activity occurs, ceasing however during the alternate hours when the light goes off [11]. The conditions under which these experiments were performed do not enable any conclusion to be drawn about whether the intense activity during the light period is a vigorous escape reaction, or whether such activity would normally occur in natural conditions. Should the latter be the case then the light-sensitive phase may play an important role in the life of nocturnal animals in temperate climates, causing them to be more active during the early morning hours than during the night.

It seems likely that at least some of the rhythms which become biphasic under certain conditions do so because of a similar

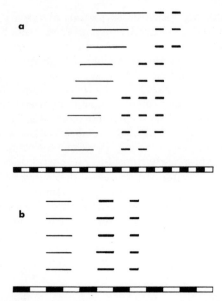

Figure 6.1. Diagrammatic representation of the activity of *Leucophaea* in (a) LD 1 : 1; (b) LD 2 : 2. The LD regime is indicated by black and white enclosed bars.

light-sensitive phase, and it would be of interest to know whether in these cases the two peaks keep the same phase-relationships under different light intensities.

Although the effect of the intensity of the light, in both LL and LD, has been studied extensively in other animals, few such studies have been made with insects. Lohmann [13] has, however, shown that the period of the free-running rhythm of the nocturnal beetle *Tenebrio* increases as the light intensity increases, in accordance with Aschoff's Rule. He has also made the interesting observation that when the light intensity is increased by a factor of 100 (within the range 0.01-100 lux) the period lengthens by about 50 min; on the other hand when the light intensity is decreased by the same factor the period decreases by only about 30 min. Lohmann [20] has also found that when a light-step, from lower to higher intensity, is given to a free-running *Tenebrio* the magnitude of the consequent change in period length is related to the circadian time at which the intensity change took place. Maximum changes in period occur when the increase in intensity is given 2-3 h after the onset of activity, and minimum changes in period occur when the light-step comes about 12 h later than this. A similar phase-dependent shift-response was shown earlier in cockroaches [21]. Lohmann points out, however, that since both light-on and light-off stimuli were included in the latter experiments the "two directional" response curve found for cockroaches may be the product of a combination of the two stimuli: only a light intensity increase was given to *Tenebrio* and a "one directional" response curve was obtained.

Lohmann found similar effects at much lower intensities, that is when the light-steps were only from 0.01 to 2 lux, and an animal in its natural environment must frequently experience intensity changes of this order. Normally, however, an animal would also be under the much more effective stimulus of the change from day to night, and in such conditions the phase-relationship between Zeitgeber and the response-curve might act towards stabilizing the system.

The intensity of the light in LD 12 : 12 also has a clear effect on the timing of the activity peak in *Tenebrio* [13]. Activity normally starts before the light-off signal, and with low light intensity during the light phase the time-difference between this onset of activity and the light-off signal is greater than it is at higher intensities. This intensity effect may be quite significant in a natural environment; observation of nocturnal moths, for example, suggests that they become active earlier after an overcast day.

The effect of alterations in the timing of the LD cycle on the activity rhythms of cockroaches has been studied by a number of authors [4, 5, 9, 11]. Some differences appear in their results, partly because of the different types of recorder used in the different experiments. Harker [4, 5] showed that when the timing of an LD 12 : 12 cycle was altered so that the onset of darkness came 7 h earlier than usual, the major peak of activity in *Periplaneta* did not phase-shift immediately to the new time of onset of darkness, but that after small phase-shifts had brought the activity peak to within about 5 h of the beginning of darkness a rapid phase-shift occurred. Brady [9] observed that, out of 7 animals given an environmental phase-shift of more than 5 h, possibly three showed what he calls "slight sign of an acceleration in phase-shifting" when the activity peak came to within about 5 h of the new LD transition. Examination of his text-figures however, shows that at this stage there is at least a three hour jump, followed by another on the following day, in contrast to shifts of considerably less than an hour at other times. However his point that the final steady state is reached by gradually decreasing phase-shifts, when the onset of activity is within about an hour of the LD transition, is clearly supported by his figure. It is here that once again the advantages of the method of recording in a running wheel and the use of direct traces can be seen, for with the methods used originally [4, 5], involving the summation of the activity within each hourly period, such small shifts could not be seen. It is also clear that the use of the running wheel, which, as discussed earlier, appears to induce a very sharp onset of activity, allows for a different type of measurement of phase-shifts. This can also be seen clearly in Brady's figures, for when he recorded from a photo-cell box, although he states that no jump can be seen in the phase-shift, such a jump can be seen if the results are summed in hourly groupings. Phase-shifts showing jumps at the time when onset of activity comes to within about 4 h of the new LD transition can also be seen in records published by Roberts [11] showing the resetting of phase by *Leucophaea*.

Environmental phase-shifts in which the onset of darkness comes later than previously cause an immediate, or in the case of very large shifts a very rapid, phase-shift in the activity rhythm of both *Periplaneta* and *Leucophaea*. In another cockroach, *Blaberus,* the rate of phase-shift has been shown to depend on the light-intensity during the light phase [22, 23].

As has been mentioned earlier the reaction of an insect to both LL and LD cycles is related to the intensity of the light, and further studies like those made of *Blaberus* are badly needed, particularly in relation to phase-setting.

One interesting aspect of phase-shifting has hardly been considered, and that is the question of stability. When the peak of some particular activity has been phase-shifted the question must arise as to whether the *total* rhythmicity of the animal has also been phase-shifted, or whether it is possible that different processes phase-shift at different rates.

Although dissociation of different physiological processes has been shown in the case of man [24] it has not, to my knowledge, been closely studied in insects, although some of the evidence on the instability of phase-shifts suggests that it might occur. Warnecke [25], for example, has shown that when the beetle *Geotrupus* is placed in a reversed LD cycle the activity cycle is phase-shifted in about 8 days (depending on the light-intensity in the light-phase), but if the LD cycle is reversed again a very rapid phase-shift back to the old phase-setting occurs. Such a rapid reversal takes place even after a new cycle has been maintained for a considerable time, and only ceases after about 4 weeks of the new conditions.

Since most of the work described in this section deals with nocturnal animals it is of interest that Warnecke has also studied three closely related species of *Geotrupus; G. silvaticus* and *G. vernalis* being diurnal in habit, while the third *G. stercorarius* is nocturnal. When kept in DD from the egg onwards all three show the same pattern of activity, with a bimodal rhythm in the adult stage. However the free-running frequency shows a basic difference, being longer than 24 h in the diurnal species, and shorter in the nocturnal. This is an interesting example of a fundamental difference between the biological clock of closely related species, a phenomena which will be discussed further in a later section.

Temperature

In common with other organisms the period of circadian rhythms in insects is only very slightly, if at all, affected by temperature. On the other hand fluctuating temperature may act as a Zeitgeber in the same way as does light. For most insects, however, temperature is a very weak Zeitgeber compared with light, and may act only in the absence of any fluctuation in other environmental factors.

Roberts [11] has shown that the locomotory activity rhythm of *Leucophaea* may be entrained by a 24-h sinusoidal temperature cycle with a 5°C range, the onset of activity occurring at the high point of the temperature cycle. It is interesting that the peak of activity should be phased to the time of maximum temperature since in a nocturnal animal activity would normally come at a time of falling temperature, although Roberts states that sunset and the high point of the temperature cycle are roughly coincident in nature.

Single low temperature "pulses" will cause a phase-shift in the activity rhythm of both *Leucophaea* [11] and *Periplaneta* [5, 26], the magnitude of the phase-shift being related to the circadian time at which the low temperature treatment occurs. Roberts found that the maximum phase-shift resulting from a 12-h temperature drop from 25° to 7° occurred when the return to the higher temperature came 14 h after the previous peak of the activity rhythm, and the minimum shift occurred when the temperature rise came 18 h after the previous activity peak. When the low temperature was continued for 48 h, however, the phase-shift which followed was strictly correlated with the time of the temperature rise.

Insects living in areas with large temperature differences between day and night experience the range used by Roberts and for a nocturnal insect the day-time increase in temperature would come about 14 h after its activity peak. Field observations relating to this aspect would be of interest.

In contrast to the effect of low temperature steps high temperature "pulses", during which the temperature is increased from 21°C to 31°C for 4 or 8 h, do not induce phase-shifting in *P. americana* [26].

The temperature range to which an insect is subjected may affect the phase-setting of an activity rhythm in relation to the LD cycle, and this type of effect must be of great importance in natural conditions. Edwards [27] has made a very careful study of the activity rhythm of two Lepidoptera from hatching through to the adult stage, and finds that in LD the larva of one, *Halisidota argentata*, undergoes a temperature-related phase-reversal, becoming diurnal at low temperatures and nocturnal at higher temperatures. A similar phase-reversal is seen in the ant *Messor semirufus*; it too is diurnal at low and nocturnal at high temperatures [28]. Some insects show unimodal activity rhythms at one temperature and bimodal rhythms at others; for example *Calliphora stygia* shows a single-peak curve in winter, but has a morning and an evening peak in summer [29].

6.1.2 Flight Activity

Studies on flight activity have been made mainly in the field rather than the laboratory, in contrast to those on walking activity. The results from field work on flight periodicities are so extensive that it is not possible to discuss them here in any detail: Lewis and Taylor [30], for example, analysed collections of about 5 million insects from 46 habitats, Williams [31, 32] trapped insects over a 4-year period, and many other very detailed studies have been made of particular genera. Laboratory studies have been related to field studies in a few cases, in particular with reference to *Drosophila*, aphids and mosquitoes: discussion will be confined here to the first two, since mosquitoes are dealt with in a separate section (p. 000).

Aphids

In a series of papers Johnston, Taylor, and their colleagues, have analysed the factors involved in the flight activity patterns of *Aphis fabae* [33, 34].

The number of aphids in flight above a summer breeding site shows a diurnal fluctuation which is bimodal in form (Fig. 6.2). The bimodal curve must result from single acts by a succession of individuals, since the winged aphids fly away from the crop a few hours after ecdysis to the adult stage and do not return: the curve therefore represents a true population periodicity.

Analysis of the sequence of events leading to flight has shown that the length of the teneral period (when the wings are expanding and hardening) is dependent on temperature. Once this developmental period is over flight itself is affected by temperature, light and the presence of adjacent insects. The afternoon peak is thus produced by a combination of two factors: a shortening of the teneral period as the temperature increases, and an increase in the rate of take-off. The morning peak appears to be due to aphids which have matured in the late evening, or overnight, all of which start to fly when the rising temperature makes this possible. If the light-intensity remains high late in the evening the number of flying insects still drops because the falling temperature lengthens the teneral period, although the number flying the next morning is lower than usual because all the flight-mature insects will have

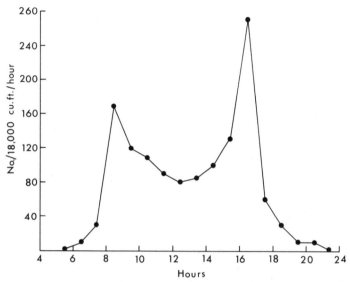

Figure 6.2. Aerial density of aliencolae of *Aphis fabae* flying above a bean crop on which they were produced, on a July day in England (after Johnson and Taylor [34]).

flown the previous evening. By contrast, if the temperature remains high, but light-intensity decreases, then take-off is inhibited but large numbers of flight-mature aphids are available for the next morning peak.

It seems from these results that the entire sequence of events leading to flight periodicity is under the control of exogenous environmental factors, and that circadian rhythmicity is not involved. Yet by comparison with other studies a circadian rhythm of ecdysis might be expected, and would underlie the subsequent events. Haines [35] studied the periodicity of moulting in constant light and temperature, and found that although peaks of ecdysis were maintained during the morning and afternoon these were not statistically significant unless the population had been maintained in LL for weeks beforehand. This of course still suggests a true rhythmicity. It should be noted that these experiments were not however confined to the imaginal ecdysis, but dealt with a population of mixed age groups, so that unless moulting has the same phase-setting in all instars any periodicity would be blurred. In further experiments a very clear periodicity appeared in LD 12 : 12, but this virtually disappeared when the light cycle was reversed; again since nothing is known about how different instars react to phase-shifting it is difficult to draw any conclusions. It might, however, be unwise to rule out any circadian rhythmicity in the flight of aphids.

This species has been discussed as an example of the way in which an interrelation of exogenous factors may produce a clear rhythm, and to show that even if there is an underlying endogenous rhythm it may have little obvious influence in the field.

Drosophila

Many species of *Drosophila* show a bimodal dawn and dusk flight pattern [36, 37, 38]. The time of flight is closely related to light-intensity, and indeed Pavan *et al.* have shown that this bimodality disappears in rain forests, particularly on cloudy days. *D, subobscura* also shows a bimodal periodicity in open areas of meadows and open woodland, but in dense woodland the closely related *D. obscura* shows only a unimodal peak [39]. It has been suggested [40, 41] that the dawn and dusk flight of *Drosophila* is an ecological adaptation to avoid dessication, but humidity is not a factor governing its flight in nature. Lewis and Taylor [30] postulate that this bimodal flight is in fact an adaptation to the type of feeding rather than due to susceptibility to dessication. Taylor and Kalmus [41] earlier suggested that *Drosophila* has become adapted to dry conditions, not by an increased resistance to dessication, but by the increased visual efficiency in low light intensities, thus allowing for a crepuscular burst of activity.

It is perhaps curious that despite the very detailed analysis of the periodicity of adult eclosion in *Drosophila* there does not seem to have been a related study on the time of flight, although Roberts [43] has shown that there is unimodal rhythm of flight activity in the laboratory, in contrast to the bimodal rhythm described above.

6.1.3 The Influence of Non-physical Factors on Locomotory Rhythms

The biotic, as distinct from the physical, environment has received relatively little attention in either field or laboratory studies on circadian rhythmicity. This is hardly surprising in view of the complexity of the results obtained after varying even the relatively easily controlled physical factors, yet if the role of rhythms in the life of animals is to be considered we cannot ignore the interaction between individuals of the same species, let alone interaction between species.

Competition

In equable climates, in which the environment is most predictable, and most permissive for many activities, there is a high adaptive value in a rhythm which can reduce inter-specific competition by temporal diversification, as Corbet has pointed out [44]. It is in such situations that laboratory and field studies tend to reveal similar activity patterns, whereas in harsher climates laboratory and field studies often reveal dissimilar patterns. Perhaps this discrepancy arises because in the field the requirements for an immediate response by the animal to any permissive physical condition when it occurs, regardless of the time of day, may override endogenous rhythms which may indeed even have a rather low adaptive value.

Differences in phase-setting of activity rhythms of individuals of the same species living in adjacent habitats has frequently been recorded. Light trap catches, for example of male *Chrysops centuriones* and *Tabanus thoracinus*, have revealed a biphasic activity rhythm in open woodland, with peaks in the early night and before sunrise, whereas in forests the activity is confined to a single hour before sunrise [45]. Complete reversal of phase-setting in the beetle *Feronia madida* occurs between open ground and adjacent woodland, the insect being diurnal in the former and nocturnal in the latter [46]. Changes in pattern in *Drosophila* flight rhythms have been mentioned earlier, those occurring in mosquitoes are discussed in the last section, and many other instances could be quoted.

It is probable that in many cases the differences in phase-setting are due to physical factors, in particular to light-intensity, operating at different levels in the different environments. Yet when individuals remain within the particular habitat, and do not pass in and out of adjacent areas, it is possible to distinguish local populations with differences in phase-setting in the laboratory under identical physical conditions.

The number of animals in the population may also affect the phase-setting, both of individuals and of the population as a whole. In some cases this may be due to changes in the physical environment brought about by the animals themselves; for example the eclosion of *Drosophila* adult occurs later in the day as the population increases [47] and it is possible that this is due to changes in food, or gas, concentration with increasing numbers. On the other hand this type of competition may not always be a major factor, and it would be interesting to investigate the role of mutual stimulation, or stress, in such situations, particularly in view of the results obtained from the use of different types of activity recorder in the laboratory.

Closely related species living in the same habitat frequently show specific differences in their time of activity, a feature of obvious adaptive value. A very clear example of such a species difference has been shown in the males of Doryline ants [48] (Fig. 6.3), and the extensive work on the eclosion of *Drosophila* has revealed characteristic emergence patterns, both for different species and for mutant strains. Surprisingly little study has been made of the genetical control of patterns of phase-setting, particularly since such detailed information on each of

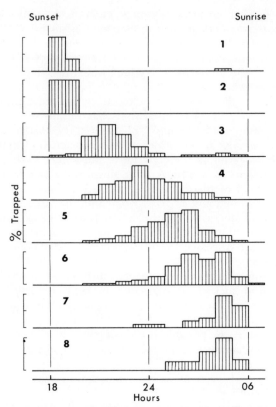

Figure 6.3. The entry of Doryline male ants into light traps 1. *Dorylus moestus* 2. *D. nigricans* 3. *D. affinis exilis* 4. *D. fulvus* 5. *D. alluaudi* 6. *D. katanensis* 7. *D. burmeisteri* 8. *D. fimbriatus laevipodex* (after Haddow, Yarrow, Lancaster and Corbet [48]).

these aspects is available for *Drosophila;* the results of one such investigation are however considered in a later section (p. 206).

On the whole the two sexes are active at the same time of day in many animals, and Lewis and Taylor [30], in their large-scale study of flight times, found this to be so for the vast majority of insects. There are, however, some exceptions. The male of the moth *Lithocolletis messaniella* is found predominately in the morning flight peak, and the female in the sunset peak; a similar dissociation occurs in another moth *Anagasta kuhniella* [49]. The hymenopteran *Neuroterus* shows even more dissociation, with a dawn peak consisting entirely of males and a day time peak entirely composed of females. The males of *Drosophila subobscura* predominate in the dawn peak and the females in the dusk, despite the fact that both sexes respond to the same light intensity. On the other hand there is some evidence that in the laboratory groups of *Drosophila,* in which only one sex is present, show a bimodal rhythm, whereas when both sexes are present only an evening activity peak occurs [50].

Activity records of cockroaches show differences between adult males and females [3, 17], but to what extent these differences are related to reproductive activity is not yet entirely clear.

Further study of behaviour patterns during reproductive cycles is needed in relation to all those rhythms which show sex differences, but it does seem that it is not only the reproductive cycle which is responsible for the sexual distinction in phase-setting, for it is also found in the eclosion rhythm of various species [51, 52, 53].

Age

Very different patterns of activity may be shown by insects at different stages in their life cycle, and although this is certainly not unexpected considering the great difference between adult and larval habits and habitats in so many insects, it is perhaps less expected in the exopterygotes in which all stages may occupy the same environmental niche. Remarkably little attention has been paid to this changing pattern, and yet it poses some fascinating problems, for it implies that the alteration in phase-setting is produced by some process within the developing animal.

The commonest type of change in pattern is probably a "sharpening" of the rhythmicity, involving a change from a rather broad peak of activity to perhaps a clear single peak or to even a biphasic pattern. Such a change is seen in *Carausius* which is active over most of the 24 h in the first instar with only a slight increase around midnight, whereas in later instars activity during the day virtually ceases and a peak of activity occurs at dawn and dusk [54, 55]. The moth *Halisidota,* when in a regular LD cycle and constant temperature, shows in the second and third instars a peak of activity around the time of light-off, and another small peak just after light-on. In the fourth and fifth instars the evening peak moves earlier and activity declines more sharply after light-on, and by the eighth instar the activity peak is occurring during the afternoon. After emergence as an adult the activity peaks again occur near the light-on and light-off signals.

In some insects the activity rhythm may become very weak, or even disappear altogether at times. This can be seen to happen in the last nymphal instar of *Gryllus domesticus;* since the rhythm reappears in the adult it seems likely that it is also maintained, but is not overt, at the earlier stage [12].

Even within the same instar the pattern of activity may change with age; for instance the adult dragonfly *Anax imperator,* in addition to the sunset activity peak, shows a peak at dawn when it is newly emerged, this peak moving to midday by the time it is more than a week old [56], and several species of Ephemeroptera are active at different times of day according to their age [57].

Effect of Feeding

Relatively few experimental studies have dealt with the question of whether feeding, or the presence of food, can act as a Zeitgeber. It is, however, well known that bees can be trained to feed at certain times of day, and the presence of food is undoubtedly a Zeitgeber in this insect. For bees, and carnivorous insects, the presence of food is dependent on the rhythmicity of other species, and there may be an adaptive value in a close relationship between feeding and the phase-setting of activity rhythms. On the other hand the food of many insects is constantly present; yet it is to such insects that experimental work has been almost entirely confined.

Green [58, 59] has made an extended study of the relationship between feeding and the locomotor rhythm of the fly *Phormia regina.* In the absence of food and water the amount of activity increases as a function of the deprivation time, but when sucrose solutions are supplied the amount of activity is immediately reduced. Further operational procedures give results suggesting that the amount of activity is controlled by a hormone released from the corpus cardiacum, the release of the hormone being under the control of receptors in the foregut. The circadian rhythm however overrides these effects, so that only the amount and not the timing of the rhythm is affected. A similar effect on the degree of activity of food-deprived *Drosophila* has been observed by Connolly [42]. The beetle *Geotrupus silvaticus* also becomes increasingly active in the absence of food [25], but in this species it is interesting that only one peak of the bimodal rhythm is affected. In all these examples it appears that food does not have any affect on the entrainment of the rhythm, but the results are worth noting in relation to weak or obscure rhythms in the field, for it may be that the presence of abundant food may reduce activity to a level at which a rhythm is difficult to measure.

The effect of feeding on activity rhythm of the cockroach *P. americana* has been studied by Harker [3] and Lipton & Sutherland [17]. The latter measured the feeding activity by recording the movement of a pendulum on which the food was suspended; use of this method reveals a feeding rhythm which is very similar to the locomotor rhythm in terms of time of onset and period length, in its ability to be reset by a change in the LD cycle, and most important, in its persistence in DD. Unfortunately these authors did not test the effect of giving food at only one time of day during DD to establish whether a phase-shift could be effected, so that the question of whether the presence of food can act as a Zeitgeber is still an open one. In the earlier experiments [3] an attempt was made to answer this question by restricting food availability to a few hours during the light phase of an LD cycle. Although the insect became active in the presence of food the normal activity rhythm continued, and the activity during the light phase appears to be only an immediate response to the presence of food; when food was not presented on any particular day there was no increase in activity at that time of day. The experiments were repeated in LL with the same result, but in view of the results of further work in recent years it is now apparent that the presentation of food during a free-running rhythm in DD might have been a more appropriate test, particularly as an activity recorder of the confined type was used.

An interesting observation has been made concerning the relationship of feeding behaviour to locomotion in the cricket *Gryllus domesticus* [12]. Feeding activity reaches a maximum about 4 h before the onset of darkness in LD, the number feeding then drops, but rises again about an hour before dusk; thus the major feeding activity comes before a minor activity peak in the afternoon, and only a minor feeding peak comes at the time of the major activity peak.

Feeding behaviour of the milkweed bug *Oncopeltis fasciatus,* again an insect whose food is always available, follows the LD cycle with maximum feeding occurring at the end of the light period. This rhythm only persists for one cycle in LL, and not at all in DD [60]. It is not possible to tell from these results whether this is really a persistent rhythm, for the one cycle shown in LL could be a direct hunger effect.

Information is still needed about the phase-shifting of feeding rhythms in insects, particularly in those like bees for which the presence of food appears to be a strong Zeitgeber. A valuable extension of the work on feeding cycles of bees might come from a study of the effect of a phase-shift in the feeding rhythm on its phase-relations with some other function; the work of Bennett & Renner [61], on the use of a bee laboratory, suggests that it might now be possible to make this type of study under controlled conditions.

An effect of the time of feeding of the larvae on the actual period of the rhythm of pupation of the mosquito *Aedes taeniorhynchus* has been shown by Nayar [62]; it is possible however that this results from an effect on the synchronization of the population rather than an effect on the periodicity of the individual larva.

6.2 RHYTHMS OF PUPATION AND ECLOSION
TO THE ADULT STAGE

Many insects, particularly those with aquatic larval stages, emerge as adults in very large numbers over a limited area, and during only a rather confined time of day. It is not surprising therefore, that many field studies have been made of this event [51, 52, 63-67].

The most extensive of the laboratory studies are those made of the eclosion of *Drosophila*. Kalmus [68] first showed that a definite peak of emergence occurred just after the change from darkness to light; this was later confirmed by other workers, and a careful study by Brett [69] showed the effect of light and temperature on the timing of eclosion. Pittendrigh and his associates, in a long series of papers (see [70]) have since then explored this phenomenon in depth, particular attention being paid to experiments designed to give information about the fundamental properties of the circadian system: experiments which include studies of the effect of "skeleton" photoperiods and of single light periods on the timing of emergence some days later. In the final analysis all information about the circadian system must come into our picture of the role it plays in the life of the animal, but while our understanding of this system is still not clear it is not always possible to consider it in perspective against the rhythms we observe. Since this chapter concerns the role of rhythms in the normal life of insects, not all of Pittendrigh's results will be discussed here, valuable though they may be in the interpretation of the properties of biological clocks.

Although in LD 12 : 12 the eclosion peak occurs just after the beginning of the light period the position of the peak changes with different photoperiods. For example, in LD 4 : 20 the peak occurs during, although towards the end of, the dark period [69], whereas in LD 14 : 10 it occurs some hours after light on, and in LD 18 : 6 there are two emergence peaks, one soon after dawn and the other about 9 h later [71,72].

The timing of the eclosion peaks is virtually temperature-independent, having a Q_{10} of 1.02 [73], despite the fact that the actual length of the developmental period of the pupa is of course very temperature sensitive. The rhythm of eclosion can also be entrained by a temperature cycle when the insects are kept in LL, in which they do not otherwise reveal an eclosion rhythm [72].

By following the development of over 16,000 individual pupae it has been shown that the timing of the eclosion peak is related to a circadian rhythm which affects the development of the pupal stage, and which results in the synchronization of the population [71]. Pittendrigh & Skopik [74] have however also followed the development of three species of *Drosophila,* using a different procedure, and have concluded that there is no circadian factor in the development of the pupa which can affect the eclosion. A detailed analysis of Pittendrigh & Skopik's paper in relation to the previous papers will be made elsewhere.

A clear rhythm of puparium formation has been described in *D. victoria,* and a less distinct one in *D. hydei* and *D. melanogaster* [76, 77]; Pittendrigh & Skopik, however, state that there is no rhythm in *D. melanogaster* although no evidence is cited to support this and they accept Rensing's results for *D. victoria.* The two strains of *D. melanogaster* used in Harker's study underwent pupation at all times of day, but it occurred very rarely at some times and most commonly at others.

Rensing & Hardland [77] have also shown that the number of pre-pupae formed reaches a maximum at dusk, and a minimum in the morning, and that this rhythm is controlled by the LD cycle: the rhythm of puparium formation is at least partly due differences in the length of the developmental time, for pre-pupae formed in the morning pupate 18 h later at $18.5°C$, whereas those formed in the afternoon take 19-20 h to pupate. It is interesting to note, in relation to the studies on the development of the pupa itself, that Rensing also finds that in LD 20 : 4 puparium formation occurs earlier in the day than it does in LD 12 : 12 [76].

A phase-relationship has been shown between the puffing-pattern of the salivary gland chromosomes in the 3rd instar larva, and the time of puparium formation, the pattern associated with puparium formation appearing 4-6 h earlier than the actual formation of the puparium [78]. On the day before puparium formation takes place the nuclear size of the cells in three endocrine systems show a rhythmical variation, and the size of the cells of the corpus allatum, prothoracic gland, and the neurosecretory brain cells all reach a maximum, in LD 12 : 12, three hours before light-on and three hours before light-off. On the day of puparium formation the maximum occurs 3-4 h *after* light-on with the change in the puff-pattern occurring immediately afterwards [79], suggesting that ecdysone release has taken place [80, 81].

Several authors have shown that the emergence rhythm of the mosquito *Aedes taeniorhyncus* is dependent on the pupation rhythm, and that the two events are separated by an interval affected by temperature and not by photoperiod [62, 82, 83].

6.3 CUTICULAR DAILY GROWTH LAYERS

One of the most fascinating manifestations of a circadian rhythm has been revealed by the work of Neville on the growth layers in the cuticle of a variety of insects. This work does not seem to have attracted the attention it deserves by workers on circadian rhythmicity, despite the fact that it now makes available a near-ideal system for fundamental work on the biological clock.

Neville [84, 85] first showed that the endocuticle of *Schistocerca gregaria,* which is deposited after each ecdysis, can be seen to display daily growth layers when viewed in section with a polarizing light microscope. The layer which is deposited during the night has the chitin organized into several lamellae, whereas that deposited during the day is non-lamellate. More recent work [86], using electron microscopy, has shown that the night-deposited layer appears lamellate under polarized light because of the helicoidal orientation of the planes of the microfibrils. The day-deposited cuticle appears non-lamellate since all the microfibrils run in one direction.

The deposition of these growth layers in *Schistocerca* continues when the locust is kept in DD and constant temperature, so that it can be regarded as showing a true circadian rhythmicity. The free-running rhythm in DD has a period of about 23 h [87], and as with other rhythms the free-running period is virtually temperature-independent. In LL the rhythm disappears, only non-lamellate cuticle being produced.

One of the most interesting results from this work has been the discovery that in the locust, epidermal cells in different parts of the insect differ in their response to LL. Even cells which are spatially very close may show quite different responses; for instance in constant low temperature and high intensity LL the endocuticle of the thickened proximal region of the hind tibia is laid down continuously in the lamellate form, despite the fact that the endocuticle of the rest of the hind tibia is, under these conditions, non-lamellate. This suggests that either each epidermal cell is controlled by its own biological clock, and the clocks differ in their response to LL, or, as Neville suggests, that many cells are controlled by a single clock, but the way in which individual epidermal cells respond to it, or are coupled with it, differs.

Although it has previously been suggested that different processes may be differently affected by environmental phase-shifts [21], this very clear demonstration at the level of closely related epidermal cells is a striking development, particularly when the pleuripotency of the insect epidermal cell is considered.

In this respect it is interesting that the deposition of endocuticle layers in the beetle *Oryctes rhinocerus,* which is rhythmical, although perhaps not truly circadian, shows a variation in the *frequency* of deposition of the different layers in different parts of the body [88], again suggesting some degree of autonomy for different epidermal cells.

Light does not act through either the compound eyes or the ocelli of the locust, nor does it appear to act directly on the neurosecretory cells of the brain, since layering of the cuticle is lost in LL even after cauterization of eyes and ocelli and the blackening of the entire head capsule. Neither is there an indirect effect via the food plant. Neville [89] concludes that these results provide evidence against neural timing of this rhythm, but it could be argued that the clock itself continues to function under these conditions, but that the epidermal cells themselves are sensitive to light, and in its presence cease to react to fluctuations in the internal environment. Neville suggests this, in effect, when he speaks of uncoupling the epidermal cell from the clock, and he concludes that different epidermal cells by reacting differently to LL are showing different degrees of uncoupling from the clock. Furthermore the way in which epidermal cells react to the environment may show a time-sequence; the same epidermal cells which produce non-lamellate endo-cuticle in LL will, before ecdysis, produce lamellate exocuticle, so that there is evidence here that the cell itself may change in its reactions.

There does not, however, appear to be any direct control of the rhythm by the nervous system, or by neurosecretion being passed directly to the cells by axons. This is clearly shown when small cylinders of hind tibia are implanted into the haemocoele; after an initial period, during which the epidermal cells of the implant migrate and encapsulate the whole implant, new endocuticle is laid down and appears in the banded form. This banding continues in DD.

Not all insects lose their rhythmicity in LL, both *P. americana* and the pharate adult of the bug *Oncopeltis fasciatus* continue to show a rhythm in both LL and DD [90, 91].

Very small changes in light-intensity may be sufficient to entrain a rhythm, as has been suggested by Neville [89] in the cave cricket *Dolichopoda linderi*. This cricket lives in an environment of darkness and almost constant temperature, yet still shows a rhythm of lamellogenesis; it is possible that its nocturnal excursions outside the cave expose it to sufficient moonlight (or temperature change?) for this to act as a weak Zeitgeber.

The effect of light intensity on the loss of rhythmicity is marked. Those epidermal cells of the locust which do lose their rhythmicity in LL take longer to do so the lower the light intensity; at intensities below 1 ft candle it may take up to 6 days before lamellate cuticle production ceases, whereas at 75 ft candles lamellate cuticle is only produced during the first experience of LL. As Neville points out this suggests that there is an accumulation of a hormone or metabolite which is regulated by light intensity.

The adaptive features of the system, whereby a regime which induces non-lamellate endocuticle in the major part of the locust but does not prevent lamellate endocuticle production in some specific regions, is of interest. It seems likely that the unaffected parts are those in which lamellae are functionally indispensible, including as they do the ends of the tibia which are involved in the stresses of jumping, the edge of wing veins, the rubber-like cuticle of the pre-alar arm and wing hinge ligaments, and the cuticle of the median ocellus and compound eye.

6.4 PHYSIOLOGICAL RHYTHMS

Insects showing clear activity rhythms might also be expected to show rhythmical variation in at least some metabolic substances. Nowosielski and Patton [92] have measured the blood sugar concentration in the cricket *Gryllus,* and find a circadian rhythm in the total blood sugar concentration which can be related, in the main, to a fluctuation in trehalose. In LD 12 : 12 the concentration shows a peak about 3 h before light-on, but surprisingly this peak is not related to the feeding rhythm, which has its peak in mid-afternoon, nor does it appear to be dependent on locomotor activity since the blood sugar rhythm is maintained during the last larval instar in which there is no overt locomotory rhythm. By contrast, no circadian fluctuations appear in the protein, amino acid or lipid fraction of the haemolymph [93]. Glycogen levels in adults of the mosquito *Culex tarsalis* fluctuate rhythmically, with the highest levels occurring towards the end of the light period [94]. The relative levels of the principal hydrocarbons in the cockroach *P. americana* are also said to fluctuate in a circadian manner, with greater fluctuations in the male than in the female [95].

Haemolymph potassium and sodium concentration has been measured in *P. americana* at intervals covering about 9 h spanning the time of onset of darkness in an LD 12 : 12 cycle, and the results suggest that there is a change in concentration over the day. Analysis of 166 samples showed a fall of about 10% in the potassium concentration during the first hour of darkness and a gradual decline of 2% in the sodium concentration, although when successive samples were taken from individual cockroaches no daily change was recorded [96]. Wall [97] has shown that the volume of fluid secreted by the malphigian tubules in *P. americana* is greater in the dark period, although there is a good deal of variation, and suggests that the drop in haemolymph potassium might be related to increased tubular secretion, since tubules secrete a fluid that contains about six times as much potassium as that in the haemolymph.

The part played by a circadian rhythm in the behavioural response of male noctuid moths to the sex pheromone of the female has been described by Shorey and co-workers [98, 99]. The female releases the pheromone rhythmically, and the male responds to it rhythmically being maximally responsive 6 to 12 h after light-on in LD 12 : 12, but electroantennograms from pherome-stimulated antennae of males show no change in olfactory response related to the LD cycle. In this case, therefore, the rhythmical behaviour must be related to a central inhibitory mechanism.

A bimodal rhythm of oxygen consumption has been shown to occur in *Drosophila* throughout larval, pupal and adult life, although the phase-setting of the peaks changes, the maxima moving earlier in the day [100] with age. There is also a difference between the sexes, the male maxima being about equal in morning and evening, but the female showing a smaller morning than evening peak.

Rensing *et al.* [101] have found that, in some mutant stocks, the female shows only a single, evening, peak. By a series of experiments, in which mutants with different ratios of X chromosomes to autosomes were used, they showed that as the ratio of X chromosomes to autosomes decreases so the oxygen consumption curve shows a relative decrease in the evening maximum and an increase in the morning maximum. From the results of crossing experiments they suggest that there may be two circadian oscillations, with a phase-difference of 12 h, and that these are quantitatively dependent on the X chromosome and the autosomes respectively.

Rhythmical changes in nuclear, nucleolar and cell volumes have been investigated by a number of workers. Klug [102] found a rhythmical fluctuation in nuclear volume in the cells of the corpora allata of carabiid beetles, and in the relative numbers of two types of neurosecretory cells in the brain. Rensing has also found a rhythmical change in the nuclear size of the corpora allata and the brain neurosecretory cells of adult female *Drosophila*. In larval *Drosophila* a bimodal rhythm has been found in the nuclear size of corpora allata, brain neurosecretory cells, prothoracic gland cells and the fat body cells, and in the nuclear and nucleolar volume of salivary gland cells [78], although not all the curves appear to be statistically significant. Brady [103] found a statistically significant increase in nuclear size of the neurosecretory cells of the suboesophageal ganglion of cockroaches during the peak of the loco-motor activity rhythm. Associated with these observations on volume changes attempts have also been made to estimate the amount of neurosecretory material in brain and suboesophageal ganglion neuro-secretory cells, and some suggestion of a rhythmical fluctuation has been reported from the suboesophageal ganglion cells in the phasmid *Clitumnus* [104], and the cockroaches *Leucophaea* [105] and *Periplaneta* [103]. In *Periplaneta* there is some suggestion that the neurosecretory granules may be slightly more aggregated peripherally before the onset of darkness, although no obvious migration of neurosecretory material has been seen. More recently the conclusion that the fluctuations in nuclear volume reflect the activity of neurosecretory cells has been questioned, for fluctuations in RNA synthesis are not always related to fluctuations in nuclear size [106].

By means of autoradiography, estimates of the level of RNA synthesis in the neurosecretory cells of the male cricket *Acheta domesticus* have been made at different times in its locomotor activity cycle [106]. The highest level of RNA synthesis in the pars intercerebralis cells occurs just after the light-on signal, this is followed by a decrease in intensity until another peak appears at about the middle of the light period, coinciding with the time of a small peak of locomotor activity. During the period of maximum locomotor activity, that is, after the onset of darkness, the lowest level of RNA synthesis is seen. The level of RNA synthesis in the suboesophageal ganglion follows a different pattern, with a low level after the light-on signal, and with the maximum occurring half way throught the light period. At the time of peak locomotor activity no synthesis was observed, but another high level was reached in the middle of the dark period. Insects cultured in LL, which show no activity rhythm, also show no fluctuation in the level of RNA synthesis. Unfortunately it is not clear exactly which cells in the suboesophageal ganglion were concerned in these experiments, for both ventral cells and neurosecretory cells are mentioned and pictured separately, but the calculated figures are for ventral cells, so that it is not possible to know whether any, or how many, neurosecretory cells are included.

In a later paper [107], although no figures are given, it is said that in the female the most intensive RNA synthesis in the suboesophageal ganglion neurosecretory cells occurs during the dark period, with the maximum about 3 h after the onset of darkness (i.e. at the time when no synthesis took place in the male), and that during the light period there is a far lower level of synthesis. The authors do not comment on this apparent reversal of the cycle compared with the male, nor have they measured RNA synthesis in the brain of the female but they assume that it follows the same pattern as in the male.

Cymborowski and Dutkowski [106] argue from the results obtained from the male that the sudden increase in RNA synthesis in the pars intercerebralis after the onset of light, which is followed by an increase in synthesis in the suboesophageal ganglion 6 h later, suggests a relationship between the two systems; they also suggest that the suboesophageal ganglion synthesis is being stimulated by the passage of neurosecretory material down a nerve to that ganglion, the combination of these two events leading to the locomotor activity rhythm (Fig. 6.4). However, since the relationship between the two events appears to be quite different in the female, until further details are published it is difficult to draw any conclusion. Furthermore the increase in RNA synthesis in the pars intercerebralis, after the onset of light, would seem to point to this onset of light as the Zeitgeber for locomotor activity, yet the onset of activity is entrained to the onset of darkness.

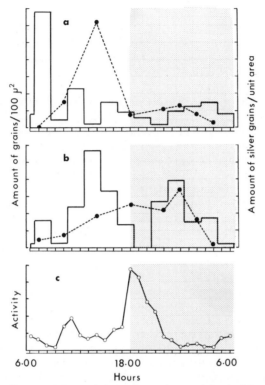

Figure 6.4. House crickets, *Acheta domesticus*, originating from LD 12 : 12. a Intensity of RNA synthesis in neurosecretory cells of the pars intercerebralis—histograms; intensity of protein synthesis from same cells − − ● − −. b Intensity of RNA synthesis in the ventral cells of the suboesophageal ganglion—histograms; intensity of protein synthesis in the neurosecretory cells of the same ganglion − − ● − −. c Locomotor activity expressed in number of movements per hour (after Cymborowski and Dutkowski [106, 108]).

After cauterization of the neurosecretory cells of the pars inter-cerebralis no RNA synthesis was observed in the suboesophageal ganglion neurosecretory cells on the third day after the operation [107]. Dutkowski and Cymborowski conclude that there can be no further RNA synthesis in these cells at any time once the pars intercerebralis has been destroyed. It is, perhaps, premature to draw this conclusion however, since trauma may suppress RNA synthesis, and after many different types of operation a cessation of rhythms has been observed for a period of some days before they reappear [8, 28]. Indeed even in normal conditions rhythms may cease to be overt for a period of time, although when they reappear their phase-setting is such that it is apparent that they have continued at a subthreshold level.

In contrast to the post-operative lack of RNA synthesis in the suboesophageal ganglion, synthesis in the follicular epithelial cells of egg vesicles is increased, although here too the rhythmical fluctuations cease, as they do in the fat body cells.

A further study has been made on the normal fluctuations in protein synthesis in the same nerve cells [108]. In the pars intercerebralis cells maximum protein synthesis occurred about 7 h after both RNA synthesis and accumulation of neurosecretion reach their maxima. In the suboesophageal ganglion, in which protein synthesis does not reach such a high level, a gradual rise continues during the light period and, after a slight drop at the beginning of the dark period, reaches its maximum about half way through the dark period. The fluctuation in the number of cells filled with neurosecretion is not as distinct as in the pars intercerebralis, and the results are difficult to interpret, but the accumulation reaches its maximum towards the end of the dark period, and its minimum before the onset of darkness.

In view of the number of types of cell which have been shown to vary in a rhythmical manner in respect of size, content, or synthetic activity, it is difficult to draw any conclusions about the sequence of events, since all are likely to be affected by the total physiological and behavioural rhythmicity of the animal. Indeed Rensing [109] has found that the volume of the nuclei of the salivary glands of *Drosophila* shows rhythmical fluctuations in culture solution, for up to ten days after the removal from the insect, suggesting that a metabolic circadian rhythm may indeed be present in all cells [110], for it seems extremely unlikely that the salivary gland cells would be the basic biological clock responsible for maintaining the circadian rhythmicity of the entire organism.

Perhaps the rhythmical changes which have been observed are more significant in relation to the way in which insects, like other organisms, show differing responses to various stimuli at different times of day. Rensing [109], for example, has found that the rhythm of the cultured salivary gland cells is affected by pulses of ecdysone, but when this is added to the medium at one time of day a phase-shift of between 3 and 9 h occurs, whereas its addition at other times of day is not followed by any phase shift. This clearly has implications for the way in which insects may react to ecdysone during moulting, and an extension of this type of study might give information about moulting rhythms. However it must be emphasized that it seems very likely that a multiplicity of cycles may be involved in any physiological rhythm, and that different cells may be "sensitive" at different times, so that only certain combinations, or sequences, of sensitivity may give any particular physiological event.

6.5 THE CONTROL OF THE LOCOMOTOR ACTIVITY RHYTHM

The search for a specific system which might control locomotor activity rhythms has been diligently pursued for a number of years, and considerable controversy has arisen over the results.

The first suggestion that hormones might be involved in the control of these rhythms came from experiments in which apparently arrhythmic cockroaches, after being joined in parabiosis with rhythmic cockroaches, took up a clear locomotory rhythm [110]. It now appears possible that this result may have been affected by the conditions under which the activity of the insects was recorded, for while the insects appear to have become arrhythmic in LL in the type of recorder then used, rhythmicity is maintained in LL in running-wheel recorders. It might thus be argued that while the cockroaches before the operation showed no overt rhythm sub-threshhold rhythms become overt under the stimulus of wounding. Little is known about the effect of wounding; Brady [8] has found a tendency for a decrease in the amount of activity at the peak of the circadian rhythm after one type of wounding, although other types of stress are followed by an increase in the activity at the peak period. However nothing is known about the particular set of conditions relevant to parabiotic experiments, and further work is clearly needed. On the other hand rhythms free-run in LL, so that the phase-setting of the rhythms of animals maintained in LL would not always be identical with that of animals maintained in LD 12 : 12, yet in the parabiosis experiments the arrhythmic animals took up a rhythm in phase with that of their parabiotic partner. In any case over the following 16 years a number of hormones have been found to affect activity rhythms.

To verify the assumption that an endocrine system is involved in the control of activity, the main endocrine systems were removed by beheading the insect; cockroaches can still walk after this operation, but are arrhythmic [3], as has since been confirmed by a number of authors [8, 111, 112, 113, 114]. Transplantation of the brain, corpora cardiaca, corpora allata, or thoracic or abdominal ganglia together with their associated neurohaemal organs, from rhythmic donors into headless cockroaches is not followed by any rhythmic activity [3, 5, 8, 115]. On the other hand the transplantation of a suboesophageal ganglion from a rhythmical donor into a headless cockroach is followed by an overt activity rhythm, in phase with the previous rhythm of the donor.

This latter result has been particularly controversial. Roberts [112] attempted to repeat these experiments, using a running-wheel recorder, and in 19 trials reported no successes, although Brady [9] comments that the method of presenting the results might obscure some rhythms. A film has since been taken of the movements of headless cockroaches in a running-wheel, similar to that used by Roberts, and it can be seen that, owing to the postural stance of a beheaded animal, the insect tends to tip forward on a curved surface, pushing down on to the severed neck and frequently struggling violetly. Brady [9] also repeated the experiments, using a number of different types of recorder, and concluded that in 2 out of 29 trials suboesophageal ganglion implants appeared to induce rhythms in their hosts, with a further eight possibly becoming rhythmic. While this does not provide certain confirmation it does not rule out this method of control, for *any* positive result is important in showing whether a rhythm can be induced by a particular method.

However the question remains as to what differences in method have been involved in the operations performed by the different authors. Apart from the fact that the present author has always left the nerves of the ganglion as long as possible for easier handling, and that when Brady did this he observed some evidence of an induced rhythm in four out of nine trials, no obvious difference in technique has been discovered.

Nishiitsutsuji-Uwo & Pittendrigh [114] claim to have found positive evidence that the suboeosphageal ganglion cannot control the activity rhythm by a hormonal agent in the cockroach *Leucophaea*. They found that "severance of the ventral cords between the first thoracic and the suboesophageal ganglion caused nearly total inactivity"; they also found that "the animal is so inactive that the presence or absence of rhythmicity cannot be resolved". However they later state that the arrhythmia of 10 out of 14 animals in which the ganglion was present, but in which the nerve cord had been severed anterior to either the first or second thoracic ganglion, is incompatible with any humoral control from the suboeosphageal ganglion. It is difficult to reconcile all these statements.

In another experiment these authors cut the circum-oesophageal connectives between the brain and the suboesophageal ganglion, and found that five out of six animals immediately resumed their normal rhythm of activity. This result rules out the control of rhythmicity by a nervous pathway from the brain. Brady [116] has questioned this result; he finds that in *Periplaneta* this operation is followed by such a high level of activity that any rhythmicity would be masked, he also quotes a personal communication from Roberts stating that rhythmicity is lost in *Leucophaea,* and Brady suggests that Nishiitsutsuji-Uwo & Pittendrigh had in fact cut the maxillary nerves rather than the circum-oesophageal connectives. It is difficult to see how such skilled operators could, however, have made such an error. Furthermore in the cricket *Acheta domesticus* the entire brain can be removed and rhythmical activity is still maintained [118]. In this latter insect not even hormonal, let alone nervous, control from the brain can be involved, and since it has been confirmed that the thoracic and abdominal ganglia play no part in the control of the rhythm only the suboesophageal ganglion can be doing so.

The possible control of rhythmicity by a humoral agent from the brain has been explored by a number of workers, and again the results are contradictory. Roberts [112] found that surgical lesions bisecting the pars intercerebralis of *Leucophaea* produced arrhythmicity, although lesions lateral to this area did not do so, and in a few insects rhythmic activity reappeared some weeks after the former operation. These results were later questioned by Nishiitsutsuji-Uwo and Pittendrigh [114], who found that many animals remained rhythmic after protocerebral bisection, or even when one whole lobe of the brain was removed. In an earlier paper [119] it was claimed that removal of the neurosecretory cells of the protocerebrum caused loss of rhythmicity. On the other hand only 19 out of 47 animals became arrhythmic in these circumstances, and the authors concluded that in the remaining animals some neurosecretory cells must have remained intact. Brady [9] cauterized the medial neurosecretory cells *in situ* and found that the animal remained rhythmically active thereafter.

From the results of later experiments Nishiitsutsuji-Uwo & Pittendrigh [114] conclude that the lateral neurosecretory cells are involved, for when a lateral-sagittal cut has been made through both protocerebral lobes all of the seven operated animals failed to recover any rhythmicity before they died. This conclusion must be based on the assumption that the critical affect of this operation is the cutting of the axons of the neurosecretory cells. The authors note however that the brain of the one animal examined histologically was completely deformed, and that it was impossible to recognize normal tissue relationships, so that the question remains as to what effect such an operation has on the total nervous control of the animal: from the results of this experiment alone it would not be possible to conclude that it must be the cutting of the neurosecretory axons which stops the locomotor rhythm. Indeed it is well known that destruction of the pars intercerebralis is followed by degeneration of the central body, and that the central body is the co-ordinating centre for pre-motor interneurones, so that such degeneration interferes with movements of all kinds. The endocrine role has, however, been argued by the authors because they have shown that cutting the circum-oesophageal connectives does not stop the rhythm, and they therefore assume that hormonal control must be derived from the brain, since they have also assumed that it cannot come from elsewhere. As noted above there is still no certainty about any of these assumptions, and the question has also been raised as to whether the connectives had actually been cut.

In yet another set of operations the optic nerves were cut between the compound eyes and the optic lobes, and although the animals remained rhythmic the rhythm could no longer be entrained by LD. When the optic tract between the optic lobes and the rest of the brain was cut on both sides, thus isolating the optic lobes, 14 out of 18 animals became permanently arrhythmic; when the optic lobes were removed completely all six animals became arrhythmic. From this interesting result it has been concluded that the circadian rhythmicity of locomotor activity is ultimately caused by an autonomous self-sustaining oscillator in the nervous output of the optic lobes, and that this causes a non-autonomous circadian rhythmicity of secretion by the pars inter-cerebralis, which in turn imposes a circadian periodicity on the activity of the thoracic ganglia.

This attractive theory must still be questioned however, for, as has been pointed out, it is still not proven that the secretion from the pars intercerebralis controls the activity of the thoracic ganglia, and although removal of the optic lobes caused arrhythmia in all animals, the severance of the nervous connection between the brain and the optic lobes did not prevent three animals from retaining a weak rhythm and one from retaining a perfectly normal rhythm. No nervous control from the optic lobes could have been involved in these rhythmical animals, and for them the optic lobe could not be the primary clock.

Despite the divergence of opinion about which regions are involved in the control of activity rhythms it seems very likely that at least two systems are involved, one endocrine and one nervous, as was originally proposed by Harker [5], but whether either, or both, are autonomous for any length of time, or whether there is considerable feedback from other systems remains to be seen.

The questions which have been raised by the different types of activity shown in different types of recorder may in the end lead to a further understanding of the whole problem. Already the results from two types of recorder have brought to light an endocrine system which, although not controlling the rhythm, plays a part in the intermediary processes. In a photocell recorder cockroaches became arrhythmic after the nerve carrying neurosecretion between the corpora cardiaca and the suboesophageal ganglion is cut [115]. Cockroaches in a running-wheel, in which there is considerable feedback from the environment, do not however become arrhythmic even when the corpora cardiaca are removed altogether [8]. These results suggest that the corpus cardiacum hormone is involved in the actual control of the level of activity in the absence of other types of stimulation, but that in the presence of other types of feedback (perhaps nervous) the corpus cardiacum system is bypassed.

Curiously little attention has been paid to the role of the endocrine system in relation to the metabolic processes in the insect. As has been noted earlier, the trehalose concentration in crickets fluctuates rhythmically, but the peak concentration does not coincide with the peak of either the locomotor or feeding rhythm. In general carbo-hydrates are transported in the blood of insects in the form of trehalose, and it is utilized as a source of energy during flight and almost certainly during other locomotor activity. An independently maintained rhythm of trehalose concentration would seem to be an excellent example of the way in which rhythmical activities could aid in the preparation for events which may be triggered by predictable fluctuations in the environment.

The conversion to trehalose of the glycogen stored in the fat body is under the control of the corpus cardiacum hormone, the major effect being produced by the secretion from the glandular lobe (that is, the region of the corpus cardiacum which is intrinsically secretory, and not concerned with the storage of hormones coming from the brain neurosecretory cells). The secretion acts on two enzyme systems, converting the fat body phosphorylase to its active form, and effecting the breakdown of glucose-6-phosphate to trehalose-6-phosphate. The secretion also seems to effect the actual transfer of trehalose across the cell membrane into the haemolymph [120, 121]. The neurosecretory storage region of the corpus cardiacum also contains a hyperglycaemic factor, and it is well known that stress causes release of both secretions: these hormones also affect the rate and amplitude of heart beat, gut peritalsis and the movements of the malphigian tubules. When an insect is in a running-wheel it may be under a certain amount of stress once it begins to move; perhaps in these circumstances the corpus cardiacum is stimulated and its secretions heighten the level of activity, thus producing the very intense activity so often seen in these recorders. In a cardiectomized insect under stress the secretion which would normally pass from the brain down the axons to the cardiaca may still be released from the ends of the cut axons, and it may be in this way that the high level of activity is maintained in such animals.

The rhythmical fluctuations of a number of other physiological systems have been noted earlier in this chapter, some of which do not appear to be directly related to the rhythm of locomotor activity, but it would be surprising if they too did not affect such activity, which must in the final analysis be a function of the overall condition of the total animal. In addition the role of proprioceptive stimulation has not yet been considered at all. Perhaps the rather simple systems which have so far been postulated have only touched the surface of the problem.

6.6 CIRCADIAN RHYTHMICITY AND MOSQUITO BEHAVIOUR

Now that many of the major lines of work on circadian rhythmicity in insects have been outlined, it is interesting to take one group, with complex circadian behaviour patterns, and consider how far we have proceeded towards an understanding of this rhythmicity.

Adult mosquitoes have been chosen, despite the multiplicity of species and their vast geographical range, for this group has probably been more extensively studied in the field than any other (over 60 papers have been published by Haddow and his colleagues alone), and very careful attention has been paid to their different behavioural activities.

The flight activity of the adult, from the time of eclosion, can be divided into four categories: (a) flight associated with host-seeking; (b) flight associated with mating; (c) flight associated with oviposition after a blood meal has been taken; (d) what may be loosely termed "free flight", that is flight which has not yet been correlated with any other activity.

Haddow [122] suggests that the different types of biting behaviour fall into two main classes: (i) that in which there is a single, very short and very pronounced wave of biting activity, although there is always scattered biting at other times; (ii) that in which biting activity goes on over a prolonged period. This latter type is strikingly irregular apart from being either diurnal or nocturnal, and comparison of counts from different days shows very marked differences in timing. Because of the considerable differences between the two types of biting cycle the published records need careful attention for, as Haddow points out, where observations have been grouped into longish periods, for example 4-h periods, not only may the sharpness of a pronounced wave of biting activity be concealed, but minor peaks may be completely eliminated.

By analogy with both laboratory experiments and field studies in other animals it would be expected that changes in light intensity might play a major role in defining the timing of an activity peak. It is difficult to relate changing light intensity to biting cycles in forests, however, because of the continuous intensity fluctuations caused by movement of the vegetation. To avoid this difficulty a study has been made on a tower in Zika Forest at a height of 80 ft, above the surrounding low canopy, but no relationship between the onset of biting time and absolute light intensity was revealed: on the other hand a clear relationship appears between the onset of biting and the angle of the sun [123]. When Crep Units [124] (a measure of the length of time between sunset and civil twilight, when the sun is 6° below the horizon) are plotted against biting times the form of the waves of activity are clearly revealed. As the authors comment these results strongly suggest that the *rate* of change of light intensity is the critical factor which determines the onset of crepuscular biting. For the seven species observed in this study onset of biting falls within the very short period when the rate of change is reaching its maximum.

Few experimental studies have been made with insects on the effect of rate of change of light intensity, but it has been established that the threshold control of swarming of *Culex pipiens fatigans* is not simply one of light intensity, but requires a gradual change of intensity towards the permissive range [44].

Clear as these results are, however, it is difficult to correlate them with records from a tower in Mpanga Forest [125] which show that although at the canopy level the main wave of biting of *Mansonia fuscopennata* begins when the light intensity falls to about 0.5 ft candles and ends when it drops to 0.002 ft candles, the biting activity at ground level continues throughout the day although the light intensity is at times as high as 300 ft candles.

These results suggest that the level at which mosquitoes bite may also be related to light intensity, and indeed very clear vertical movements are seen in, for example, *Aedes ingrami,* and in *Taeniorhychus* which bite almost exclusively at ground level by day and in the trees at night [122].

The preferred level at which biting takes place has not been studied in the laboratory, yet this aspect must have a marked effect on laboratory results, for in some species the precision with which levels are maintained is extraordinary. *Eretmapodites chrysogaster,* for example, bites from 6 to 18 in above the forest floor; if a human bait lies down flat hardly any of the mosquitoes will bite, whereas increasing the bait level by lying on the side brings immediate biting [126]. Clearly here again further studies are badly needed.

A further complexity revealed by studies of biting cycles is the change in biting pattern with the type of habitat. McClelland [127] for instance, found that the domestic population of *A. aegypti* on the Kenya coast shows a very ill-defined biting cycle, whereas an outdoor population of the same species in Uganda has a very sharply defined cycle. Furthermore in the latter area the time of biting by the same species found inside huts differs by 12 h from the biting time outside the huts. Although light intensity must be a factor involved in such phase-differences there must also be many other factors concerned, and again there is little experimental evidence on this point. The recent concern with variation between activity records of insects in different types of recorder in the laboratory suggests that many significant variables may have been ignored in other laboratory experiments in the past.

Flight activity apparently unassociated with biting cycles has also been studied, and it is useful here to consider whether biting activity can be associated with a general activity rhythm. Haddow [125] has examined the relationship between a number of activities of *M. fuscopennata* (Fig. 6.5). It can be seen that the biting patterns vary from level to level in the forest, but at all levels in the forest they differ from the pattern in huts; the general flight activity pattern in the forest also differs from that outside the forest. It might also be expected that the flight activity of males would influence that of females, but this does not appear to happen, and the swarming of males is confined to the hour after sunset

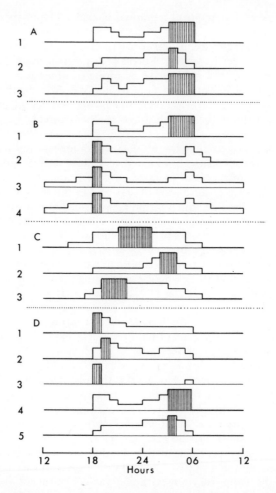

Figure 6.5. Synoptic figure of the circadian rhythms shown by *Mansonia fuscopennata* in the Entebbe area. Hatched areas represent maximal activity of each function. *A* (1) Attraction of females to mercury-vapour light in forest; (2) Attraction of females to light in open; (3) Oviposition. *B* (1) Oviposition; (2) Biting activity above forest canopy; (3) Biting activity in the canopy and understory; (4) Biting at ground level in forest. *C* (1) Biting activity in banana plantation; (2) Biting activity in huts; (3) Probable entry into huts. *D* (1) Attraction of males to light in forest; (2) Attraction of males to light in the open; (3) Swarming activity of males; (4) Attraction of females to light in the forest; (5) attraction of females to light in the open (after Haddow [125]).

and the hour before sunrise. Within the forest, apart from a close correlation between oviposition and flight activity, which will be discussed later, there does not seem to be any relationship between the biting cycle and the general activity rhythm: this is rather surprising since it might be expected that when the insect is most active there would be more frequent contact with the host.

Haddow in discussing these rather unexpected findings points out that Harker's [110] theory of multiple clocks, each of which may be phase-set to a different time in relation to an entraining factor, does not explain the difference in timing of a single activity in different habitats, and he postulates strong environmental influences as the causative factor. Environmental factors must of course be setting the phase in any case, but it is possible that the expression of some rhythms is inhibited by exogenous factors, and peak activity occurs only when the inhibitory factor ceases to act; what is then seen as a peak would not appear as such if the true circadian peak had not been suppressed. Corbett [128] has, for example, shown that in exposed sites outside a forest *A. africanus* oviposits at evening twilight, 4 h after the time dictated by the rhythm, because exogenous factors do not become permissive until then.

We still really need to know whether the effect of an environmental factor on one activity affects all other rhythmical processes in the same way, or whether the phase-setting of different activities can be dissociated. Examination of Fig. 6.5 suggests that although there is a major phase-shift in biting time between forest, banana plantation and huts, there is no such major shift in time of flight activity: perhaps there is here some dissociation of the rhythms. Another possibility is that small differences in genotype control the phase-setting in relation to the Zeitgeber; little is known of this aspect, although it is known that females of the butterfly *Colias*, differing by one sex-linked gene, fly at different times of day [21]. It is clear that we need to know more about the exogenous factors which operate in different types of locality, and their effects on the actual expression of each type of activity before the problems mentioned above can be solved.

Laboratory experiments on mosquito flight have shown that in LD 12 : 12 individual sugar-fed female *Anopheles* show a peak of activity after both light-on and light-off, but in DD only that peak associated with light-off persists. Light itself appears to have an inhibitory effect on this endogenous activity peak, and if the insects are exposed to light at this time virtually no activity takes place [130]. This finding is particularly relevant for many field studies in which flight activity is measured by the attraction of insects to light, for if the presence of light actually inhibits flight at the usual time of maximum activity then totally false conclusions could be drawn from collecting records. A high-intensity light may also cause an abnormal degree of activity at certain sensitive times in the circadian cycle, as has been discussed in relation to cockroach activity. Nothing is known about this in relation to the attraction of insects to light-traps, but again it is possible that such an effect may give erroneous ideas about the patterns of normal flight activity.

Flight activity of individual *Aedes aegypti* in the laboratory under LD 12 : 12 shows a main peak of activity 1 to 2 h before the onset of darkness. Although the amount of activity is decreased in darkness the rhythm is maintained in DD, as it is in LL [131]. It was earlier suggested that the rhythm was not overt in LL [132], but examination of the figures suggests that a free-running rhythm with a period of about 26 h is present. In this connection it is interesting that the period is about 22.5 h in DD, so that this insect does not follow Aschoff's Rule, which demands that the period would be shorter in LL than in DD, although the total activity is in this case higher in DD in accordance with Aschoff's Rule.

Aedes aegypti shows a bimodal rhythm of flight in certain photo-periods: in LD 4 : 20, although activity is concentrated within a short light period, the phase-setting appears to be achieved by the light-off signal, although a very small peak may occur in the dark, being phased by the light-on signal. Both peaks are maintained on transfer to DD. In LD 20 : 4, however, although the rhythm is still bimodal, the minor peak which appears 22 to 23 h after light-off, and so here occurs during the next light period, is not maintained when a transfer is made to DD [131]. It is possible that this minor peak may represent a phase of the rhythm in which the insect is very light-sensitive, as we saw in the cockroach.

It is interesting that a LD 12 : 12 regime which is advanced by 6 h, through shortening one dark period by 6 h, causes the flight activity of *Anopheles* to be entrained very rapidly to the new cycle, peak activity shifting after just one exposure to the new light-off signal. By contrast if the LD regime is advanced 6 h by shortening one light period, several cycles of the new regime are necessary before the steady state is reached [131].

In natural conditions it is likely that heavy cloud may cause a decrease in light intensity unexpectedly early in the day; a circadian system which allowed for immediate entrainment so that high activity occurred early on the following afternoon would be clearly disadvantageous. On the other hand it is very unlikely that light intensity would increase hours earlier than normal in the day, and therefore the possibility of immediate entrainment is immaterial.

The adaptive value of peak flight activity in the early hours of morning or evening may possibly be related to the lower temperature at these times of day, at least in very hot climates, for although temperature itself is not the primary entraining factor Rowley and Graham [133, 134] have shown that flight in virgin female mosquitoes is more readily undertaken, and sustained, at temperatures below 27° C.

An experimental approach to biting cycles has proved to be more difficult, and virtually nothing is known about this aspect in experimental conditions.

The common occurrence of bimodal biting cycles in the field has led to the suggestion that there may be two sections of the population involved, each showing different behaviour patterns according to age [122, 135, 136]. Several early studies [137, 138, 139] revealed no significant differences in the biting pattern of young and old females, although there appears to be a significant drop in biting "drive" during ovarian development [140, 141]. Corbet [142] however, found that although the biting cycle of M. fuscopennata is not related to the age-composition of the population, there are significant differences in the average age of females biting at different *levels* at different times. At ground level the mosquitoes biting at night are younger than those biting during the day: above the canopy the average age increases from sunset to sunrise, with a corresponding drop below the canopy. Corbet suggests tentatively that these age changes may be related to the fact that nullipara may have just mated before biting, whereas parous females will have oviposited before biting and their flight from oviposition sites may affect the position at which they bite.

The pattern of oviposition appears to be related to the biting cycle in some species, as has already been seen (Fig. 5). On the other hand although A. africanus shows a single oviposition peak followed by a biting peak, it also shows another general activity peak related to neither biting or oviposition [143]. McClelland [144] studied the oviposition periodicity of A. aegypti at 12 different heights above ground level, and with different exposures to sunlight, and in this species found a pattern resembling that of biting and flight activity.

A number of workers have explored the oviposition rhythm of captive populations of A. aegypti, and find that it can be entrained by a light period of as short a length as 5 min. When 5 min light is given every 24 h two oviposition peaks occur, one taking place before the onset of light and the other 4 to 8 h after onset of darkness [145-148]. The authors concluded that no oviposition cycles occurred in either LL or DD. In a later study Gillett [149] points out that ovarian development takes about 55 h from the time of the blood meal, on which ovarian development depends. The problem therefore arises as to how this developmental cycle contributes to a circadian rhythm of oviposition.

Gillett has used an ingenious experimental procedure which enabled him to follow the ovisposition cycle of individual mosquitoes. He found that in LD 12 : 12 a mosquito which has completed ovarian development withholds its eggs until the first "available" laying period, and should all the eggs not be laid during that period at least some are retained for a further 24 h. In LL however the eggs are laid as soon as development is completed, and oviposition continues until all the eggs are laid. If there is a change to DD during ovarian development oviposition becomes rhythmical, and any mosquito which has already completed egg development close to the time when the light is switched off delays laying until 24 h after the onset of darkness. Although Gillett does not specifically make this point these results must imply that the rhythm is indeed a truly circadian one which can be maintained in DD.

McClelland [144] points out that testing oviposition rhythms in the laboratory may have drawbacks, in that confined mosquitoes are continuously exposed to some of the stimuli which would be received in the field only after appetitive behaviour, and the act of oviposition might occur in the laboratory in conditions which might not stimulate the act in the field. However oviposition rhythmicity of *A. aegypti* appears to show the same general characteristics in both field studies and in the laboratory study.

Just what selective advantage there might be in the synchronization of oviposition is difficult to assess, but it should be kept in mind that the actual act of oviposition may not be the major factor on which selection acts, for closely related to oviposition must be the timing of the flight activity which precedes egg laying. Those mosquitoes which oviposit early in the evening, and have a peak of biting activity later in the night, are thus ensured of having time for another blood meal after egg-laying and so can begin the sequence again in the shortest possible time [150].

Taylor and Jones [131] have made the interesting suggestion that the geographical range of *A. aegypti* is not, as previously thought, limited by the range of summer temperature, but may be limited by a range of summer day-length which will permit a physiological balance between factors controlling flight activity and oviposition rhythms.

This brief survey of the rhythmical activities of just one group of insects indicates the complexity and diversity of the problems which still await solution. Even should the nature of the biological clock itself be finally resolved we shall still be very far from understanding the full range of the inter-relationship between the clock, the physiological and behavioural systems, and the biotic and physical environment.

REFERENCES

1. J. E. Harker, Diurnal rhythms and homeostatic mechanisms. *Symp. Soc. Exp. Biol.*, **18**, 283-300 (1964).

2. J. E. Harker, Control of diurnal rhythms of activity in *Periplaneta americana* L., *Nature, Lond.*, 175, 733 (1955).
3. J. E. Harker, Factors controlling the diurnal rhythm of activity of *Periplaneta americana* L., *J. exp. biol.*, 33, 224-234 (1956).
4. J. E. Harker, The effect of perturbations in the environmental cycle on the diurnal rhythm of activity of *Periplaneta americana* L. *J. exp. biol.*, 37, 154-163 (1960).
5. J. E. Harker, Internal factors controlling the suboesophageal ganglion neurosecretory cycle in *Periplaneta americana* L. *J. exp. Biol.*, 37, 164-170 (1960).
6. J. L. Cloudsley-Thompson, Studies in diurnal rhythms VI. *Ann. Mag. Nat. Hist.*, 9, 305-309 (1956).
7. D. L. Gunn, The daily rhythm of activity of the cockroach, *Blatta orientalis. J. exp. Biol.*, 17, 267-277 (1940).
8. J. Brady, Control of the circadian rhythm of activity in the cockroach. I. The role of the corpora cardiaca, brain and stress. *J. exp. Biol.*, 47, 153-163 (1967).
9. J. Brady, Control of the circadian rhythms of activity in the cockroach. II. The role of the sub-oesophageal ganglion and ventral nerve cord. *J. exp. Biol.*, 47, 165-178 (1967).
10. S. K. Roberts, Circadian activity rhythms in cockroaches. I. The free-running rhythm in steady-state. *J. cell. comp. Physiol.*, 55, 99-110 (1960).
11. S. K. Roberts, Circadian activity rhythms in cockroaches. II. Entrainment and phase shifting. *J. cell. comp. Physiol.*, 59, 175-186 (1962).
12. J. N. Nowosielski and R. L. Patton, Studies on circadian rhythm of the house cricket, *Gryllus domesticus* L. *J. Insect Physiol.*, 9, 401-410 (1963).
13. M. Lohmann, Der Einfluss von Beleuchtungsstärke und Temperatur auf die tagesperiodische Laufaktivität des Mehlkäfers *Tenebrio molitor* L. *Z. vergl. Physiol.*, 49, 341-389 (1964).
14. J. E. Harker, In discussion J. L. Cloudsley-Thompson, Adaptive functions of circadian rhythms. *Cold Spring Harb. Symp. Quant. Biol.*, 25, 354 (1960).
15. J. D. Palmer, A persistent, light-preference rhythm in the fiddler crab, *Uca pugnax* and its possible adaptive significance. *Am. Nat.*, 98, 431-434 (1964).
16. J. Aschoff, U. Saint-Paul and R. Wever, Circadiane Periodik von Finkenvögeln unter dem Einfluss eines Selbstgewählten Licht-Dunkel-Wechsels. *Z. vergl. Physiol.*, 58, 304-321 (1968).
17. G. R. Lipton and D. J. Sutherland, Activity rhythms in the American cockroach, *Periplaneta americana. J. Insect Physiol.*, 16, 1555-1566 (1970).
18. J. E. Harker, Diurnal rhythms in the animal kingdom. *Biol. Rev.*, 33, 1-52 (1958).
19. J. L. Kavanau, Compulsory regime and control of environment in animal behaviour. I. Wheel running. *Behaviour*, 20, 251-281 (1963).
20. M. Lohmann, Phase dependent changes of circadian frequency after light steps. *Nature, Lond.*, 213, 196-197 (1967).
21. J. E. Harker, *The Physiology of Diurnal Rhythms.* Cambridge Univ. Press (1964).
22. U. Wobus, Der Einfluss der Lichtintensität auf die Resynchronisation der circadian en Laufaktivität der Schabe *Blaberus craniifer* Burm. *Z. vergl. Physiol.*, 52, 276 (1966).
23. U. Wobus, Der Einfluss der Lichtintensität auf die circadiane Laufaktivität der Schabe *Blaberus craniifer* Burm. *Biol. Zentr.*, 85, 305-323 (1966).
24. P. R. Lewis and M. C. Lobban, Dissociation of diurnal rhythms in human subjects living on abnormal time routines. *Q. J. exp. Physiol.*, 42, 371-386 (1957).

25. H. Warnecke, Vergleichende Untersuchungen zur tagesperiodischen Aktivität von drei *Geotrupes*—Arten. *Z. Tierpsychol.*, 23, 513-526 (1966).
26. E. Bünning, Zur Analyse des Zeitsinnes bei *Periplaneta americana. Z. Naturforsch.*, 14B, 1-4 (1959).
27. D. K. Edwards, Activity rhythms of Lepidopterous defoliators. II. *Halisidota argentata* Pack. (Arctiidae) and *Nepytia phantasmaria* Stkr. (Geometridae). *Canad. J. Zool.*, 42, 939-958 (1964).
28. F. S. Bodenheimer and H. J. Klein, Über die Temperaturabhängigkeiten von Insekten. *Z. vergl. Physiol.*, 11, 345-385 (1930).
29. K. R. Norris, Daily patterns of flight activity of blowflies in the Canberra district as indicated by trap catches. *Australian J. Zool.*, 14, 835-854 (1966).
30. T. Lewis and L. R. Taylor, Diurnal periodicity of flight by insects. *Trans. R. ent. Soc. Lond.*, 116, 393-479 (1964).
31. C. B. Williams, The times of activity of certain nocturnal insects, chiefly Lepidoptera, as indicated by a light trap. *Trans. R. ent. Soc. Lond.*, 83, 523-562 (1935).
32. C. B. Williams, An analysis of four year captures of insects in a light trap. *Trans. R. ent. Soc. Lond.*, 89, 79-131 (1939).
33. C. G. Johnson, L. R. Taylor and E. Haine, The analysis and reconstruction of diurnal flight curves in alienicolae of *Aphis fabae* Scop. *Annal. appl. Biol.*, 45, 682-701 (1957).
34. C. G. Johnson and L. R. Taylor, Periodism and energy summation with special reference to flight rhythms in aphids. *J. exp. Biol.*, 34, 209-221 (1957).
35. E. Haine, Periodicity in aphid moulting and reproduction in constant temperature and light. *Z. angew. Ent.*, 40, 99-124 (1957).
36. N. W. and E. A. Timofeeff-Ressovsky, Populations-generische Versuche an *Drosophila. Z.f. indukt. Abstamm.- Verebungsl.*, 79, 28 (1940).
37. T. Dobzhansky and C. Epling, Contributions to the genetics, taxonomy and ecology of *Drosophila pseudo-obscura* and its relatives. *Publ. Carneg. Instn.*, 544, 1-46 (1944).
38. C. Pavan, T. Dobzhansky and H. Burla, Diurnal behaviour of some Neotropical species of *Drosophila. Ecology*, 31, 36-43 (1950).
39. V. R. D. Dyson-Hudson, The daily activity rhythm of *Drosophila subobscura* and *D. obscura. Ecology*, 37, 562-567 (1956).
40. D. F. Mitchell and C. Epling, The diurnal periodicity of *Drosophila pseudo-obscura* in Southern California, *Ecology*, 32, 696-708 (1951).
41. L. R. Taylor and H. Kalmus, Dawn and dusk flight of *Drosophila subobscura*, *Nature, Lond.*, 174, 221 (1954).
42. K. J. Connolly, Locomotor activity in *Drosophila* as a function of food deprivation. *Nature, Lond.*, 209, 224 (1966).
43. S. K. Roberts, "Clock" controlled activity rhythms in the fruit fly. *Science*, 124, 172 (1956).
44. P. S. Corbet, The role of rhythms in insect behaviour, in: *Insect Behaviour*, pp. 13-28, Haskell (ed.). *R. Ent. Soc. Sump.* (1966).
45. G. A. Lancaster and A. J. Haddow, Further studies on the nocturnal activity of Tabanidae in the vicinity of Entebbe, Uganda. *Proc. R. ent. Soc. Lond. (A)*, 42, 39-48 (1967).
46. G. Williams, Seasonal and diurnal activity of Carabiidae, with particular reference to *Nebria, Notiophilus* and *Feronia. J. Anim. Ecol.*, 28, 309-330 (1959).
47. S. Mori, Population effect on the daily periodic emergence of *Drosophila. Mem. Coll. Sci. Kyoto (B)*, 25, 49-55 (1954).

48. A. J. Haddow, I. H. H. Yarrow, G. A. Lancaster and P. S. Corbet, Nocturnal flight cycle in the males of African doryline ants (Hymenoptera: Formicidae), *Proc. R. ent. Soc. Lond.*, 41, 103-106 (1966).
49. D. K. Edwards, Laboratory determination of the daily flight times of separate sexes of some moths in naturally changing light. *Canad. J. Zool.*, 40, 511-530 (1962).
50. W. Ohsawa, K. Matutani, H. Tukuda, S. Mori, D. Miyadi, S. Yanagisima and Y. Sato, Sexual properties of the daily rhythmical activity in *Drosophila melanogaster. Physiol. Ecol.*, 5, 26-45 (1942).
51. H. Caspers, Rhythmische Erscheinungen in der Fortpflanzung von *Clunio marinus* (Dipt. Chiron.) und das Problem der lunaren Periodizität bei Organismen. *Arch. Hydrobiol.*, Suppl. 18, 415-494 (1951).
52. E. T. Nielsen and J. S. Haeger, Pupation and emergence in Aedes *taeniorhynchus* (Weid.). *Bull. ent. Res.*, 45, 757-768 (1954).
53. H. F. Barnes, On some factors governing the emergence of gall midges (*Cecidomyidae*). *Proc. zool. Soc. Lond.*, 381-393 (1930).
54. H. Eidmann, Über rhythmische Erscheinungen bei der Stabheuschrecke *Carausius morosus* Br. *Z. vergl. Physiol.*, 38, 370-390 (1956).
55. F. Steiniger, Die Erscheinungen der Katalepsie bei Stabheuschrecken und Wasserläufern. *Z. morph. Ökol. Tiere*, 26, 591-594 (1933).
56. P. S. Corbet, The life-history of the emperor dragonfly, *Anax imperator* Leach (Odonata: Aeschnidae), *J. Anim. Ecol.*, 26, 1-69 (1957).
57. A. Tjønneland, The flight activity of mayflies as expressed in some East African species. *Univ. Bergen Arb. Naturv. R.*, 1, 1-88 (1960).
58. G. W. Green, The control of spontaneous locomotor activity in *Phormia regina* I. Locomotor activity patterns of intact flies. *J. Insect Physiol.*, 10, 711-726 (1964).
59. G. W. Green, The control of spontaneous locomotor activity in *Phormia regina*. II. Experiments to determine the mechanisms involved. *J. Insect Physiol.*, 10, 727-752 (1964).
60. R. L. Caldwell and H. Dingle, The regulation of cyclic reproductive and feeding activity in the milkweed bug, *Oncopeltus*, by temperature and photoperiod. *Biol. Bull. Wood's Hole*, 133, 510-525 (1967).
61. M. F. Bennett and M. Renner, The collecting performance of honey bees under laboratory conditions. *Biol. Bull. Wood's Hole*, 125, 416-430 (1963).
62. J. K. Nayar, The pupation rhythm in *Aedes taeniorhynchus*. II. Ontogenetic timing, rate of development, and the endogenous diurnal rhythm of pupation. *Ann. ent. Soc. Amer.*, 60, 946-971 (1967).
63. E. Palmen, Diel periodicity of pupal emergence in natural populations of some chironomids. *Ann. Zool-Bot. Soc. fenn. Vanamo*, 17, 1-30 (1955).
64. E. Palmen, Diel periodicity of pupal emergence in some north European chironomids. *Int. Congr. Ent. X*, 2, 219 (1958).
65. O. Park, Nocturnalism—the development of a problem. *Ecol. Monographs*, 10, 485-536 (1940).
66. N. C. Morgan and A. B. Waddell, Diurnal variation in the emergence of some aquatic insects. *Trans. R. ent. Soc. Lond.*, 113, 123-137 (1961).
67. P. S. Corbet, The Biology of Dragonflies, Witherby, London (1962).
68. H. Kalmus, Periodizität und Autochronie als zeitregelnde Eigenschaftern der Organismen. *Biol. gen.*, 11, 93-114 (1935).
69. W. J. Brett, Persistent diurnal rhythmicity in *Drosophila* emergence. *Ann. ent. Soc. Amer.*, 48, 119-131 (1955).
70. C. S. Pittendrigh, The circadian oscillation in *Drosophila pseudoobscura* pupae: A model for the photoperiodic clock. *Z. Pflanzenphysiol.*, 54, 275-307 (1966).

71. J. E. Harker, The effect of photoperiod on the developmental rate of *Drosophila* pupae. *J. exp. Biol.*, 43, 411-421 (1965).
72. C. S. Pittendrigh, Circadian rhythm and the circadian organization of living systems. *Cold Spring Harb. Symp. Quant. Biol.*, 25, 159-184 (1960).
73. C. S. Pittendrigh, On the temperature independence in the clock system controlling emergence time in *Drosophila*. *Proc. nat. Acad. Sci. Wash.*, 40, 1018-1029 (1954).
74. C. S. Pittendrigh and S. D. Skopik, Circadian Systems. V. The driving oscillation and the temporal sequence of development. *Proc. nat. Acad. Sci. U.S.A.*, 65, 500 (1970).
75. J. E. Harker, The effect of a biological clock on the developmental rate of *Drosophila* pupae. *J. exp. Biol.*, 42, 323-431 (1965).
76. L. Rensing, Die Bedeutung der Hormone bei Steuerung circadianer Rhythmen. *Zool. Jb. Abt. allg. Zool. Physiol.*, 71, 595-606 (1965).
77. L. Rensing and R. Hardeland, Zur wirkung der circadianen Rhythmik auf der Entwicklung von *Drosophila*. *J. Insect Physiol.*, 13, 1547-1568 (1967).
78. L. Rensing, B. Tach and V. Bruce, Daily rhythms in the endocrine glands of *Drosophila* larva. *Experientia*, 21, 103-104 (1965).
79. L. Rensing, Zur circadianen Rhythmik des Hormonsystems von *Drosophila*. *Z. Zellforsch.*, 74, 539-558 (1966).
80. U. Clever, Genactivitäten in den Riesenchromosomen von *Chironomous tentans* und ihre Beziehung zur Entwicklung. I. Genaktivierung durch Ecdyson. *Chromosoma*, 12, 607-675 (1961).
81. H. J. Becker, Die Puffs der Speicheldrüsenchromosomen von *Drosophila melanogaster* Chromosoma, 13, 341-386 (1962).
82. E. T. Nielsen and D. G. Evans, Duration of the pupal stage of *Aedes taeniorhynchus* with a discussion of the velocity of development as a function of temperature, *Oikos*, 11, 200-221 (1960).
83. M. W. Provost and P. T. M. Lunn, The pupation rhythm in *Aedes taeniorhyncus*. I. Introduction. *Ann. ent. Soc. Amer.*, 60, 138-149 (1967).
84. A. C. Neville, Daily growth layers in locust rubber-like cuticle, influenced by an external rhythm. *J. Insect Physiol.*, 9, 177-186 (1963).
85. A. C. Neville, Growth and deposition of resilin and chitin in locust rubber-like cuticle. *J. Insect Physiol.*, 9, 265-278 (1963).
86. A. C. Neville and B. M. Luke, A two-system model for chitin-protein complexes in insect cuticles. *Tissue and Cell*, 1, 689-707 (1969).
87. A. C. Neville, Chitin lamellogenesis in locust cuticle. *Q. J. Micr. Sci.*, 106, 269-315 (1965).
88. B. Zelazny, Quoted in Neville, Cuticle ultrastructure in relation to the whole insect. *Symp. R. ent. Soc. Lond.*, 5, 17-39 (1970).
89. A. C. Neville, Chitin orientation in cuticle and its control; in: *Advances in Insect Physiology*, 4, 213-286 (1967).
90. A. C. Neville, Circadian organization of chitin in some insect skeletons. *Q. J. micr. Sci.*, 106, 315-325 (1965).
91. H. Dingle, R. L. Caldwell, J. B. Haskell, Temperature and circadian control of cuticle growth in the bug *Oncopeltus fasciatus*. *J. Insect Physiol.*, 15, 373-378 (1969).
92. J. W. Nowosielski and R. L. Patton, Daily fluctuations in blood sugar concentration in the house cricket, *Gryllus domesticus* L., *Science*, 144, 180-181 (1964).
93. J. W. Nowosielski and R. L. Patton, Variation in the haemolymph protein, amino acid, and lipid levels in adult house crickets, *Acheta domesticus* L. of different ages. *J. Insect Physiol.*, 11, 263-270 (1965).

94. S. Takahashi and R. F. Harwood, Glycogen levels of adult *Culex tarsalis* in response to photoperiod. *Ann. ent. Soc. Amer.*, **57**, 621-623 (1964).

95. R. B. Turner and F. Acree, The effect of photoperiod on the daily fluctuations of haemolymph hydrocarbons in the American cockroach. *J. Insect Physiol.*, **13**, 519-522 (1967).

96. J. Brady, Control of the circadian rhythm of activity in the cockroach. III. A possible role of the blood-electrolytes. *J. exp. Biol.*, **49**, 39-47 (1968).

97. B. J. Wall, Effects of dehydration and rehydration on *Periplaneta americana*. *J. Insect Physiol.*, **16**, 1027-1042 (1970).

98. R. J. Bartell and H. H. Shorey, A quantitative bioassay for the sex pheromone of *Epiphyas postvittana* (Lepidopt.) and factors limiting male responsiveness. *J. Insect Physiol.*, **15**, 33-40 (1969).

99. T. L. Payne, H. H. Shorey and L. K. Gaston, Sex pheromones of noctuid moths: factors influencing antennal responsiveness in males of *Trichoplusia* N.I. *J. Insect Physiol.*, **16**, 1043-1055 (1970).

100. L. Rensing, Zur circadianen Rhythmik des Sauerstoffverbrauches von *Drosophila*. *Z. vergl. Physiol.*, **53**, 62-83 (1966).

101. L. Rensing, W. Brucken and R. Hardeland, On the genetics of a circadian rhythm in *Drosophila*. *Experientia*, **24**, 509-510 (1968).

102. H. Klug, Histo-physiologische Untersuchungen über die Aktivitätsperiodik bei Carabiden, Wiss. *Z. Humbolt-Univ.* Berlin, **8**, 405-434 (1958).

103. J. Brady, Histological observations on circadian changes in the neurosecretory cells of cockroach sub-oesophageal ganglia. *J. Insect Physiol.*, **13**, 201-213 (1967).

104. M. Raabe, Recherches sur la neurosecretion dans la chaine nerveuse ventrale du Phasme, *Clitumnus extradentatus*: Variations d'activité des differentes elements neurosecreteurs. *C.r-hebd. Séanc. Acad. Sci. Paris*, **262**, 303-306 (1966).

105. N. de Besse, Recherches histophysiologiques sur la neurosecretion dans la chaine nerveuse ventrale d'une blatte, *Leucophaea maderae*. *C.r-hebd. Séanc. Acad. Sci. Paris*, **260**, 7014-7017 (1965).

106. B. Cymborowski and A. Dutkowski, Circadian changes in RNA synthesis in the neurosecretory cells of the brain and suboesophageal ganglion of the house cricket. *J. Insect Physiol.*, **15**, 1187-1197 (1969).

107. A. B. Dutkowski and B. Cymborowski, Role of neurosecretory cells of pars intercerebralis in regulating RNA synthesis in some tissues of *Acheta domesticus*. *J. Insect Physiol.*, **17**, 99-108 (1971).

108. B. Cymborowski and A. Dutkowski, Circadian changes in protein synthesis in the neurosecretory cells of the central nervous system of *Acheta domesticus*. *J. Insect Physiol.*, **16**, 341-348 (1970).

109. L. Rensing, Circadiane Rhythmik von *Drosophila*—Speisheldrüsen *in vivo*, *in vitro* und nach Ecdysonzugabe. *J. Insect Physiol.*, **15**, 2285-2303 (1969).

110. J. E. Harker, Diurnal rhythms. *Ann. Rev. Ent.*, **6**, 131-146 (1961).

111. S. K. Roberts, Significance of endocrines and central nervous system in circadian rhythms, in: *Circadian Clocks*, J. Aschoff (ed.) (1965).

112. S. K. Roberts, Circadian activity rhythms in cockroaches. III. The role of endocrine and neural factors. *J. cell. comp. Physiol.*, **67**, 473-486 (1966).

113. J. Nishiitsutsuji-Owu and C. S. Pittendrigh, Central nervous control of circadian rhythmicity in the cockroach. II. The pathway of light signals that entrain the rhythm. *Z. vergl. Physiol.*, **58**, 1-13 (1968).

114. J. Nishiitsutsuji-Owu and C. S. Pittendrigh, Central nervous control of circadian rhythmicity in the cockroach. III. The optic lobes, locus of the driving oscillation? *Z. vergl. Physiol.*, **58**, 14-46 (1968).

115. J. E. Harker, Endocrine and nervous factors in insect circadian rhythms. *Cold Spring. Harb. Symp. Quant. Biol.*, 25, 279-288 (1960).

116. J. Brady, How are insect circadian rhythms controlled? *Nature, Lond.*, 223, 781-784 (1969).

117. M. Fingerman, A. D. Lago and M. E. Lowe, Rhythms of locomotor activity and oxygen consumption of the grasshopper *Romalea microptera. Am. Midl. Nat.*, 59, 58-67 (1958).

118. B. Cymborowski, Investigations on the neurohormonal factors controlling circadian rhythm of locomotor activity in the house cricket (*Acheta domesticus* L.). I. The role of the brain and the suboesophageal ganglion. *Zool. Poloniae*, 20, 103-126 (1970).

119. J. Nishiitsutsuji-Owo, S. F. Petropulus and C. S. Pittendrigh, Central nervous control of circadian rhythmicity in the cockroach. I. Role of the pars intercerebralis. *Biol. Bull. mar. biol. hab. Wood's Hole*, 133, 679-696 (1967).

120. J. E. Steele, The action of the insect hyperglycaemic hormone. *Gen. comp. Endocrinology*, 3, 46-52 (1963).

121. G. J. Goldsworthy, The action of hyperglycaemic factors from the corpus cardiacum of *Locusta migratoria* on glycogen phosphorylase. *Gen. comp. Endocrinology*, 14, 78-85 (1970).

122. A. J. Haddow, Studies of the biting habits of African mosquitoes. An appraisal of methods employed, with special reference to the twenty-four-hour catch. *Bull. ent. Res.*, 45, 199-242 (1954).

123. A. J. Haddow, D. J. L. Casley, J. P. O'Sullivan, P. M. L. Ardoin, Y. Ssenkubuge and A. Kitma, Entomological studies from a high steel tower in Zika Forest, Uganda. II. The biting activity of mosquitoes above the forest canopy in the hour after sunset. *Trans. R. ent. Soc. Lond.*, 120, 219-236 (1968).

124. E. T. Neilsen, Twilight and the 'crep' unit. *Nature, Lond.*, 190, 878 (1961).

125. A. J. Haddow, The biting behaviour of mosquitoes and tabanids. *Trans. R. ent. Soc. Lond.*, 113, 315-335 (1961).

126. A. J. Haddow, Observations on the biting-habits of African mosquitoes in the genus *Eretmapodites* Theobald. *Bull. ent. Res.*, 46, 761-771 (1956).

127. G. A. H. McClelland, Observations on the mosquito *Aedes* (*Stegomyia*) *aegypti* (L) in East Africa. II. The biting cycle in a domestic population on the Kenya Coast. *Bull. ent. Res.*, 50, 687-696 (1960).

128. P. S. Corbet, The oviposition-cycles of certain sylvan culicine mosquitoes (Dipter: Culicidae) in Uganda. *Ann. trop. Med. Parasit.*, 57, 371-381 (1963).

129. W. Hovanitz, Differences in the field activity of two female colour phases of *Colias* butterflies at different times of day. *Contrib. Lab. Vertbr. Zool. Michigan* 41, 1-37 (1948).

130. M. D. R. Jones, M. Hill and A. M. Hope, The circadian flight activity of the mosquito *Anopheles gambiae*: Phase setting by the light regime. *J. exp. Biol.*, 47, 503-511 (1967).

131. B. Taylor and M. D. R. Jones, The circadian rhythm of flight activity in the mosquito *Aedes aegypti* (L); The phase setting effects of light-on and light-off. *J. exp. Biol.*, 51, 59-70 (1969).

132. J. D. Gillett, A. J. Haddow and P. S. Corbet, The sugar-feeding cycle in a cage-population of mosquitoes. *Entomologia exp. appl.*, 5, 223-232 (1962).

133. W. A. Rowley and C. L. Graham, The effect of temperature and relative humidity on the flight performance of female *Aedes aegypti. J. Insect Physiol.*, 14, 1251-1257 (1968).

134. W. A. Rowley and C. L. Graham, The effect of age on the flight performance of female *Aedes aegypti* mosquitoes. *J. Insect Physiol.*, 14, 719-728 (1968).

135. J. A. Downes, Habits and life cycle of *Culicoides nubeculosis* Mg. *Nature, Lond.*, 166, 510-511 (1950).

136. W. H. R. Lumsden, The crepuscular biting-activity of insects in the forest canopy in Bwamba, Uganda; a study in the relation to the sylvan epidemiology of yellow fever. *Bull. ent. Res.*, **42**, 721-760 (1952).
137. R. A. Senior White, On the biting activity of three neotropical *Anopheles* in Trinidad, British West Indies. *Bull. ent. Res.*, **43**, 451-460 (1953).
138. J. D. Gillett, Hormonal mechanisms involved in the reproductive cycle of mosquitoes. *Rep. E. Afr. Virus Res. Inst.*, 39-45 (1957).
139. M. T. Gillies, Age groups and the biting cycle in *Anopheles gambiae*. A preliminary investigation. *Bull. ent. Res.*, **48**, 553-559 (1957).
140. M. M. Lavoipierre, Presence of a factor inhibiting biting activity in *Aedes aegypti*. *Nature, Lond.*, **182**, 1567 (1958).
141. A. N. Clements, *The Physiology of Mosquitoes*, Pergamon Press, London (1963).
142. P. S. Corbet, The age-composition of biting mosquito populations according to time and level. *Trans. R. ent. Soc. Lond.*, **113**, 336-345 (1961).
143. J. D. Gillett and A. J. Haddow, Laboratory observations on the oviposition-cycle in the mosquito *Aedes (Stegomyia) africanus* Theobald. *Ann. trop. Med. Parasit.*, **51**, 170-174 (1957).
144. G. A. H. McClelland, Field observations on periodicity and site preference in oviposition by *Aedes aegypti* (L) and related mosquitoes (Diptera: Culicidae) in Kenya. *Proc. R. ent. Soc. Lond.* (A) **43**, 147-154 (1968).
145. A. J. Haddow and J. D. Gillett, Observations on the oviposition-cycle of *Aedes aegypti* (L). *Ann. trop. Med. Parasit.*, **51**, 159-169 (1957).
146. A. J. Haddow and J. D. Gillett, Laboratory observations on the oviposition-cycle in the mosquito *Taeniorhynchus (Coquillettidia) fuscopennatus* Theobald. *Ann. trop. Med. Parasit.*, **52**, 320-325 (1958).
147. J. D. Gillett, A. J. Haddow and P. S. Corbet, Observations on the oviposition-cycle of *Aedes (Stegomyia) aegypti* (L). *Ann. trop. Med. Parasit.*, **53**, 35-41 (1959).
148. J. D. Gillett, P. S. Corbet and A. J. Haddow, Observations on the oviposition-cycle of *Aedes (Stegomyia) aegypti* (L) III. *Ann. trop. Med. Parasit.*, **53**, 132-136 (1959).
149. J. D. Gillett, Contributions to the oviposition-cycle of the individual mosquitoes in a population. *J. Insect Physiol.*, **8**, 665-681 (1962).
150. A. J. Haddow and Y. Ssenkubuge, Laboratory observations on the oviposition-cycle in the mosquito *Anopheles (Cellia) gambiae*. *Ann. trop. Med. Parasit.*, **56**, 352-355 (1962).

CHAPTER 7

Circadian Rhythms in Plants

Malcolm B. Wilkins

Regius Professor of Botany
The University
Glasgow, W.2, Scotland

7.1 INTRODUCTION

The existence of circadian rhythms in living organisms was first established during a detailed study of leaf movement in plants more than 200 years ago. The movement of leaves had attracted the attention of Pliny in the first century A.D. and their complex patterns were being noted by Albertus Magnus in the thirteenth century. The Dutch astronomer, de Mairan [1], made a discovery of fundamental importance in 1729 in finding that the regular pattern of leaf movement (Fig. 7.1) continued after plants had been transferred to the relatively uniform environment of his cellar. The persistence of rhythmic leaf movement after plants were isolated from the periodic variation of environmental parameters such as light intensity and temperature was later confirmed both by Duhamel [2] in 1758 and Zinn [3] in 1759. These investigations therefore clearly focused attention upon the fact that the rhythms of leaf movement were not merely the response of the plant to

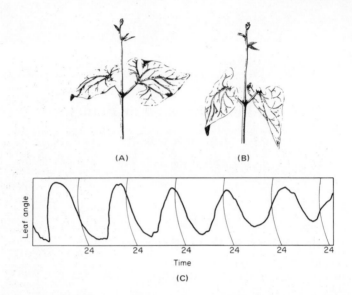

Figure 7.1. The position of the primary leaves of seedlings of *Phaseolus multiflorus* during the day (A) and at night (B), and the persistence of these up and down movements in the form of a circadian rhythm after seedlings are placed in continuous dim light at a uniform temperature (C). The reference lines are 24 h apart. Note that the period is about 27 h. (A and B, original; C, from Bünning and Tazawa [59])

periodic variation in environmental parameters, but, rather, were controlled by an unknown mechanism within the plant. It is this mechanism which is now termed the "biological clock".

The observation of a persistent circadian rhythm in some aspect of the physiology, chemistry or behaviour of a living organism is not the only evidence for the existence of a "biological clock". In 1920 and 1923 Garner and Allard [4, 5] published systematic and critical evidence that plants had the capacity to respond to changes in day-length. The onset of flowering (Fig. 7.2) and tuberization, and indeed the dormant state in many trees, is controlled by the relative lengths of the day and night. This phenomenon, known as photoperiodism, is found in a large number of different physiological responses in both plants and animals. An important point about photoperiodism in plants and poikilothermic animals is that the "critical" day or night length, that is the length at which a changeover in developmental pattern is induced, is scarcely affected by ambient temperature within the range 10-25°C [6]. Thus, in such organisms having a temperature close to that of the environment, there is the capacity to detect changes in day-length or night-length of a

A   B

Fig. 7.2. Photoperiodic control of flowering in the short-day plant *Kalanchoë blossfeldiana*. One plant (A) has been subjected to 24-hour cycles comprising 8 hours light and 16 hours darkness and the other plant (B) to cycles of 16 hours light and 8 hours darkness. (Reproduced by kind permission of Prof. W. W. Schwabe, Wye College, Ashford, Kent, U.K.)

[*To face page 236*

few minutes from a critical value. This degree of precision can only be achieved if the organism possesses a reference mechanism against which day-length or night-length can be compared, and the mechanism must be able to operate at a relatively uniform rate despite changes in ambient temperature. In other words, the organisms must possess a biological clock.

A third line of enquiry which has led to the inescapable conclusion that living organisms possess some form of temperature compensated time measuring mechanism stems from the pioneering work on animal navigation and orientation by Kramer [7] and von Frisch [8]. Details of this work can be found elsewhere [6], see also Ch. 8, but it must be mentioned that there is now unequivocal evidence from animals as diverse as bees, fish, lizards and birds that a biological clock is used to compensate for the relative movement of the sun across the sky during the day; without such a clock, accurate "solar compass" direction computation is impossible.

This chapter is primarily concerned with circadian rhythmicity in plants. The separation of plant and animal rhythms for discussion is rather arbitrary since their characteristics are astonishingly similar. At the present time, there is no evidence that the chemical reactions comprising the biological clock are different in plants and animals. The observed rhythms are, of course, different in plants and animals, but the basic circadian system may be identical, and might be located at a key level of cell chemistry common to all living organisms—nucleic acid metabolism or protein synthesis—or at the site of a particular membrane phenomenon. This question will be discussed in detail later in this chapter. It must be stressed, however, that the ultimate aim, the location and identification of the clock system in chemical terms, has not yet been achieved in any organism. In this chapter interest will be centred on the physiological and biochemical investigations which have attempted to establish the nature of the biological clock in plant tissues. It should be remembered, however, that in all these studies the behaviour of the clock system is inferred from the changes induced in some observable rhythm which may be several steps removed from the basic mechanism.

Research on biological clocks in both plants and animals has only recently emerged from the descriptive stage. Much of this descriptive work provided an insight into the properties of the clock system and in some cases a general impression of the kind of mechanism involved. However, little further progress will be made in establishing the fundamental nature of the circadian system until the problem is attacked much more intensely at the biochemical and biophysical level. Plants and unicellular organisms offer great advantages as experimental material for such studies.

7.2 DISTRIBUTION OF CIRCADIAN RHYTHMS IN PLANTS

Persistent circadian rhythms have been observed in all the major groups of plants with the exception of the *Bryophyta* (mosses and liverworts). However, since a number of Bryophytes have been shown to have photoperiodic responses, they must have some form of time measuring mechanism. At present, this mechanism has not been shown to regulate any physiological or biochemical processes in a rhythmic manner.

Rhythms are found in a large number of different physiological, chemical, or behavioural processes in plants, from the unicellular algae to the most complex, multicellular, higher plants. Strangely, a persistent rhythm in a particular aspect of behaviour in one species may not be found in a closely related species. The rate at which sporangia are discharged from the sporangiophores of the dung fungus *Pilobolus sphaerosporus* shows a most distinct rhythm which persists for several days in constant conditions. In contrast, while cultures of the closely related species *Pilobolus crystallinus* show a rhythm of sporangium discharge in a natural environment, all trace of the rhythm disappears abruptly after the onset of uniform environmental conditions [9].

Studies of rhythms have been made with a number of plants representative of the major groups. A selection of the many papers is given in Table 7.1. Undoubtedly most of the early studies were made on the leaf movements of higher plants because of the ease with which this aspect of plant behaviour could be recorded. Over the past 40 years, Bünning [6] and his co-workers have made a most detailed study of the rhythmic movement of the primary leaves of the runner bean plant, *Phaseolus multiflorus*. The advent of new physical techniques such as stable photomultiplier tubes, and infra-red gas analysis, has facilitated the continuous recording of many other physiological processes in plants. The rhythm of luminescence in the photosynthetic, armoured marine dinoflagellate *Gonyaulax polyedra* has been studied by Hastings and Sweeney [10] and their co-workers, and the rhythm of carbon dioxide metabolism in the succulent plant *Bryophyllum fedtschenkoi* by Wilkins [11, 12].

No attempt will be made to provide an encyclopaedic coverage of the studies of plant rhythms. The books by Bünning [6] and Sweeney [13], and the reviews by Hastings [14] and Sweeney [15] provide sources of much further information on plant rhythms. In addition, listings in the Environmental Biology Handbook edited by Altman and Dittmer [16] give valuable tabulated information, and the proceedings of the Cold Spring Harbor Symposium [17] and of the NATO Summer School on Circadian Clocks [18] are sources of more detailed treatments of various aspects of this subject. This chapter aims at providing a general insight into the characteristics of plant rhythms and to indicate the areas where more detailed investigation might profitably be undertaken.

Table 7.1

Some examples of circadian rhythms in plants

Group	Organism	Rhythm	Reference
Photosynthetic Protozoans	*Gonyaulax polyedra*	luminescence, photo-synthetic capacity, growth	Hastings and Sweeney[10]
		luciferin activity	Bode, De Sa and Hastings[19]
	Euglena gracilis	phototaxis	Bruce and Pittendrigh[20]
Algae	*Hydrodictyon reticulatum*	photosynthesis, respiration	Pirson, Schön and Döring[21]
	Chlamydomonas reinhardi	phototaxis	Bruce[22]
	Oedogonium cardiacum	sporulation	Bühnemann[23]
	Acetabularia major	photosynthetic capacity	Sweeney and Haxo[24]
	Acetabularia mediterranea	RNA synthesis	vanden Driessche and Bonotto[25]
		chloroplast ATP	vanden Driessche[26]
Fungi	*Daldinia concentrica*	spore discharge	Ingold and Cox[27]
	Pilobolus sphaerosporus	sporangium discharge	Uebelmesser[9]
	Neurospora crassa	growth	Pittendrigh *et al.*[28]
	Neurospora crassa	conidiation	Sargent, Briggs and Woodward[29]
			Sargent and Briggs [30]
Pteridophyta	*Selaginella serpens*	plastid shape	Busch[31]
Spermatophyta	*Phaseolus multiflorus*	leaf movement	Bünning[32]
	Kalanchoë blossfeldiana	petal movement	Bunsow[33]
	Avena sativa	growth rate of coleoptile	Ball and Dyke[34]
	Bryophyllum fedtschenkoi	CO_2-fixation in darkness	Warren and Wilkins[35]
	Bryophyllum daigremontianum and *Coffea arabica*	CO_2 compensation point	Jones and Mansfield[36]
	Lemna perpusilla	CO_2 output	Hillman[37]
	Coleus blumei × *C. frederici*	leaf movement	Halaban [38, 39, 40, 41]

7.3 THE PROBLEM OF ESTABLISHING THE ENDOGENOUS NATURE OF THE RHYTHM

One of the major controversies in this field over the past two decades has been the question of establishing unequivocally whether living organisms really possess a clock system capable of sustained temperature-compensated oscillation, or whether the oscillations observed in the so-called uniform environment really reflect the periodic stimulation of the organism by some uncontrolled, subtle environmental parameter, such as the intensity of cosmic ray bombardment, or of the earth's magnetic field. The latter view has been propounded principally by Brown [42]. Most investigators favour the former view that the rhythmic systems are endogenous. This view is based upon the rhythm conforming with five conditions proposed by Pittendrigh [43]. First, the physiological rhythm, which presumably reflects the activity of the basic clock system, must continue for several days in an environment in which as many parameters as possible are kept uniform. Second, the phase of the rhythm must be able to be shifted, and the new phase maintained under uniform environmental conditions. Third, the rhythm should be able to be initiated in aperiodic organisms raised under constant conditions by a *single* stimulus. Fourth, the phase of the rhythm should be delayed when organisms are deprived of oxygen or exposed to low temperatures, thus indicating a dependence of oscillation upon the availability of metabolic energy. Fifth, the period of the rhythm must not be exactly 24 h, and should be capable of being varied slightly by changing the level at which environmental parameters such as light intensity and temperature are held. Indeed the term circadian (about-a-day) was introduced because the free-running period of most circadian rhythms lies at a value between 22 and 27 h.

If a rhythm has these five characteristics it is generally agreed to have arisen from within the organism. Brown [42, 44], however, disputes this conclusion and maintains that these characteristics do not eliminate the possibility that the rhythm could arise from the effects of subtle 24-h rhythms in the physical environment of the earth. There seems to be no means of finally resolving these conflicting views. The possibility of ascertaining whether rhythms would persist in organisms dispatched in a deep space probe would, at first sight look a promising approach, but, quite apart from the expense and technical difficulties, it would be absolutely necessary to ensure that the space craft was not subjected to *any* physical rhythms of *any* frequency, since both plants and animals appear to have the capacity for a specialized type of entrainment known as frequency demultiplication, that is to accept non-24-h cycles of, for example, light and darkness, and exhibit a 24-h physiological rhythm [45].

7.4 THE CHARACTERISTICS OF CIRCADIAN RHYTHMS IN PLANTS

Circadian rhythms in plants persist in a uniform environment for varying lengths of time depending upon the organism. In the fungi for example, the rhythm of spore discharge from the stroma of *Daldinia concentrica* will continue for about 14 days in darkness [27], as shown in Fig. 7.3, and a similar degree of persistence has been found in the rhythm of leaf movement of *Phaseolus multiflorus* [6]. In *Bryophyllum fedtschenkoi*

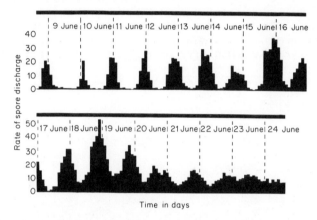

Figure 7.3. The persistence of the rhythm of spore discharge from the stroma of the fungus *Daldinia concentrica* for 17 days under uniform environmental conditions of temperature and darkness. (From Ingold and Cox [27])

the rhythm of CO_2 metabolism will last for between 3 and 8 days, the actual time apparently being related to the light intensity to which the plant was subjected previously, and the physiological state of the plant. In the unicellular organism *Gonyaulax polyedra* the rhythm persists for only a few days in darkness, but for several weeks in a constant low intensity of light. Obviously, the length of time a rhythm will persist in a darkened plant will be markedly affected by the depletion of food reserves. The persistence of the *Gonyaulax* rhythm for many days in constant dim light may reflect the capacity of this organism to photosynthesize carbohydrates under these conditions.

When a rhythm persists in an environment uniform with respect to temperature and darkness the observed period is said to be the free-running or natural period of the endogenous oscillating system. In such circumstances the effects of variation of environmental parameters upon the phase and period of the rhythmic system can be established, and it is from such data that our present ideas of the nature of the biological clock in plants are based. Discussion will be confined to relatively few organisms which have been thoroughly studied, and which may be regarded as showing responses reasonably typical of those observed in other plants.

7.4.1 Entrainment

Circadian rhythms in plants, as in animals, can be entrained to periodicities other than their natural frequency by exposure to cycles in light intensity or temperature [45]. At a uniform temperature, cycles of light and darkness can entrain the rhythms to periodicities ranging from about 6 h to 36 h. An example of the entrainment of the rhythms of leaf movement in *Canavalia ensiformis* [46] and of CO_2 metabolism in *Bryophyllum fedtschenkoi* [47] are shown in Fig. 7.4. In most plants,

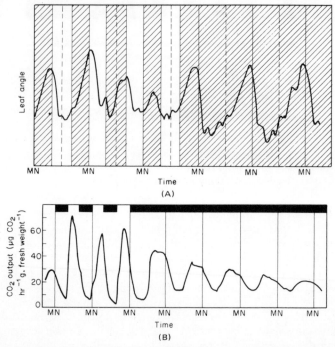

Figure 7.4. Entrainment of the rhythms of leaf movement in *Canavalia ensiformis* (A) and of carbon dioxide emission by leaves of *Bryophyllum fedtschenkoi* (B) on exposure to 16-h cycles comprising 8 h light and 8 h darkness, and the immediate reappearance of a circadian period with the onset of continuous darkness. MN indicates the time of midnight. The times of darkness are shown by shading (A) or by the black area of the bar across the top of the figure (B). (A, from Kleinhoonte [46]; B, from Wilkins [47])

however, there appears to be a limit to the rapidity with which the plant's circadian system can be forced to oscillate by imposed cycles of light and darkness. Bünning [48] has for example found that the rhythm of leaf movement in the runner bean seedling *(Phaseolus multiflorus)* can be entrained easily to an 18-h period by exposure to LD 9 : 9. On the other hand, exposure to LD 8 : 8 did not entrain the circadian rhythm to a period of 16 h. Instead, the rhythm reverted to its natural frequency. These limits of entrainment are not, however, a fixed characteristic of the circadian oscillating system in an organism, but are a function of the energy input into the system. Wilkins [47] has shown that the rhythm of carbon dioxide metabolism in leaves of *Bryophyllum* could be entrained to LD 3 : 3 when the intensity during the photoperiod was 1,000 lux, but to only LD 6 : 6 with an intensity of 500 lux, and to LD 8 : 8 with 100 lux (Fig. 7.5).

In all except two of the cases reported in the literature the period of an entrained circadian rhythm reverts to its natural value immediately the plant is returned to a uniform environment, as illustrated in the cases of *Canavalia* and *Bryophyllum* in Figs 7.4 and 7.5. This is what would be expected from both a physical and biological standpoint; there is no evidence in the vast majority of organisms, plant or animal, that a new periodicity can be acquired from environmental experience. Two organisms which appear to be exceptions to this general rule are the fungus *Pilobolus sphaerosporus* and the alga *Hydrodictyon reticulatum.* In *Pilobolus* the period of the entrained rhythm of sporangium discharge appears to be retained for a limited time after a uniform environment is restored, and then the natural period abruptly reappears [9]. The apparent retention of the entrained frequency has been discussed in detail by Wilkins [11] and attributed entirely to the development and maturation times of the sporangiophore. Trophocysts, from which the sporangiophores develop, are initiated in the cultures by a change from darkness to light, and require approximately 28 h to develop a mature sporangium at room temperature. Detailed examination of Üebelmesser's published data reveals that the last peak to occur on the entrained frequency after a uniform environment was restored was always approximately 28 h after the culture last experienced a change from darkness to light. This was so irrespective of the entraining frequency. The rhythm of photosynthesis and respiration in *Hydrodictyon* is not very marked but the data of Pirson, Schön and Döring [21] appear to indicate that an entrained periodicity is retained for a few days in a uniform environment. After exposure to cycles comprising 10.5 h light and 7 h darkness, a 17.5 h rhythm persisted for 2-3 days in darkness. In so far as the author is aware, this finding in *Hydrodictyon* is unique, and the question it poses undoubtedly merits further investigation.

Circadian rhythms can also be entrained by exposing plants to

BACR–9

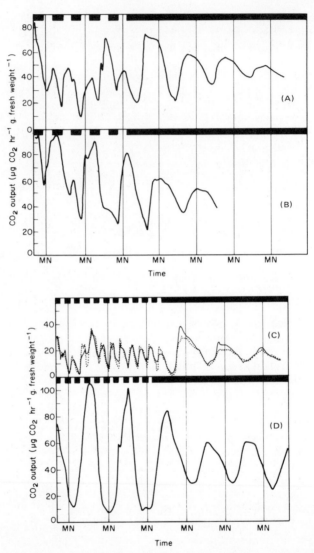

Figure 7.5. The effect of light intensity on the extent to which the circadian system in leaves of *Bryophyllum fedtschenkoi* can be entrained by cycles of light and darkness. The system is entrained by 6 h light and 6 h darkness when the intensity is 500 lux (A), but not when it is 100 lux (B). Entrainment occurs to 3 h light and 3 h darkness when the intensity is 1000 lux (C), but not when it is 500 lux (D). Programmes of light and darkness (black) are shown by the bars above each figure. Note the reappearance of the circadian rhythm after continuous darkness begins. (From Wilkins [47])

periodic fluctuations in temperature. The rhythm of sporulation in the alga *Oedogonium cardiacum* can be entrained to temperature cycles having a period of 18 and 24 h, but not to those having periods of 12, 30 and 48 h. A periodic temperature variation of only 2.5°C is sufficient to achieve this entrainment [23]. In the succulent plant *Kalanchoë blossfeldiana* entrainment of the rhythm of petal movement has been achieved with a temperature fluctuation of only 1°C [49].

The effect of subjecting plants simultaneously to cycles of light and darkness and of temperature in different phase relationships to one another has not been systematically investigated in plants. Preliminary investigation of this question in the author's laboratory several years ago utilized a temperature variation of 10°C, and a light intensity of approximately 1,000 lux, both environmental parameters being varied on a 24-h cycle in as near a sinusoidal oscillation as was technically possible In almost all cases the phase of rhythm of carbon dioxide metabolism in *Bryophyllum* leaves was entrained to the cycle of light intensity rather than the cycle of temperature, though at some phase angle relationships between the entraining environmental rhythms, the physiological rhythm in the plant became too indistinct for reliable deductions to be made. Whether larger variations in temperature could reverse this situation is not known, and further study of these effects might indeed be worthwhile.

7.4.2 Frequency Demultiplication

Frequency demultiplication is a special case of entrainment which has been observed in a number of plants. In contrast to normal entrainment, where the driven oscillator acquires the precise period (or frequency) of the driving oscillation, frequency demultiplication involves the driven oscillator acquiring a period which is a multiple of the periodicity of the entraining cycles. The rhythm of transpiration in lemon cuttings had an *exactly* 24-h period when exposed to LD 4 : 4 [50], and the rhythm of leaf movement in *Phaseolus* seedlings showed periods of 28 h and 32 h under LD 7 : 7 and LD 8 : 8 respectively [51]. Other examples of frequency demultiplication, including one in the rhythm of phototactic response in the photosynthetic protozoan *Euglena gracilis,* are fully discussed by Bruce [45].

7.4.3 Initiation and Inhibition of Rhythms

Circadian rhythms in plants can be initiated by a single stimulus. In some cases, for example, cultures of the fungus *Pilobolus sphaerosporus* which have been raised under conditions of total darkness, a rhythm can be initiated with a flash of light of only 0.02 sec [9]. A rhythm can be induced in the growth rate of the coleoptiles of *Avena sativa* (oat) seedlings which have been grown in continuous red light merely by transferring them to darkness [34], and the leaf movement rhythm in *Phaseolus multiflorus* can be started by transferring the seedlings from continuous light in which they have been grown to darkness, or vice versa [52, 53]. An arhythmic culture of *Gonyaulax polyedra* which has been maintained in continuous light for three years shows a circadian rhythm in luminescence after a single change in light intensity [54]. The rhythm of carbon dioxide metabolism in *Bryophyllum* leaves is inhibited by light but begins again immediately the light is extinguished [55, 56].

The environmental parameters which have the most marked capacity to inhibit circadian rhythms in plants are light, high and low temperatures, and anaerobic conditions. In the Crassulacean plant *Bryophyllum fedtschenkoi* and the photosynthetic marine dinoflagellate *Gonyaulax polyedra,* two plants at widely different levels of organization, the circadian rhythms are inhibited by prolonged exposure to light of high intensity (Fig. 7.6). That the basic circadian oscillator is inhibited, and not just the observed physiological process, is shown by the fact that oscillation begins again when darkness is restored and that the phase of the new rhythm is set by this change, the first peak occurring a specific time after the restoration of darkness and subsequent ones at approximately 24-h intervals [55, 11, 56, 10]. Similar reports have been made for the rhythm in the growth rate of *Avena sativa* coleoptiles [34]. The fact that in these plants the first peak of the rhythm occurs a specific time after the light is extinguished strongly suggests that light actively inhibits the basic oscillators by forcing them to a particular phase of their cycle, and holding them in that phase for as long as the exposure continues. The oscillations would thus always begin again from a given phase of the cycle regardless of the length of time for which the inhibition had lasted.

In *Bryophyllum,* exposure to high temperature has an effect upon the system almost identical to that of high intensity light [57]. When leaves are kept in darkness at $25°C$ a rhythm of carbon dioxide output persists with a period of about 22.5 h. If the temperature is increased to $35°C$ rhythmicity stops, though a high level of CO_2 output remains. When the temperature is lowered once again to $25°C$ a rhythm reappears in the carbon dioxide output of the leaves, the first peak occurring a specific time (18-19 h at $25°C$) after the original temperature is restored regardless of the length of exposure to $35°C$. In the case of leaves kept at a lower temperature, for example $15°C$, the inhibition resulting from exposure to $35°C$ is closely similar except that on the restoration of the lower temperature ($15°C$) the first peak of the rhythm occurs about 24-25 h after the temperature reduction (Fig. 7.7). In these respects, the effect of exposing leaves to $35°C$ is identical to that achieved by exposing the leaves to light. At $25°C$ and at $15°C$ the first peak of the

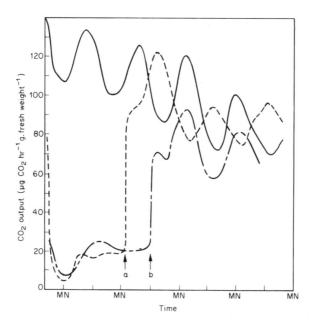

Figure 7.6. Inhibition of the rhythm of carbon dioxide emission in small pieces of mesophyll of *Bryophyllum* leaves from which the epidermis had been removed. Continuous line shows the rhythm in tissue placed in darkness at the onset of the experiment (1600 h). The dashed and dotted lines show the rhythm induced in the tissue on transfer to darkness after continuous exposure to white light (1000 lux) for a further 33 h (arrow, a), and 44 h (arrow, b) respectively. Note the phases of these rhythms in relation to that of the control rhythm. MN indicates time of midnight. (From Wilkins [55])

rhythm occurs 18-19 h and 24-25 h respectively after darkness is restored. These findings clearly imply that on exposure to high temperature the oscillator is driven to, and held at, the same phase in the cycle as that at which it is held by light.

Reduction of metabolism by exposure to low temperature or to anaerobic conditions also inhibits the circadian oscillations in plants. Prolonged exposure of seedlings of *Avena sativa* [58] and *Phaseolus multiflorus* [59], and leaves of *Bryophyllum fedtschenkoi* [57] to low temperature delays the phase of the circadian rhythms, the magnitude of the delay being related to the length of the chilling period. Exposure to anaerobic conditions has also been shown to inhibit the circadian oscillations in *Avena* [58, 60], *Phaseolus* [61] and *Bryophyllum* [62].

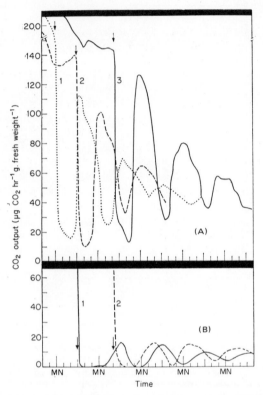

Figure 7.7. The inhibition of the rhythm of carbon dioxide emission in leaves of
Bryophyllum fedtschenkoi by prolonged exposure to a temperature of 35°C and the
onset of rhythmicity again when the temperature is reduced to 26°C (A) at three
different times shown by the arrows 1, 2 and 3 or to 16°C (B) at two different times
shown by the arrows 1 and 2. Leaves were in darkness throughout. Note how the
phase of the rhythm is determined by the time at which the temperature reduction
occurs. (From Wilkins [57] by kind permission of the Royal Society of London)

7.4.4 Control of Phase

The factors which control the phase of the circadian oscillations of the
biological clock have been extensively investigated in both plants and
animals. This is an important aspect of research on biological clocks,
since in organisms having more than one clock mechanism, and there is
evidence that this is so in man, the co-ordination of the different
circadian systems, and their synchronization with external *Zeitgebers*,
appear to have important consequences which are reflected in the health
and proficiency of the organism.

Phase control of circadian oscillators is usually studied by examining the effects of a change in one environmental factor on the phase of a rhythm persisting in organisms maintained in a uniform environment. For example, leaves held at a constant temperature and in darkness may be exposed for a few minutes or hours to a particular wavelength of light, then returned to darkness, and the characteristics of any subsequent phase change that occurs determined. Of the environmental parameters examined, light, temperature, and the partial pressure of oxygen have been found to be capable of adjusting and resetting in a predictable manner the phase of circadian oscillations in plants.

It has already been mentioned that in such plants as *Avena* and *Bryophyllum* the phase of the rhythm is set by the time they are transferred from continuous light to darkness, and similar findings have been reported for the rhythm of growth rate in the fungus *Neurospora crassa* [28] and of luminescence in *Gonyaulax polyedra* [10]. Decreasing the light intensity, rather than an absolute change from light to darkness, suffices both to initiate the rhythm and determine the phase in *Bryophyllum* [56] and in *Gonyaulax* [10]. A decrease of about 80% of the initial intensity is required in the case of *Bryophyllum* as shown in Fig. 7.8. On the other hand, a change from prolonged darkness to

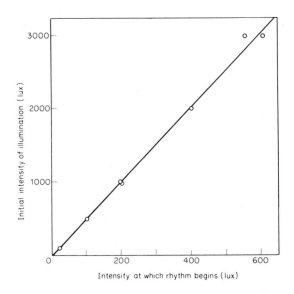

Figure 7.8. The relationship between the initial intensity of light to which leaves of *Bryophyllum fedtschenkoi* are subjected and the intensity at which the circadian rhythm of carbon dioxide metabolism begins when the intensity is gradually reduced over a period of 6 h. (From Wilkins [56])

continuous light can initiate the rhythm and set the phase in *Phaseolus* seedlings [52] and in cultures of the alga *Oedogonium cardiacum* [23].

When a plant is kept in continuous darkness, the phase of the circadian oscillating system can be reset by a brief exposure to light. The magnitude of this phase shift depends, however, upon a number of factors such as (1) the position in the cycle at which the light treatment is given, (2) the length of the treatment, (3) the radiant flux density, and (4) the wavelength of radiation.

In *Bryophyllum, Phaseolus* and *Gonyaulax,* the phase of the rhythm is shifted by an exposure to light for a few hours at one position in the cycle but not by an exposure given at another position. The effectiveness of a 5-h light treatment in shifting the phase of the rhythm of carbon

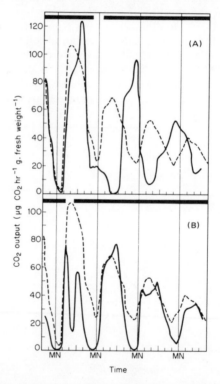

Figure 7.9. The induction of a phase shift in the rhythm of carbon dioxide emission in leaves of *Bryophyllum fedtschenkoi* kept in continuous darkness and at 26° C by a 5-h exposure to white light (3000 lux) between the peaks (A), but not by an identical exposure at the apex of a peak (B). The rhythm in control leaves kept throughout in darkness is shown by the broken lines, and the position of the light treatment by the bar above each figure. MN indicates midnight. (From Wilkins [56])

dioxide metabolism in *Bryophyllum* when given in the trough, and at the apex of the peak, is shown in Figs. 7.9A & B respectively. In the former case the phase was reset, the first peak occurring some 18-19 h after the end of the treatment, whereas in the latter case no phase shift was induced [56]. By scanning the whole cycle with light treatments it is possible to derive the detailed relationship between the time in the cycle when the treatment is given and the magnitude of the phase shift induced. Such response curves have been determined for a number of organisms [63, 64]; one published for the leaf movement rhythm in *Phaseolus multiflorus* is shown in Fig. 7.10. In this case the plants were exposed to a constant low intensity of background light and a uniform temperature. The light intensity was then raised to 15,000 lux for 3 h at various positions in the cycle. Clearly the precise timing of the treatment determines whether or not a phase shift is induced, its magnitude and whether it is in a positive or negative sense [63, 65].

When a light treatment is given in a part of the cycle where a phase shift can be induced, the duration of the treatment can affect the magnitude of the phase shift. In the case of *Bryophyllum*, Wilkins [56] could detect no significant phase shift after a 1-h exposure to light

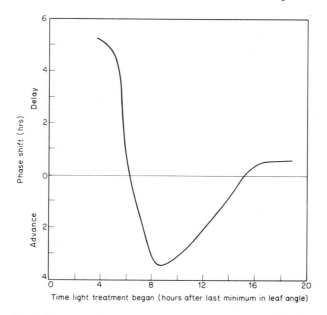

Figure 7.10. The magnitude and direction of the phase shifts induced in the rhythm of leaf movement in *Phaseolus multiflorus* seedlings kept at a constant temperature and a uniform light intensity of 150 lux, by 3-h exposures to an intensity of 15,000 lux at different times in the cycle. (From Moser [63])

between the peaks of the rhythm of carbon dioxide output, whereas 3-h and 6-h treatments gave rise to most marked phase shifts. The magnitude of the phase shift was not a function of the duration of the light treatment for exposures of at least 3-h, the first peak following the treatment always occurred 18-19 h after darkness was restored. The amount of phase shift is thus determined by the moment the light treatment ends.

The radiant flux density to which the plants are exposed during a light treatment is also important in determining the extent of the phase shift induced. Again in the case of *Bryophyllum* leaves, at 8 lux and above the flux density during a 6-h treatment given between the peaks of the carbon dioxide output rhythm had little effect on the magnitude of the phase shift. At 2 lux however the phase shift was substantially less.

The relative effectiveness of different regions of the spectrum in inducing phase shifts in circadian rhythms has been investigated in a number of plant systems, and will be discussed later (p. 260) when the whole question of the photoreceptors involved in the induction of phase shifts and in the initiation and inhibition of rhythms by light will be considered in details.

When a plant is kept in continuous dim light the phase of the rhythm can be reset by exposing the plant to darkness for a few hours. Such a response has been observed in *Bryophyllum, Phaseolus* and *Gonyaulax*. In *Bryophyllum* the effectiveness of the dark treatment is dependent upon the time in the cycle at which it is given [56].

Phase shifts can also be induced in circadian systems in plants under uniform environmental conditions by raising or lowering the temperature for a few hours. Once again whether or not the phase is changed depends upon the position in the cycle at which the temperature change occurs. In *Bryophyllum* leaves, increasing the temperature from $15°$ to $35°C$ for 3 h between the peaks of the rhythm of carbon dioxide output results in a predictable phase shift in which the first peak occurs 24-25 h after the original temperature of $15°C$ is restored, and subsequent peaks at about 24-h intervals. An exactly similar treatment given at the apex of a peak does not influence the phase of the rhythm at all (Fig. 7.11). The effect of increasing the temperature for 6 h between the peaks is exactly the same as that found with the 3-h treatment providing the treatments end at the same time. Thus, as was the case with the light treatments, the magnitude of the phase shift induced by increasing the temperature from $15°C$ to $35°C$ for 3 or 6 h is not related to the length of the treatment, but to the moment when the treatment ends. In the previous section it was mentioned that both light and high temperature appear to drive the *Bryophyllum* oscillator to a particular phase of the cycle. This concept is fully supported by the findings just discussed. The induction of phase shifts by light and high temperature treatments given between the peaks,

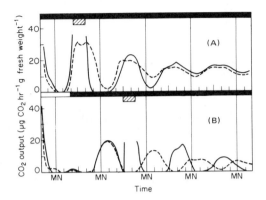

Figure 7.11. The induction of a phase shift in the rhythm of carbon dioxide emission of *Bryophyllum fedtschenkoi* leaves kept in darkness and at 16°C by a 6-h exposure to a temperature of 35°C between the peaks (B), but not by an identical exposure at the apex of a peak (A) of the rhythm. Control rhythms in untreated leaves are shown by the broken line. The phases of both rhythms in B were set for convenience of treatment by an initial exposure to light as shown by the bar above the figure. MN indicates midnight. (From Wilkins [57] by kind permission of the Royal Society of London)

but not by either treatment given at the apex of the peaks, is most easily explained on the basis that light and high temperature drive the oscillating system to the phase of the cycle corresponding to that normally found at the apex of the peak of the rhythm of carbon dioxide output. When treatments are applied at times in the cycle so that they end just when the system would be leaving this point in the cycle in normal oscillation, no phase shift is induced. In a number of other plants the induction of a phase shift by raising the temperature has also been shown to depend upon the point in the cycle at which the change is made. Results of the detailed study by Moser [63] in which the cycle of the leaf movement rhythm in *Phaseolus* seedlings was scanned with a treatment consisting of increasing the uniform background temperature of 20°C to 28°C for 4 h are shown in Fig. 7.12. In *Phaseolus* the phase of the rhythm can clearly be advanced, delayed or unaffected by the temporary increase in temperature depending upon the precise timing of the perturbation.

Exposure to low temperature and to anaerobic conditions for several hours might be expected to induce a phase shift in plant rhythms since metabolism would be severely if not totally inhibited. The circadian oscillators in both *Bryophyllum* and *Phaseolus* can be arrested, and a phase shift induced both by depriving the tissues of oxygen or by exposing them to low temperature, but the effectiveness of either

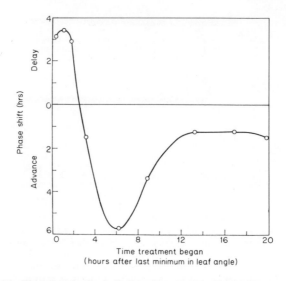

Figure 7.12. The magnitude and direction of the phase shifts induced in the rhythm of leaf movement in *Phaseolus multiflorus* seedlings, kept at a temperature of 20°C and a uniform low light intensity, by raising the temperature to 28°C for 4 h at various times in the cycle. (From Moser [63])

treatment again depends upon its precise time of application in the circadian cycle.

In *Bryophyllum* leaves exposure to low temperature for 4 h near the apex of a peak of the carbon dioxide output rhythm induced a phase shift approximately equal to the length of treatment. An identical exposure given between the peaks does not induce a phase delay at all [57]. Longer treatments of 8 and 12 h duration caused phase shifts even when given between the peaks. However, these shifts were always much less than those induced by the treatments given at the apex of the peak where the phase delay was always equal to the duration of the treatment (Fig. 7.13A). Virtually identical results reported for *Phaseolus* [66] are shown in Fig. 7.13B. No phase shift is induced when chilling takes place at point E in the cycle unless the treatment is longer than 5-6 h, whereas at point B the phase delay is always equal to the duration of the treatment. In both *Bryophyllum* and *Phaseolus* the phase delay is clearly related to the length of the chilling treatment, a finding in marked contrast to the effect of exposure to high temperature in *Bryophyllum*.

Depriving *Bryophyllum* leaves of oxygen for several hours between the peaks causes a phase delay approximately equal to the length of the treatment, whereas a similar deprivation at the apex of the peak does not

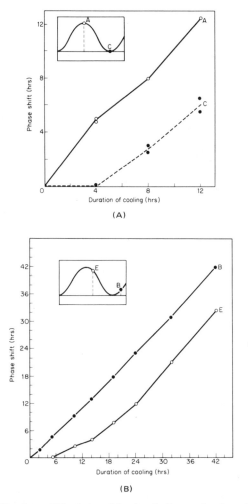

Figure 7.13. The phase shifts induced in the rhythms of carbon dioxide emission in leaves of *Bryophyllum fedtschenkoi* (A) and of leaf movement in seedlings of *Phaseolus multiflorus* (B) by exposures to temperatures of 0-1°C for various lengths of time beginning at two positions in the cycle as shown by the inset diagrams. (A, from Wilkins [57] by kind permission of the Royal Society of London; B, from Wagner [66])

affect the phase (Fig. 7.14 [62]). Similar findings have been reported for the rhythms of leaf and petal movement in *Phaseolus* and *Kalanchoë blossfeldiana* respectively [61]. In both *Bryophyllum* [57, 62] and *Phaseolus* [66, 61] the positions in the cycle at which exposure to low

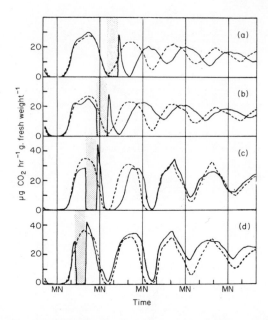

Figure 7.14. The phase shifts induced in the rhythm of carbon dioxide emission in leaves of *Bryophyllum fedtschenkoi* by exposure to anaerobic conditions for 6 h at different times in the cycle. Times of treatments are shown by the shaded areas of the figure. The leaves were kept in darkness and at 26° C throughout the experiments. MN indicates midnight. (From Wilkins [62])

temperature is effective in re-setting the phase are not identical with those in which anoxic conditions are most active. Similarly in *Bryophyllum,* the point in the cycle at which a 4-h exposure to 0° C is without effect on the phase is not the same as that at which exposure to anaerobic conditions are without effect. Since both anoxic conditions and low temperatures might have been expected to have similar effects in reducing metabolism, it would have been anticipated that both treatments would have been effective in inducing a phase shift at similar points in the cycle. On the basis of treating plants with low temperature for several hours at different points in the cycle, Bünning [32] suggested that the circadian oscillating system had the characteristics of a relaxation oscillator, that is, one in which there is an energy requirement for operation only over a portion of the cycle. Whilst this view seems to be acceptable if the data for low temperature treatments and for anaerobic treatments are viewed separately, it appears to become untenable when the data for the two types of treatment are considered together in view of the lack of coincidence between the points of maximum and minimum effectiveness in the cycle of the two types of treatment.

In summary, the induction of predictable phase shifts in circadian systems undergoing steady-state oscillation can be achieved with a variety of stimuli. The effectiveness of all the stimuli depends upon the time in the cycle at which the stimulus occurs. The relationship between the magnitude of a stimulus given at a time in the cycle when the system is responsive and the magnitude of the phase shift induced depends upon a number of factors such as the nature of the stimulus, its duration, intensity, etc.

7.4.5 Control of Period

For an oscillating system to form the basis of an accurate time measuring mechanism, precise regulation of the period is of the utmost importance. The period of a biological block is essentially circadian, that is, it has a value of between about 22 and 28 h, the actual value differing from organism to organism, and to a lesser extent upon the environmental conditions. The one feature of the circadian oscillating system, apart from its period of about 24 h, which distinguishes it from all other types of oscillators in living organisms is the fact that the period is remarkably unaffected by changes in ambient temperature. In other words, the circadian system appears to incorporate some kind of temperature compensating device.

In some species of plant, notably the runner bean *Phaseolus multiflorus*, strains have been selected having significantly different natural periods [67] under standardized environmental conditions. Crosses between these strains led to the F_1 generation having an intermediate natural period, while the F_2 generation showed evidence of a segregation of the natural period into two groups having values similar to those of the original parents. This kind of evidence, and the fact that oscillation with a definite natural period can be initiated by a single stimulus, show that an oscillating system having a natural period within the circadian range is an inherited characteristic of the organism.

The steady-state natural period of a circadian oscillator is, however, not absolutely fixed genetically, but can vary within certain limits with changes in environmental parameters. The constant level of illumination to which an organism is subjected at a uniform temperature can modify the free-running period of a circadian oscillating system. This effect is more often encountered in animals than in plants. Despite some exceptions, the changes in period of the animal systems occur in a predictable way: in diurnal animals the period becomes shorter, and in nocturnal animals it becomes longer, as the level of illumination is increased [68]. As a general rule, plants do not show such a response although in *Gonyaulax polyedra* the period of the rhythm of luminescence appears to become shorter with increasing light intensity [10]. On the other hand, exposure of plants to constant illumination in different spectral bands has a quite marked effect upon the period. In the case of *Phaseolus multiflorus,* the spectral band between 610 and 690 nm increases, and that between 690 and 850 nm decreases, the steady-state period of the rhythm of leaf movement, as compared with the value recorded in darkness, or in other bands of the spectrum [52].

Since it is reasonable to suppose that the basic oscillating system in a plant comprises a number of interlinked chemical reactions, it might be expected that the frequency of oscillation would be a function of temperature. In most organisms that have been studied, this is in fact so, but to a much smaller extent than would have been anticipated. Some measure of the change in the period of oscillation can be obtained by dividing the period at temperature $t° C$ by the period at $t + 10° C$. For almost all organisms, plant or animal, the resulting quotient is between 0.8 and 1.03. The rate of a normal dark chemical reaction would increase by a factor of between 2 and 4 after a temperature increase of $10° C$.

Some examples of the natural period of circadian rhythms in plants at different temperatures are shown in Table 7.2. These values refer to the steady-state period, that is the period of oscillation after the organism has been kept in the stated conditions for a sufficient length of time for the period to have attained a stable value. In a number of plants, *Bryophyllum, Oedogonium* and *Gonyaulax,* the period shows a slight but significant dependence upon ambient temperature, while in a few others, for example *Phaseolus* and *Avena,* no significant difference could be observed in the period of oscillation at temperatures between $15° C$ and $25-28° C$. One or two plants show a quite marked variation in the period with temperature; the rhythm of flower opening in *Cestrum nocturnum* is reported to have a period of 27 h at $17° C$ and one of 31 h at $13° C$ [71].

The smallness of the change in the period of most circadian systems at different ambient temperatures clearly implies that the systems must incorporate some form of temperature compensating mechanism. The nature of this mechanism is at the present time totally unknown. It is clear that the temperature compensating mechnism is not by any means perfect, since slight but significant variation in the period does occur. Nevertheless, the mechanism does dramatically reduce changes in the period with temperature and in a few organisms the variation is reduced to values below the level of statistical significance.

The rapidity with which the temperature compensating mechanism operates appears to vary between different plants. When *Phaseolus multiflorus* seedlings are kept at $25° C$ the period of the leaf movement rhythm is about 28 h. If the plants are transferred to $15° C$ the period at

Table 7.2

Variation of the period of circadian oscillating systems with ambient temperature

Organism	Temp °C	Period hours	Reference
Oedogonium cardiacum	17.5 25.0 27.5	20 22 25	Bühnemann[23]
Gonyaulax polyedra	15.9 21.0 26.6	22.5 24.4 26.8	Hastings and Sweeney[70]
Bryophyllum fedtschenkoi	16 26	23.8 ± 0.3 22.4 ± 0.4	Wilkins[57]
Avena sativa	15-17 26-28	24 24	Ball and Dyke[34]
Phaseolus multiflorus	15 20 25	28.3 ± 0.4 28.0 ± 0.4 28.0 ± 1.0	Leinweber[69]
Neurospora crassa	24 31	22 22	Pittendrigh, Bruce, Rosenweig and Rubin[28]

first increases to 33-34 h but then gradually decreases so that it finally attains a new steady-state value of somewhat less than 28 h after about 2 or 3 cycles [69]. On the other hand, the rhythm of carbon dioxide metabolism in leaves of *Bryophyllum* shows no such prolonged transient effects after a change of temperature. A new steady-state period is established with 12-22 h (that is, within the first cycle) following a single step-type change of temperature [57].

Modification of the period of circadian oscillating systems can be achieved with several other types of treatment. In a strain of *Phaseolus multiflorus* the period was increased from the normal value of about 24 h to one between 30 and 35 h when the atmospheric pressure to which the plant was subjected was reduced to 30-44 mm of mercury (ca 4-6 kPa) [72]. Lengthening of the period has also been achieved by chemical treatments such as the application of colchicine, phenylurethane [73, 48], theobromine and thophylline [75] to *Phaseolus* seedlings. Even the application of 2% ethanol increased the period of oscillation in *Phaseolus* by 5 h [75].

The increase in period resulting from lowered atmospheric pressure is consistent with the results of exposing organisms to anaerobic conditions as described in the previous section. The mechanism whereby the various hemical substances lengthen the period is quite unknown at the present time.

7.5 PHOTORECEPTION MECHANISMS

Attempts have been made to identify the photoreceptors responsible for bringing about the light induced responses in a number of organisms. Most studies of the effects of wavelength of light on plant rhythms have been made with spectral bands which are too broad to allow the determination of an accurate action spectrum, but they have provided a general indication of the most active regions of the spectrum. In higher plants the red end of the spectrum is undoubtedly the most active in controlling the circadian oscillating systems. The rhythm in the growth rate of coleoptiles of *Avena sativa* is initiated by a change from continuous red light, in which the seedlings were germinated, to darkness [34]. The rhythm of leaf movement of *Phaseolus multiflorus* can be initiated by exposure of dark-grown plants to irradiation in the spectral band between 600 and 700 nm [52]. On the other hand, while irradiation of plants with the band between 700 and 800 nm does not initiate the rhythm, it does have the effect of cancelling the inductive effect of a previous exposure to the 600-700 nm band. This cancellation can itself be overcome by a further exposure to the 600-700 nm band. A reversible photo-response of this type to the red and far-red spectral bands is characteristic of those mediated by the plant chromoprotein phytochrome. There is other evidence that these two spectral bands are the most active in controlling the circadian system in *Phaseolus*. It will be recalled that the period of oscillation can be lengthened or shortened by continuously exposing the seedlings to irradiation in these regions of the spectrum [52]. The circadian oscillating system in the succulent plant *Bryophyllum* seems also to be affected only by longer wavelengths of visible radiation, and interruption of continuous darkness by several hours of red light induces a phase shift whereas blue light does not [11, 56].

On the other hand, recent experiments by Halaban [41] have shown that in *Coleus* plants, the circadian rhythm of leaf movement is markedly affected by both the blue and red regions of the spectrum, but not by the far-red region. In darkness the normal value of the period was 22.5 h. Continuous irradiation with blue light lengthened the period to 24 h, whereas continuous irradiation with red light shortened the period to 20.5 h. Neither green nor far-red irradiation had any effect on the period. An 8-h exposure of darkened seedlings to red light advanced the phase of the rhythm, whereas a similar exposure to blue light delayed the phase. Irradiation with green and far-red did not induce phase shifts.

Far-red radiation did not reverse any of the effects of either red or blue light. These results, which are rather different from those obtained with *Phaseolus* seedlings [52], make it unlikely that phytochrome is the photoreceptor in *Coleus*.

The experiments described hitherto reveal prominent activity in one or other regions of the spectrum, but they do not provide unequivocal evidence for the identification of the photoreceptor molecule. This identification can only be achieved when an action spectrum for the light induced response has been properly determined, and in a strict sense this has not been attained with any circadian system. However, the relative activity of several hours exposure to equal quantum flux densities in different narrow spectral bands has been compared in three different organisms, and in each case a quite different action spectrum has been found (Fig. 7.15A, B & C).

The action spectrum for shifting the phase of the rhythm of the circadian system in *Gonyaulax polyedra* shows two peaks (Fig. 7.15A), a major one in the blue region and a minor one in the red region [76]. The peaks of the action spectrum do not coincide with those of the absorption spectra of any of the pigments extracted from *Gonyaulax,* but Hastings and Sweeney [76] suggest that one of the chlorophylls may be the primary photoreceptor.

An action spectrum for photo-inhibition of the rhythm of conidiation in *Neurospora crassa* has been determined by Sargent and Briggs [30]. Activity is wholly confined to the spectral region below 540 nm (Fig. 7.15B). There is a strong peak in the blue region of the spectrum and a broad shoulder in the near ultra-violet. While this spectrum again does not allow the identification of the photoreceptor, Sargent and Briggs [30] suggest that it might be either a carotenoid or a flavin compound. However, they advance two lines of evidence which rather rule out the possibility of a carotenoid being involved.

An action spectrum has also been determined for the induction of phase shifts in the rhythm of carbon dioxide metabolism in *Bryophyllum* (Wilkins, unpublished, Fig. 7.15C). Activity in this case is wholly confined to the spectral band between 600 and 700 nm. Although the action spectrum rather indicates that phytochrome may be the photo-receptor in this case, it must be stated that despite several attempts red/far-red reversibility has not yet been demonstrated.

Thus, almost every organism investigated with a view to establishing the nature of the photoreceptor mechanism has given rather different information. In general, while red light is undoubtedly active in all higher plants, blue light is also active in *Coleus*. In only one plant *(Phaseolus)* has the red/far-red reversibility characteristic of the phytochrome system been demonstrated. In *Gonyaulax* the major peak of activity is in the blue region with another smaller peak in the red region of the spectrum,

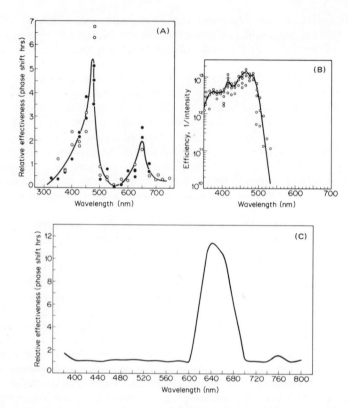

Figure 7.15. Action spectra for the induction of phase shifts in the rhythms of luminescence in *Gonyaulax polyedra* (A) and of carbon dioxide emission in leaves of *Bryophyllum fedtschenkoi* (C) by exposure for several hours to equal quantum flux densities in different narrow bands of the spectrum, and for the inhibition of the rhythm of conidiation in *Neusospora crassa* (B) by continuous irradiation with equal quantum flux densities in different spectral bands. (A, from Hastings and Sweeney [76]; B, from Sargent and Briggs [30]; C, Wilkins original unpublished)

and in the fungus *Neurospora* activity appears to be confined to the blue end of the spectrum. The latter finding is probably of general application in the fungi, since, as Wilkins [100] pointed out, a number of them respond to white light stimuli but, so far as is known, no red light responses have yet been found. Totally different pigments thus appear to be involved in mediating the light induced effects on circadian systems in *Neurospora*, *Gonyaulax* and the higher plants, and within the latter group different species may use different photoreceptors. In no case has a photoreceptor been unequivocally identified.

Ultra-violet radiation has been shown to induce stable phase shifts in the circadian system in *Gonyaulax* [77]. As with other light induced phase shifts, the magnitude of the shift depends upon the time in the cycle at which irradiation occurs. No effect of ultra-violet radiation could be detected on the phase of the rhythm of carbon dioxide metabolism in *Bryophyllum* (Wilkins, unpublished).

7.6 LOCATION AND IDENTIFICATION OF THE CIRCADIAN OSCILLATING SYSTEM

In unicellular organisms such as *Gonyaulax polyedra* each cell must have a circadian system controlling the rhythms manifest in its own physiological and chemical activity. Indeed in this organism, as in *Acetabularia*, rhythms have been observed in individual cells [78]. The phase of the circadian oscillation in an individual cell in a *Gonyaulax* culture is unaffected by the surrounding cells. A mixture of two cultures in which the circadian rhythms have been set in quite different phases by exposure to different cycles of light and darkness shows a persisting rhythm which is exactly that predicted on the basis of a summation of the rhythms of the two independent cultures [93]. In multicellular organisms, however, the question arises as to whether the circadian system is confined to specialized cells, or whether the activity of each cell in the organism is controlled by its own independent circadian system. Evidence obtained from the rhythm of carbon dioxide metabolism in leaves of *Bryophyllum* suggests that each cell has its own circadian system. Even the smallest pieces of leaf mesophyll from which the epidermis has been removed exhibit circadian rhythms of carbon dioxide metabolism, regardless of the region of the leaf from which they are taken [55]. Furthermore, tissue cultures of leaf mesophyll cells of *Bryophyllum* also show circadian rhythmicity in carbon dioxide metabolism (Fig. 7.16A, Wilkins and Holowinsky [79]). So in this plant at least, the circadian system is neither confined to particular regions of the leaf nor does it depend upon the organization of the leaf.

In other higher plants it is not clear whether the controlling circadian system is localized or not. In *Phaseolus*, for example, the periodic changes in the turgidity of the cortical cells of the pulvinus give rise to the rhythm of leaf movement. These changes in cortical cell turgidity appear to be dependent upon the presence of the lamina [80], but it is not known whether they are due to a rhythm in the synthesis of a hormone in the lamina, to a rhythm in the transport of the hormone from the lamina to the pulvinus or to periodic changes in the sensitivity of the pulvinus cells to the hormone. All three possibilities, and more, could, of course, be involved simultaneously, and further study of the

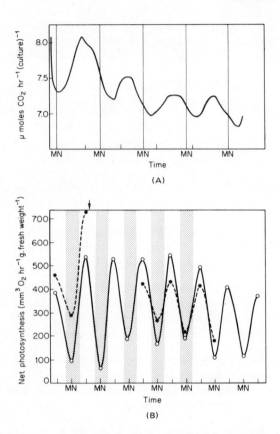

Figure 7.16A. The rhythm of carbon dioxide emission from a callus culture of leaf mesophyll cells of *Bryophyllum fedtschenkoi* under conditions of darkness and a uniform temperature. MN indicates midnight. (From Wilkins and Holowinsky [79])

B. The persistence of a rhythm in photosynthetic capacity in cells of *Acetabularia major* both in cycles of light and darkness and in constant dim illumination after the nucleus has been removed at the time indicated by the arrow. Darkness is indicated by the shading. The dotted line refers to juvenile cells without caps and the continuous line to cells with caps. (From Sweeney and Haxo [24])

Phaseolus rhythm might profitably seek the answer to this question.

The most important problem in this field of research is the location of the circadian oscillating system within the cell. The solution of this problem is of fundamental importance since it requires the identification of the system's chemical mechanism.

Several organisms have been investigated with a view to ascertaining the intra-cellular location of the circadian oscillating system. Two rather different approaches have been used. On the one hand, attempts have been made to isolate the circadian system by microsurgical techniques in which organelles have been removed from cells. On the other hand, biochemical approaches have attempted either to trace the sequence of chemical events back from the observed rhythm, or to deduce the chemical events involved in the circadian system from the effects of inhibiting substances on the phase and period of the rhythms. Results obtained from inhibitor experiments must be viewed with extreme caution since the specificity of action of inhibitors of nucleic acid and protein synthesis in plant tissues must be open to very considerable doubt [81, 82, 83].

The umbrella-shaped, giant unicellular alga *Acetabularia* is a very convenient experimental organism for the microsurgical approach since its nucleus is conveniently located at the base of the stalk and can be removed relatively easily. It has been found that when *Acetabularia* cells have been enucleated the rhythm of photosynthetic capacity continues normally both in cycles of light and darkness, and in continuous dim light (Fig. 7.16B). In addition, the phase of the rhythm can be reset by light signals as readily in enucleated cells as in normal cells [84]. Obviously, a normal circadian system capable of persistent oscillation exists somewhere in these cells other than in the nucleus. The question of whether the nucleus itself has a circadian system, and whether such a system would play a dominant role when the nucleus was present in the cell is left quite unresolved by these data. The work of Schweiger and Schweiger [85] has shown, however, that the nucleus does have a circadian system and that it can dominate the other circadian systems in the *Acetabularia* cell (Fig. 7.17). Two cultures of *Acetabularia* were grown under cycles of light and darkness differing in phase by 12 h, and the phases of the rhythms of photosynthetic capacity of the two stock cultures therefore differed by 12 h. Nuclei from cells of each stock culture were transferred to enucleated cells of the other culture, and the treated cells were then placed in a constant low intensity of light. After two days, the phase of the rhythm of photosynthetic capacity was found to coincide with that of the light dark cycles to which the nuclei had been previously exposed. Thus, within two days of transfer, the nuclei had readjusted the phase of the cytoplasmic circadian systems by 12 h, and their role as the site of the master oscillator appears to be clearly established. In some way as yet unkown, the cytoplasmic oscillating systems are controlled by those of the nuclei.

The persistence of the rhythm in enucleated cells of *Acetabularia* does not rule out the possibility that the circadian oscillator involves nucleic acid and protein synthesizing systems. In *Acetabularia* large amounts of DNA have been found in the chloroplasts [86] and the synthesis of RNA has been demonstrated in cells from which the nucleus has been removed

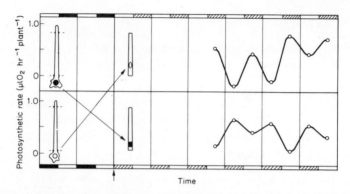

Figure 7.17. Effect on the phase of the rhythm of photosynthetic capacity in an *Acetabularia* cell in continuous dim light of the implantation of a nucleus from another cell in which the phase of the circadian rhythm differed by 12 h. For further explanation see text. The light/dark cycles to which the cultures were exposed are shown by the bars above and below the figure, and the time of nuclear implantation and transfer to uniform dim light by the arrow at the bottom of the figure. The hatched areas of the bars indicate when the cytoplasm of the cell would have experienced darkness had the original light/dark cycles been maintained. Vertical lines are 24 h apart. (From Schweiger and Schweiger [85])

[87]. The question of whether the circadian system involves nucleic acid and protein synthesis has been approached from a different point of view using inhibitory substances. Neither the phase nor the period of the rhythm of photosynthetic capacity were affected by the high concentrations of Actinomycin D, puromycin or chloramphenicol which drastically reduced RNA and protein synthesis in *Acetabularia* [88]. At first sight, results such as these point to circadian rhythmicity being independent of nucleic acid and protein synthesis, but only a small fraction of the total RNA metabolism might in fact be involved in the circadian oscillating system. Since a considerable fraction of RNA metabolism appears to be unaffected by Actinomycin D [88], the absence of an effect of this substance may be due to it having little or no inhibiting effect on a specific fraction of nucleic acid and protein metabolism which is involved in the circadian system or to only a small amount of the total general RNA synthesis being necessary for operation of the system. Experiments involving the use of inhibitors of nucleic acid and protein synthesizing mechanisms have not, therefore, provided evidence which can resolve the question of the extent to which these mechanisms are involved in the circadian oscillating system in *Acetabularia* cells.

The second approach to the question of identifying the circadian system has been to determine whether or not rhythmic variation occurs in the amount and the specific activity of various enzymes under uniform environmental conditions. In the case of the unicellular dinoflagellate *Gonyaulax,* the rhythm of luminescence appears to involve periodic changes in the levels of both the enzyme and substate involved in the phenomenon. The specific activity, and the total activity of luciferase has been shown to vary markedly during a 24-h period [89] (Fig. 7.18A). In addition, Bode, DeSa and Hastings [19] have shown that there is a daily rhythm in the activity of luciferin, the substrate for the enzyme luciferase, in cells of *Gonyaulax* (Fig. 7.18B). Similarly, early assays of ribulosediphosphate carboxylase in homogenates of *Gonyaulax* cells showed a cycle of activity, with approximately the same amplitude and phase as those of the rhythm in photosynthetic capacity in the cell suspensions from which the samples for enzyme assay were obtained [90]. Such findings as these naturally give rise to the idea that a periodic synthesis of enzyme is occurring, and that this in turn might be attributed to a periodic synthesis of specific messenger nucleic acids which code for the enzymes in question. If this idea were to be substantiated by experimental evidence, it would be difficult not to conclude that the circadian system or biological clock either involves, or exerts its controlling influence at the level of, nucleic acid metabolism. In the case of *Gonyaulax,* for example, the rhythm in luciferase and ribulose-diphosphate carboxylase activity could be attributed to the periodic synthesis of the m-RNA's coding for these enzymes. If this were the case, however, there should be a periodicity in the incorporation of P^{32} and labelled nucleotides into m-RNA fractions and of labelled amino-acids into proteins. In addition, inhibitors of nucleic acid synthesis, such as Actinomycin D, and of protein synthesis such as puromycin, cycloheximide and chloramphenicol ought to inhibit the rhythms and induce shifts in its phase. The induction of a phase shift by an inhibitor is, in fact, a most important criterion, since it clearly indicates an effect on the circadian system itself. A simple abolition of the rhythm might be attributed merely to the inhibition of the observed process which is under the control of the circadian system without there being any effect on the circadian system itself. Karakashian and Hastings [91], and Hastings [92] found that Actinomycin D abolished the rhythm of luminescence in *Gonyaulax,* but not immediately it was applied (Fig. 7.19). One peak of luminescence occurred after the treatment was begun. This finding raised the interesting possibility that, providing m-RNA synthesis in *Gonyaulax* is totally and immediately blocked by Actinomycin D, a state of affairs which does not appear to have been checked experimentally, the synthesis of the m-RNA responsible for a particular peak of luciferase activity and hence luminescence takes place some 24-h earlier. In other words, RNA synthesis occurring today is responsible for tomorrow's peak of

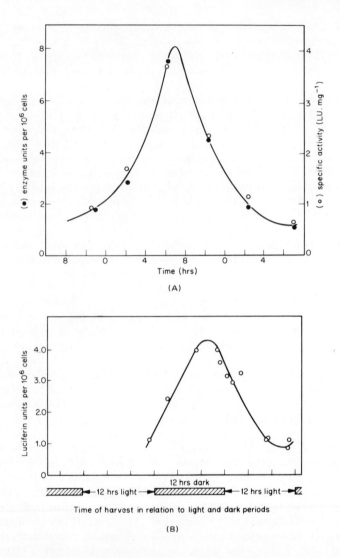

Figure 7.18A. The amounts of active luciferase extracted at different times of day from cultures of *Gonyaulax polyedra* grown at $22 \pm 2°C$ and 12 : 12 h cycles of light and darkness. The specific activity (+) and enzyme activity (●) vary in a similar way indicating that the extraction of other proteins does not change drastically at different times of day. (From Hastings and Keynan [89])

B. Variation in luciferin activity in extracts of *Gonyaulax polyedra* cells harvested at different times of day. Times of harvest in relation to the light and dark periods are shown on the ordinate. (From Bode, DeSa and Hastings [19])

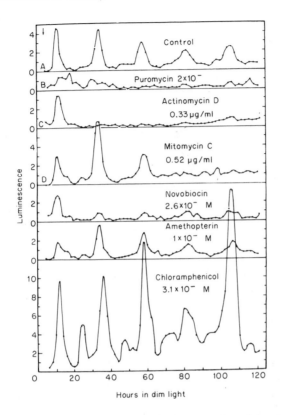

Figure 7.19. The effects of a number of inhibitors of nucleic acid and protein synthesis on the rhythm of luminescence of *Gonyaulax polyedra*. Inhibitors were applied 4 h after the onset of constant conditions as shown by the arrow. Luminescence is expressed in arbitrary units. (From Karakashian and Hastings [91])

luminescence. The action of Actinomycin D supports the idea that the peaks of luminescence are due to the rhythmic *de novo* synthesis of luciferase. Unfortunately, the effects of inhibitors of protein synthesis only partially support this idea (Fig. 7.19). Puromycin caused an almost immediate abolition of the rhythm, a finding consistent with the general scheme just outlined, but chloramphenicol had the effect of enhancing the amplitude of the rhythm which otherwise persisted normally! Substances which inhibit DNA synthesis ought to have little or no effect on the rhythm if the rhythmicity is arising at the level of RNA synthesis. Mitomycin, which inhibits DNA synthesis, has no effect on the rhythm of luminescence until about three days after application when the rhythm disappeared (Fig. 7.19). Karakashian and Hastings [91] suggest that this observation could be explained on the basis that DNA turnover occurs at a low rate and that the delayed inhibition of the rhythm is due to the lack of newly synthesized DNA to replace that broken down.

It has already been pointed out that experiments involving prolonged exposure of organisms and tissues to inhibitors are difficult to interpret. Abolition of the observed rhythm does not establish that the circadian system has been inhibited since the inhibitor might merely be inhibiting the observed process, or the link between this process and the circadian system, while the circadian system oscillates normally. The only way to assess whether an inhibitor is affecting the circadian system is to apply it for a short period of time, remove it, and then observe whether or not a significant shift in phase has been induced. There are, however, major practical difficulties with this approach in that many inhibitors are difficult if not impossible to remove adequately from the living cells if they have been applied for a period of between 4 and 12 h. Karakashian and Hastings [91] studied the effect of such pulse-type treatments of inhibitors on the rhythm of luminescence in *Gonyaulax*, but could find no indication of phase shifts being induced. Even a brief treatment with Actinomycin D inhibited the rhythm in the same way as a prolonged treatment, an observation which suggests that the inhibitor is strongly retained in the cells. This approach has not, therefore, yielded additional evidence which can help in assessing the extent to which nucleic acid metabolism is involved in the circadian system in *Gonyaulax*.

There seems to be good evidence in *Gonyaulax* that the circadian system controls both photosynthesis and luminescence, and indeed a rhythm of cell division. Over extended periods of time in constant conditions the rhythms retain precisely the same phase relationship to one another indicating that their periods are identical. The temperature dependence of all three rhythms appears to be identical. On the other hand, there is no evidence to suggest that the circadian system itself incorporates any of the chemical and other events involved in photo-synthesis, luminescence, or cell division, since there is no effect of inhibiting these processes on the phase and period of circadian system. Sweeney [13] has shown that a 3-h exposure to DCMU at a concentration which totally inhibits photosynthesis had no effect on the phase of the rhythms of photosynthesis or luminescence after the inhibitor was removed. Mechanical stimulation of *Gonyaulax* cultures to promote luminescence, a procedure which must alter the amount or state of luciferin, similarly shows no effect on the phase of the rhythms of luminescence [93], and inhibitors of cell division are also without effect on the phase and period of the rhythms of luminescence and photosynthesis. Thus, the circadian system in *Gonyaulax* appears to control the three distinct processes photosynthesis, luminescence and cell division, the component biochemistry of these processes appears not to be involved in the basic circadian system, and there is no evidence of feedback from these processes to the circadian system.

Studies of the incorporation of P^{32} and of labelled amino-acids have lent some support to the idea that both nucleic acid and protein synthesis can show circadian rhythmicity. The incorporation of P^{32} was rhythmic in seedlings of *Phaseolus multiflorus* [94] but no such rhythm could be detected in *Gonyaulax* cultures [92]. In the latter case, a rhythmic incorporation of P^{32} into a specific RNA might perhaps have been masked by the larger amounts being incorporated into a non-rhythmic fraction of RNA. A rhythm in the incorporation of labelled amino-acids into the proteins of *Euglena* under conditions where the rhythm of phototaxis persists has been reported by Feldman [94].

Several studies with the rhythm of photosynthetic capacity of *Acetabularia* and inhibitors of nucleic acid and protein synthesis have been made. Sweeney, Tuffli and Ruben [88] placed enucleated cells of *Acetabularia crenulata* in sea water in which the concentration of the inhibitor (Actinomycin D, chloramphenicol, etc.) reduced protein synthesis by 50%. No change in the period of the rhythm could be detected, though the penetration of all the substances into the cells was clearly taking place from the reduction which occurred in the overall level of photosynthesis. On the other hand, Vanden Driessche [96, 97] found the rhythms of photosynthetic capacity and of chloroplast shape in *Acetabularia mediterranea* to be inhibited by Actinomycin D at a concentration of 2.7 μg/ml^{-1} when the nucleus is present but not when it is absent. As Sweeney [13] has pointed out, the measurements were made in this investigation at only 12-h intervals and there seemed to be some doubt that the rhythmicity had been totally abolished in the intact cells. Deviation of the period from 24-h in continuous light may also be an added difficulty in interpreting the data of Vanden Driessche. Feldman [95] has shown that changes in the period of the rhythm of phototaxis in *Euglena* can be induced by the inhibitor of protein synthesis, cycloheximide. The changes are concentration dependent, but are relatively much less that the inhibition of incorporation of labelled amino-acids into protein of the cells of *Euglena*.

A rather detailed theoretical consideration of the basic circadian system has been made by Ehret and Trucco [98]. In proposing the Chronon concept, a molecular model of the biological clock, they have attempted to synthesize the diverse data on the physiology and biochemistry of the circadian rhythms into a unified scheme. They propose that the circadian rhythms in eukaryotic cells and organisms result from the recycling of a sequential process that regulates the transcription of template RNA from DNA. The eucell is regarded as an event generator (clock) whose circadian escapement consists of a sequential transcription component (Chronon) and a Chronon recycling component. The chronon is envisaged as a very long polycistronic complex of DNA, the transcription rate of which is limited by some functions of eukaryotic organization that are relatively temperature independent. Each eucell is thought to have hundreds of chronons on each of its nuclear chromosomes, and many sets of extra-nuclear chronons, free from any attachment to the euchromosomes in its different cell organelles. On a given chronon, RNA transcription proceeds unidirectionally from an initiator cistron (C_i) to the terminator cistron (C_t). Once the products of translation of the message of C_t have been formed, the point of no return is passed and the escapement mechanism operates on its chronon recycling component. This component includes, it is proposed, the post-transcriptional events of translation, end-product formation and polymer assembly, and the pre-transcriptional events that cause an initiator substance to accumulate. When the initiator substance arrives at its target cistron (C_i) the system proceeds to recycle into its next circadian phase.

This scheme clearly places the biological clock at the level of RNA transcription and if this is the case the inhibitors of RNA synthesis such as Actinomycin D should clearly induce phase shifts. In addition, rhythmic behaviour in all organisms should be dependent upon the periodic synthesis of enzymes. This seems certainly to be the case in the rhythm of luminescence of *Gonyaulax*. Recent studies by Sweeney [13] have shown that in the case of the rhythm of photosynthesis in *Gonyaulax*, only one of the several partial processes making up photosynthesis as a whole is rhythmic. While the capacity for photosynthesis is changing rhythmically, that for carrying out the Hill reaction is not. The pigment level is uniform throughout a cycle, and so is the sensitivity of photosynthesis to inhibitors of photosynthetic phosphorylation such as DCMU and n-chlorophenylhydrazone. On the other hand, as mentioned earlier, the ribulose-diphosphate carboxylase activity in crude cell extracts is very different at different phases of the rhythm, and the maximum and minimum of activity coincide with the maximum and minimum of the photosynthetic capacity rhythm in the intact cells. This variation is only observed, however, when the concentration of bicarbonate in the assay mixture is low. At high bicarbonate concentrations, the activity of the enzyme extracted when the cells show a minimum in photosynthetic capacity is as high as that in extracts of cells made when the photosynthesizing capacity is high. Furthermore, in the intact cells, increasing the bicarbonate level in the culture medium markedly raises the level of photosynthetic capacity in the troughs of the rhythm. These findings strongly suggest that the rhythm in photosynthetic capacity is not being caused, as had at first been supposed, by the periodic synthesis and destruction of enzymes, and that the rhythmic synthesis of protein is not a critical part of either the basic circadian system or the mechanism linking the circadian system to the photo-synthetic reactions. Changes which alter the activity rather than the amount of enzyme are indicated.

Although this conclusion stands in contrast, at present, to that drawn from the studies of luciferin and luciferase in *Gonyaulax* [19, 89], they are in very close agreement with conclusions drawn much earlier by Warren [99]. The rhythm of carbon dioxide metabolism in excised leaves of *Bryophyllum* has been shown to be due to the periodic activity of the mechanism responsible for the fixation of carbon dioxide in the dark. The variation in the activity of this system *in vivo* is almost absolute (Fig. 7.20), being virtually zero for several hours, then rising to a high value, then declining to zero again for several hours, in continuing darkness [35]. The sequential switching on, and switching off, of this

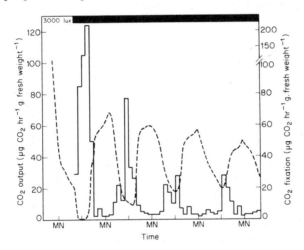

Figure 7.20. The rhythms of carbon dioxide emission (smooth curve) and of carbon dioxide ($^{14}CO_2$) fixation (histogram) in leaves of *Bryophyllum fedtschenkoi* at 26°C after being transferred from light at 3000 lux to continuous darkness at 08.00 h as shown by the beginning of the black bar above the figure. (From Warren and Wilkins [35])

dark carbon dioxide fixation system is particularly valuable for a biochemical investigation. Several enzyme systems have the capacity to bring about carbon dioxide fixation in darkness in these leaves, but the enzyme most likely to be principally involved is phosphoenolpyruvic carboxylase. Although the studies made by Warren [99] must be regarded as being of a preliminary nature, they have attempted to ascertain the mechanism underlying the rhythm of carbon dioxide fixation rhythm in the leaves. Extracts of leaves at different times in the cycle showed that there were large amounts of phosphoenolpyruvic carboxylase present, even when carbon dioxide fixation by the leaves was zero. Thus, in contrast to the findings for luciferase in *Gonyaulax,* but in agreement with the recent findings for ribulose-diphosphate carboxylase in that organism, the data for *Bryophyllum* suggest that periodic synthesis and destruction of enzyme is not involved in the origin or expression of the circadian rhythm. Some other mechanism must clearly be involved—cycles in the availability of substrates for the carboxylation reaction, or in the concentration of an inhibitor of the reaction are the two most obvious possibilities. When carbon dioxide concentration is non-limiting, infiltration of leaves with phosphenol-pyruvic acid at times in the cycle when carbon dioxide fixation was zero caused no increase in fixation. It seems unlikely, therefore, that substrate concentration is limiting when fixation by the leaves is extremely low. Possible periodic variation in the availability of co-enzymes has not yet been investigated. However, the rate of oxygen uptake by *Bryophyllum* leaves in darkness shows no rhythmic variation. The periodic appearance of an inhibitor of phosphoenolpyruvic carboxylase in the tissues has been examined. At least two inhibiting substances are present in the leaves, but they are present in substantial amounts at all phases of the cycle of carbon dioxide fixation. One of these substances appears to be citric acid.

Interpretation of these results is rather difficult at this stage of the investigation but a tentative conclusion is that at least one of the inhibitors arises as an end product of the carbon dioxide fixation reaction. Further fixation might then be prevented until the concentration of inhibitor at the site of fixation is reduced by removal of the inhibitor to a new intracellular location [99].

It must again be stressed that our knowledge of the biochemistry of the *Bryophyllum* rhythm is primitive, and the application of more sophisticated techniques will be necessary to make appreciable further progress. Nevertheless, it is of interest that neither the data available from *Bryophyllum,* nor those for the rhythm of photosynthesis in *Gonyaulax* make it necessary to postulate the involvement of nucleic acid or protein synthesis in the basic circadian system. On the other hand, the luminescence rhythm is consistent with involvement of these synthetic processes, though as Sweeney [13] has suggested, a speculative analysis of the data could lead to other conclusions. At all events, the Chronon concept of Ehret and Trucco [98], while an interesting proposition, lacks really substantial supporting evidence, and in addition it does not really deal with the problem of temperature independence. Notional models based on end-product inhibition and the subsequent removal of inhibitor to isolated cell compartments can be envisaged as the basis of the circadian system, if the correct assumptions are made about reaction rates, but these too suffer from the oversimplification of not really incorporating a mechanism to account for temperature independence. Further knowledge of the mechanism of the basic circadian system or biological clock will only stem from a more detailed biochemical approach than has been generally attempted hitherto. The unicellular organisms and plants, and especially plant tissue cultures, provide excellent material for such studies and the application of sophisticated techniques of biochemistry, biophysics and molecular biology will undoubtedly lead to real progress in the future.

REFERENCES

1. De Mairan, Observation botanique. *Histoire de l'Academie Royale des Sciences, Paris*, 35 (1729).
2. Duhamel de Monceau, *La Physique des Arbres*, 2, 158-159 (1758).
3. J. G. Zinn, Von dem Schafe der Pflanzen. *Hamburgischen Magazin*, 22, 40-50 (1759).
4. W. W. Garner and H. A. Allard, Effect of the relative length of day and night and other factors of the environment on growth and reproduction in plants. *J. Agr. Res.*, 18, 553-606 (1920).
5. W. W. Garner and H. A. Allard, Further studies in photoperiodism, the response of plants to relative length of day and night. *J. Agr. Res.*, 23, 871-920 (1923).
6. E. Bünning, *The Physiological Clock* (Longmans, Springer-Verlag, New York, 1967).
7. G. Kramer, Die Sonnenorientierung der Vögel *Verh. dtsch. zool. Geo.*, 72-84 (1952).
8. K. von Frisch, Die Sonne als Kompass im Leben der Bienen. *Experientia*, 6, 210-221 (1950).
9. E. R. Uebelmesser, Über den endonomen Tagesrhythmus der Sporangientrager-bildung von *Pilobolus*. *Archiv für Mikrobiol.*, 20, 1-33 (1954).
10. J. W. Hastings and B. M. Sweeney, The *Gonyaulax* Clock, in: *Photoperiodism and Related Phenomena in Plants and Animals*, Withrow (ed.). American Assoc. Adv. Sci. (1959).
11. M. B. Wilkins, The effect of light upon plant rhythms. *Cold Spring Harb. Symp. quant. Bid.*, 25, 115-129 (1960).
12. M. B. Wilkins, The influence of temperature and temperature changes on biological clocks, *Circadian Clocks*, pp. 146-166, Aschoff (ed.), Amsterdam: North-Holland Publishing Co. (1965).
13. B. M. Sweeney, *Rhythmic Phenomena in Plants*. Academic Press (1969).
14. J. W. Hastings, Unicellular clocks. *Ann. Rev. Microbiol.*, 13, 297-312 (1959).
15. B. M. Sweeney, Biological clocks in plants. *Ann. Rev. Plant. Physiol.*, 14, 411-440 (1963).
16. P. L. Altman and D. S. Dittmer, Environmental Biology. *Fed. Am. Soc. Exptl. Biol.*, 565-608 (1966).
17. A. Chovnick (ed.), 'Biological clocks'. *Cold Spring Harb. Symp. quant. Biol.*, 25 (1960).

18. J. Aschoff (ed.), *Circadian Clocks*. (Amsterdam: North-Holland Publishing Co., 1965).
19. V. C. Bode, R. DeSa and J. W. Hastings, Daily rhythm of luciferin activity in *Gonyaulax polyedra*. *Science*, 141, 913-915 (1963).
20. V. G. Bruce and C. S. Pittendrigh, Temperature independence in a unicellular 'clock'. *Proc. natn. Acad. Sci., U.S.A.*, 42, 676-682 (1956).
21. A. Pirson, H. Schön and H. Döring, Wachstums- und Stoffwechselperiodik bei *Hydrodictyon*. *Z. Naturforsch*, 9b, 350-353 (1954).
22. V. G. Bruce, The biological clock in *Chlamydomonas reinhardi. J. Protozool.*, 17, 328-334 (1970).
23. F. Bühnemann, Das endodiurnale System der Oedogoniumzelle. II. *Biol. Zent.*, 74, 691-705 (1955).
24. B. M. Sweeney and F. T. Haxo, Persistence of a photosynthetic rhythm in enucleated *Acetabularia. Science*, 134, 1361-1363 (1961).
25. T. vanden Driessche and S. Bonotto, The circadian rhythm in RNA synthesis in *Acetabularia mediterranea. Biochim. Biophys. Acta.*, 179, 58-66 (1969).
26. T. vanden Driessche, Circadian variation in A.T.P. content in the chloroplasts of *Acetabularia mediterranea. Biochim. Biophys. Acta*, 205, 526-528 (1970).
27. C. T. Ingold and V. J. Cox, Periodicity of spore discharge in *Daldinia. Ann. Bot.*, 19, 201-209 (1955).
28. C. S. Pittendrigh, V. G. Bruce, N. S. Rosenweig and M. L. Rubin, Growth patterns in *Neurospora. Nature*, 184, 169-171 (1959).
29. M. L. Sargent, W. R. Briggs and D. O. Woodward, Circadian nature of a rhythm expressed by an invertaseless strain of *Neurospora crassa. Plant Physiol.*, 41, 1343-1349 (1966).
30. M. L. Sargent and W. R. Briggs, The effects of light upon a circadian rhythm of conidiation in *Neurospora. Plant Physiol.*, 42, 1504-1510 (1967).
31. G. Busch, Über die photoperiodische Formänderung der Cloroplasten von *Selaginella serpens. Biol. Zbl.*, 72, 598-629 (1953).
32. E. Bünning, Cellular clocks. *Nature, Lond.*, 181, 1169-1171 (1958).
33. R. Bünsow, Endogene Tagesrhythmik und Photoperiodismus bei *Kalanchoë blossfeldiana. Planta*, 42, 220-252 (1953).
34. N. G. Ball and I. J. Dyke, An endogenous 24-hour rhythm in the growth rate of the *Avena* coleoptile. *J. exp. Bot.*, 5, 421-433 (1954).
35. D. M. Warren and M. B. Wilkins, An endogenous rhythm in the rate of dark-fixation of carbon dioxide in leaves of *Bryophyllum fedtschenkoi. Nature, Lond.*, 191, 686-688 (1961).
36. M. B. Jones and T. A. Mansfield, A circadian rhythm in the level of carbon dioxide compensation point in *Bryophyllum* and *Coffea. J. exp. Bot.*, 21, 159-163 (1970).
37. W. S. Hillman, Carbon dioxide output as an index of circadian timing in *Lemna* photoperiodism. *Plant Physiol.*, 45, 273-279 (1970).
38. R. Halaban, The circadian rhythm of leaf movement of *Coleus blumei* × *C. frederici*, a short day plant. I. Under constant light conditions. *Plant. Physiol.*, 43, 1883-1886 (1968).
39. R. Halaban, The circadian rhythm of leaf movement in *Coleus blumei* × *C. frederici*, a short day plant. II. The effects of light and temperature signals. *Plant. Physiol.*, 43, 1887-1893 (1968).
40. R. Halaban, The flowering response of *Coleus* in relation to photoperiod and the circadian rhythm of leaf movement. *Plant Physiol.*, 43, 1894-1898 (1968).
41. R. Halaban, Effect of light quality on the circadian rhythm of leaf movement of a short day plant. *Plant Physiol.*, 44, 973-977 (1969).

42. F. A. Brown, A unified theory for biological rhythms: rhythmic duplicity and the genesis of "circa" periodism. *Circadian Clocks,* J. Aschoff (ed.), Amsterdam: North-Holland Publishing Co. (1965).
43. C. S. Pittendrigh, On temperature independence in the clock system controlling emergence time in *Drosophila. Proc. natn. Acad. Sci. U.S.A.,* 40, 1018-1029 (1954).
44. F. A. Brown, Response to pervasive geophysical factors and the biological clock problem. *Cold Spring Harbor Symp. quant. Biol.,* 25, 57-72 (1960).
45. V. G. Bruce, Environmental entrainment of circadian rhythms. *Cold Spring Harb. Symp. quant. Biol.,* 25, 29-48 (1960).
46. A. Kleinhoonte, Untersuchungen über die autonomen Bewegungen der Primär-blatter von *Canavalia ensiformis. Jb. wiss. Bot.,* 75, 697-725 (1932).
47. M. B. Wilkins, An endogenous rhythm in the rate of carbon dioxide output of *Bryophyllum.* IV. Effect of intensity of illumination on entrainment of the rhythm by cycles of light and darkness. *Plant Physiol.,* 37, 735-741 (1962).
48. E. Bünning, Endogenous diurnal cycles of activity in plants, in: *Rhythmic and Synthetic Processes in Growth,* Rudnick (ed.). Princeton Univ. Press (1957).
49. O. Oltmanns, Über den Einfluss der Temperatur auf die endogene Tagesrhythmik und die Blühinduktion bei der Kurztagpflanze *Kalanchoë blossfeldiana. Planta,* 54, 233-264 (1960).
50. J. B. Biale, Periodicity in transpiration of lemon cuttings under constant environmental conditions. *Proc. Am. Soc. Hort. Sci.,* 38, 70-74 (1940).
51. A. Flügel, Die Gesetzmassigkeiten der endogenen Tagesrhythmik. *Planta,* 37, 337-375 (1949).
52. L. Lörcher, Die Wirkung verschiedener Lichtqualitäten auf die endogene Tagesrhythmik von *Phaseolus. Z. Bot.,* 46, 209-241 (1958).
53. L. Wassermann, Die Auslösung endogen-tagesperiodischer Vorgänge bei Pflanzen durch einmalige Reize. *Planta,* 53, 647-669 (1959).
54. B. M. Sweeney and J. W. Hastings, Rhythms, in: *The Physiology and Biochemistry of Algae.* Academic Press Inc. (1962).
55. M. B. Wilkins, An endogenous rhythm in the rate of carbon dioxide output of *Bryophyllum.* I. Some preliminary experiments. *J. exp. Bot.,* 10, 377-390 (1959).
56. M. B. Wilkins, An endogenous rhythm in the rate of carbon dioxide output of *Bryophyllum.* II. The effects of light and darkness on the phase and period of the rhythm. *J. exp. Bot.,* 11, 269-288 (1960).
57. M. B. Wilkins, An endogenous rhythm in the rate of carbon dioxide output of *Bryophyllum.* III. The effects of temperature changes on the phase and period of the rhythm. *Proc. Roy. Soc. Lond. B.,* 156, 220-241 (1962).
58. N. G. Ball and I. J. Dyke, The effects of decapitation, lack of O_2, and low temperature on the endogenous 24-hour rhythm in the growth rate of the *Avena* coleoptile. *J. exp. Bot.,* 8, 323-338 (1957).
59. E. Bünning and M. Tazawa, Über der Temperatureinfluss auf dei endogene Tagesrhythmik bei *Phaseolus. Planta,* 50, 107-121 (1957).
60. M. B. Wilkins and D. M. Warren, The influence of low partial pressures of oxygen on the rhythm in the growth rate of the *Avena* coleoptile. *Planta,* 60, 261-273 (1963).
61. E. Bünning, S. Kurras and V. Vielhaben, Phasenverschiebungen der endogenen Tagesrhythmik durch reduktion der Atmung. *Planta (Berlin),* 64, 291-300 (1965).
62. M. B. Wilkins, An endogenous rhythm in the rate of carbon dioxide output of *Bryophyllum.* V. The dependence of rhythmicity upon aerobic metabolism. *Planta (Berlin),* 72, 66-77 (1967).

278   MALCOLM B. WILKINS

63. I. Moser, Phasenverschiebungen der endogenen Tagesrhythmik bei *Phaseolus* durch Temperatur- und Lichtintensitätsänderungen. *Planta (Berlin)*, **58**, 199-219 (1962).
64. R. Zimmer, Phasenverschiebung und andere Storlichwirkingen auf die endogen tagesperiodischen Blutenblättbewegungen von *Kalanchoë blossfeldiana*. *Planta*, **55**, 283-300 (1962).
65. E. Bünning and I. Moser, Response-kurven bei der circadianen Rhythmik von *Phaseolus*. *Planta*, **69**, 101-110 (1966).
66. R. Wagner, Der Einfluss niedriger temperatur auf die Phasenlage der endogentagesperiodischen Blättbewegungen von *Phaseolus multiflora*. *Z. Bot.*, **51**, 179-204 (1963).
67. E. Bünning, Über die Erblichkeit der Tagesperiodizität bei den *Phaseolus*-Blättern. *Jb. wiss. Bot.*, **77**, 283-320 (1932).
68. J. Aschoff, Tierische Periodik unter dem Einfluss von Zeitgebern. *Z. Tierpsychol.*, **15**, 1-30 (1958).
69. F. J. Leinweber, Über die Temperaturabhängigkeit der Periodenlänge bei der endogenen Tagesrhythmik von *Phaseolus*. *Z. Bot.*, **44**, 337-364 (1956).
70. J. W. Hastings and B. M. Sweeney, On the mechanism of temperature independence in a biological clock. *Proc. natn. Acad. Sci. U.S.A.*, **43**, 804-811 (1957).
71. L. Overland, Endogenous rhythm in opening and odor of flowers of *Cestrum nocturnum*. *Am. J. Bot.*, **47**, 378-382 (1960).
72. E. Bünning, Zur Kenntnis der erblichen Tagesperiodizität bei den Primär-blättern von *Phaseolus multiflorus*. *Jb. wiss. Bot.*, **81**, 411-418 (1935).
73. E. Bünning, Versuche zur Beeinflussung der endogenen Tagesrhythmik durch chemische Faktoren. *Z. Bot.*, **44**, 515-529 (1956).
74. E. Bünning, Über die Urethanvergiftung der endogenen Tagesrhythmik. *Planta*, **48**, 453-458 (1957).
75. S. Keller, Über die Wirkung chemischer Faktoren auf die tagesperiodischen Blattbewegungen von *Phaseolus multiflorus*. *Z. Bot.*, **48**, 32-57 (1960).
76. J. W. Hastings and B. M. Sweeney, The action spectrum for shifting the phase of the rhythm of luminescence in *Gonyaulax polyedra*. *J. gen. Physiol.*, **43**, 697-706 (1960).
77. B. M. Sweeney, Resetting the biological clock in *Gonyaulax* with ultra-violet light. *Plant Physiol.*, **38**, 704-708 (1963).
78. B. M. Sweeney, The photosynthetic rhythm in single cells of *Gonyaulax polyedra*. *Cold Spring Harb. Symp. quant. Biol.*, **25**, 145-148 (1960).
79. M. B. Wilkins and A. W. Holowinsky, The occurrence of an endogenous circadian rhythm in a plant tissue culture. *Plant Physiol.*, **40**, 907-909 (1965).
80. L. Brauner and N. Arslan, Experiments on the auxin reactions of the pulvinus of *Phaseolus multiflorus*. *Rev. Fac. Sci. Univ. Istanbul.*, **16b**, 257 (1951).
81. R. J. Ellis, Chloroplast ribosomes: stereospecificity of inhibition by chloramphenicol. *Science*, **158**, 477-478 (1969).
82. I. R. MacDonald and R. J. Ellis, Does cycloheximide inhibit protein synthesis specifically in plant tissues. *Nature, Lond.*, **222**, 791-792 (1969).
83. R. J. Ellis and I. R. MacDonald, Specificity of cycloheximide in higher plant systems. *Plant Physiol.*, **46**, 227-232 (1970).
84. B. M. Sweeney and F. T. Haxo, Persistence of a photosynthetic rhythm in enucleated *Acetabularia*. *Science*, **134**, 1361-1363 (1961).
85. H. G. Schweiger and E. Schweiger, The role of the nucleus in a cytoplasmic diurnal rhythm, in: *Circadian Clocks*, pp. 195-197, J. Aschoff (ed.). Amsterdam: North-Holland Publishing Co. (1965).

86. A. Gibor and M. Izawa, The DNA content of the chloroplasts of *Acetabularia*. *Proc. natn. Acad. Sci. U.S.A.*, 50, 1164 (1963).
87. H. G. Schweiger and S. Berger, DNA-dependent RNA synthesis in chloroplasts of *Acetabularia*. *Biochim. Biophys. Acta*, 87, 533 (1964).
88. B. M. Sweeney, C. F. Tuffli and R. H. Rubin, The circadian rhythm of photosynthesis in *Acetabularia* in the presence of Actinomycin D, Puromycin, and Chloramphenicol. *J. gen. Physiol.*, 50, 647-659 (1967).
89. J. W. Hastings and A. Keynan, Molecular aspects of circadian systems, in: *Circadian Clocks*, pp. 167-182, J. Aschoff (ed.). Amsterdam: North-Holland Publishing Co. (1965).
90. B. M. Sweeney, Rhythmicity in the biochemistry of photosynthesis in *Gonyaulax*, in: *Circadian Clocks*, pp. 190-194, J. Aschoff (ed.). Amsterdam: North-Holland Publishing Co. (1965).
91. M. W. Karakashian and J. W. Hastings, The effects of inhibitors of macro-molecular biosynthesis upon the persistent rhythm of luminescence in *Gonyaulax*. *J. gen. Physiol.*, 47, 1-12 (1963).
92. J. W. Hastings, Biochemical aspects of rhythms: Phase shifting by chemicals. *Cold Spring Harb. Symp. quant. Biol.*, 25, 131-143 (1960).
93. J. W. Hastings and B. M. Sweeney, A persistent diurnal rhythm of luminescence in *Gonyaulax polyedra*. *Biol. Bull.*, 115, 440-458 (1958).
94. A. Ruckebeil, Untersuchungen zur Tagesperiodizität der [32] P Einlagerung in die Verbindungen der Nukleinsäurefraktionen. *Z. Bot.*, 49, 1-22 (1961).
95. J. F. Feldman, Circadian rhythmicity in amino acid incorporation in *Euglena gracilis*. *Science*, 160, 1454-1456 (1968).
96. T. vanden Driessche, The role of the nucleus in the circadian rhythms of *Acetabularia mediterranea*. *Biochim. Biophys. Acta*, 126, 456-470 (1966).
97. T. vanden Driessche, Circadian rhythms in *Acetabularia*: Photosynthetic capacity and chloroplast shape. *Exptl. Cell. Res.*, 42, 18-30 (1966).
98. C. F. Ehret and E. Trucco, Molecular models for the circadian clock. I. The chronon concept. *J. Theoret. Biol.*, 15, 240-262 (1967).
99. D. M. Warren, Endogenous rhythms in the carbon dioxide fixation of plant tissues. Ph.D. Thesis, University of London, U.K. (1964).
100. M. B. Wilkins, Circadian rhythms in plants, in: *Physiology of Plant Growth and Development*, pp. 647-671, M. B. Wilkins (ed.). London: McGraw-Hill Publishing Co. Ltd. (1969).

CHAPTER 8

Biological Clocks and Bird Migration

G. V. T. Matthews

The Wildfowl Trust, Slimbridge,
Gloucester, England

8.1 INTRODUCTION

Although invasions of such birds as the Crossbill *Loxia curvirostra* occur at intervals of several years with a regularity which is in itself an intriguing problem of timing, most bird migrations have a more-or-less annual, i.e. circennial rhythm.

On arrival at the breeding grounds the birds set up territories, mate, lay eggs, incubate and display parental behaviour. These activities are controlled by a delicate interplay of internal hormonal and external factors which have been admirably summarized [31, 39] and will not be considered further here. Breeding completed and the gonads regressed, the birds undergo a plumage moult, lay on deposits of sub-cutaneous fat and migrate to their winter quarters. After a sojourn there, fat deposits are again developed and the birds migrate back to the breeding grounds. There are a number of variants of this simple cycle of events. Basically, however, the reproductive and migratory cycles are closely inter-twined. Experimental work has concentrated more on the former and we often have to assume that similar factors are controlling them both, in the absence of further dissection.

8.2 CIRCENNIAL CLOCKS

As in the case of circadian clocks, there has been prolonged controversy as to whether there are endogenous circennial clocks governing the breeding migratory cycles or whether the control is wholly by environmental factors. Neither polar position is tenable. Certainly there are cases of regular breeding at intervals widely different from the annual. The Wideawake Tern *Sterna fuscata* breeds every 9.6 months, Audubon's Shearwater *Puffinus herminieri* every nine months, the Brown Booby *Sula leucosaster* every eight [10, 72, 13]. These, be it noted, are tropical seabirds breeding in environments where there are presumably no selective pressures favouring particular months in terms of climate or food supply. The wonder is that such populations do not show continuous breeding instead of a marked synchrony. Social stimulation has been invoked as the most likely "proximate" factor controlling the actual laying date.

Again, as with circadian clocks, the closer a rhythm approaches an astronomical interval based on the earth's rotation (24 h) or its completion of an orbit round the sun (12 months), the greater is the tendency to suspect some external Zeitgeber, Yet if accurate clocks exist with periods widely bracketing the astronomical interval, why should there not be some with intervals of exactly 24 h or 12 months? One would, indeed, expect strong selective pressure for such clocks.

One of the most remarkable examples of a rigidly annual cycle is that of the Slender-billed Shearwater *Puffinus tenuirostris*. This breeds on islands off the south-east of Australia and then undertakes vast migratory journeys up the Asiatic coast to the Aleutians, down the American coast to California then looping back across the Pacific to Australia. The single egg is laid within a period of 13 days, 20 November to 3 December, 85% of the birds laying within 3 days of the peak on 25 or 26 November [71]. Moreover, these peak dates have remained almost unchanged for over a century to judge from the records of the trading concerns that harvest the subsequent fatty young. Individual birds show an even more fantastic precision, one laying on 24 November in four successive years. While social stimulation could have played a part in telescoping the egg-laying period, this was not the whole story. When birds were taken into captivity [44] and held for a year in an entirely artificial environment with a variety of lighting schedules, the birds achieved spermatogenesis, and oocytes developed, at the same time as in the freely migrating population. Since the birds were not captured after breeding and at the southern autumn equinox, we cannot exclude the possibility that they may have been exposed to some external Zeitgeber before they were incarcerated. As we shall see, physiological events are often influenced by environmental changes some months earlier. Nevertheless these shearwaters obviously have an endogenous circennial rhythm, possibly a truly annual one, whether or not it has to be aligned to astronomical events once a year.

To clinch the evidence for the existence of endogenous circennial clocks it would be necessary to maintain birds in constant conditions for several years, monitoring their breeding condition and concomitant activities such as weight changes and migratory restlessness (the additional activity shown by caged migrants at the appropriate seasons). The method is of course fraught with logistic difficulties and many of the species most used for studies of reproductive physiology are short-lived. In Mallards *Anas platyrhynchos* held in constant darkness for three years and in constant light for two [5, 6] the reproductive cycle indeed continued to wax and wane but with increasing period. Recently Willow Warblers *Phylloscopus trochilus* and Chiff Chaffs *P. collybita* have been held in constant light conditions for 28 months [25]. The former, breeding in N. Europe and migrating trans-equatorially to S. Africa and so experiencing two springs, two autumns each year, continued their cycles of migration restlessness, fat deposition and gonad recrudescence. The latter, breeding sympatrically but wintering north of the Equator, rapidly became aperiodic in activity and physiology (Fig. 8.1).

The cycle of events in the Willow Warbler, which one would expect to be the more subject to a rigid endogenous control to prevent misplaced breeding, slid forward out of phase with the external seasons. The

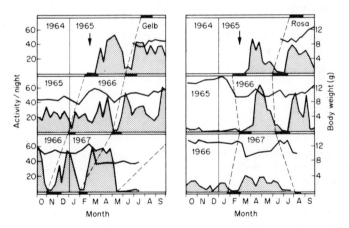

Figure 8.1. Circennial periodicity in migratory activity and associated phenomena in two Willow Warblers *Phylloscopus trochilus* maintained in constant light for 28 months. Body-weight (continuous line) migration restlessness (hatched) and moult periods (black bar and broken line). Gelb had a period of 9.0 months, Rosa of 11.8 months. (From Gwinner [25])

indications were that the circennial clock has a phase length of about ten months. It is not clear why an endogenous periodicity so widely different from the annual should have been selected. In nature the bird is not subject to years of differing length. By contrast it is subject to widely varying photoperiods both in the course of the seasons and as it migrates north/south. If it moves east/west it will encounter rapid changes in the time of dawn and dusk. It is an obvious advantage then to entrain its activity cycle to the new photoperiod. A clock whose period is different from 24 h, and thus needs to be shifted every day, might be more adaptable to such rapidly changing circumstances. It might also more easily detect and measure the changes. Thus a bird with a clock cycling at 23 h 20 min anticipates dawn by 40 min each day and has to adjust to that amount. If dawn comes 5 min earlier it would presumably be easier to detect a difference of 5 in 40 rather than 5 in 1440—which would be required if it had a clock cycling at precisely at 24 h and which normally did not want readjusting.

8.3 CONTROL OF SPRING MIGRATION AND REPRODUCTION BY PHOTOPERIOD

Since the pioneer work of Rowan [61] changes in photoperiod have been recognised as one of the main factors governing the reproductive cycle and associated activities including migration. He exposed Juncos *Junco hyemalis* to increasing photoperiods in winter and found that these induced gonad development and vernal migration. However, many birds breeding in north temperate zones winter near the Equator where the photoperiod changes only minutely, or south of the Equator where photoperiods are decreasing after 21 December. A great deal of experimental work has since been carried out that shows that a "long" photoperiod as such is the stimulus, not whether it is increasing or decreasing. Individuals suddenly subjected to days with LD 16 : 8 responded much more rapidly than birds exposed to gradual increases over a longer period. Up to LD 20 : 4 the longer the photoperiod the more stimulating it appears to be. The lower limit appears to be LD 12 : 12 which explains why equatorial conditions could provide sufficient stimulation. Indeed there are clear indications that even without long photoperiod stimulation the gonads will recrudesce to breeding condition, albeit very much more slowly.

A most interesting observation was that short photoperiods could apparently be summated. The 12 h of light required in 24 could be split up into shorter periods interrupted by darkness. Thus a 24 h schedule 5L-1D-5L-1D-5L-1D-5L-1D was just as effective as LD 20 : 4. Furthermore, a schedule LD 8 : 16 became effective in stimulating gonad development if the dark period was interrupted by one hour of light, even though the total light was only 9 h. It became apparent that interruptions at some parts of the dark period were more important than at others (Fig. 8.2).

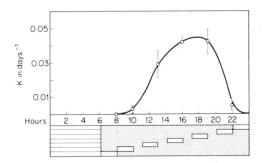

Figure 8.2. The increase in testis growth rate in White-crowned Sparrows *Zonotrichia leucophrys* subjected to cycles of LD 6:18 with 2 h light interruptions at different parts of the dark period. (From Farner [20])

The situation was probed further by setting up ahemeral light-dark cycles, i.e. combinations that did not fall into 24 h periods, such as LD 6 : 18, already known to be ineffective. The combinations LD 6 : 42 and LD 6 : 66 were also ineffective. But LD 6 : 30 and LD 6 : 54 were highly stimulating to gonad development. In these combinations the second light period fell in the middle of a subsequent 24 h period (timed from the beginning of the initial light period); in the ineffective combinations at the beginning of a subsequent 24 h period.

All this was strongly suggestive of a difference in "sensitivity" to light signals at different parts of a circadian cycle. Results of this nature have now been reported with sedentary birds such as House Finches *Carpodacus mexicanus* [29] and House Sparrow *Passer domesticus* [50]; with migratory White-crowned Sparrows [20] and Juncos [78]; and with the trans-equatorial migrant Bobolinks *Dolichonyx oryzivorus* [78].

In an elegant series of experiments [78] the activity cycles of the birds were monitored as well as their subsequent gonad development (Fig. 8.3). The activity, as would be expected, entrained to the light periods provided. When these were brief the activity continued with little change into the subsequent darkness. This indicated a circadian rhythm of locomotor and, probably neural activity. A second light signal falling within this period of activity increased the gonadal response and thus was apparently treated as an extension of the previous light signal and interpreted as a long photoperiod. If the second signal fell outside the

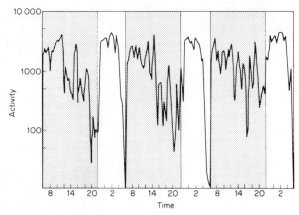

Figure 8.3. Activity patterns in White-throated Sparrows *Zonotrichia albicollis* in days 19-21 held on LD 16:8 cycles. The activity during the night is *Zugunruhe* and is correlated with fat deposition. Stippled areas represent light, white areas darkness. (From Wolfson [78])

activity period, about 15 h or more after the previous light signal, it was interpreted as the beginning of a new, separate photoperiod.

In further experiments the dark periods were extended quite considerably, with schedules of LD 16 : 56 and LD 16 : 104 giving one effective photoperiod every three days and five days. Rather surprisingly these were just as effective in stimulating gonad development as a LD 16 : 8 schedule. The birds' activity was maintained with little change in pattern during the days without light (Fig. 8.4). This suggests that an initial, stimulating light-dark experience can be stored for at least 4 days. Once the photoperiod had been measured and found to be long, the daily response to the stimulation continued to occur even in darkness.

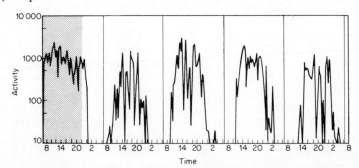

Figure 8.4. Activity patterns in White-throated Sparrow *Zonotrichia albicollis* in days 26-30 held on LD 16:104 cycles. The daylight activity pattern on day 1 is maintained through the subsequent 4 days of darkness. Stippled areas represent light, white areas darkness. (From Wolfson [78])

Indeed after an initial course of eight LD 16 : 8 days, birds maintained for 18 days in complete darkness still reached full gonad development. There were also indications that the more activity each day the greater was the degree of gonad response.

Possibly the duration of activity is only a reflection of time-measuring processes going on in the brain. It might, however, provide information needed to make decisions about the length of photoperiod. This would resurrect memories of Rowan's [62] hypothesis that the duration of activity-wakefulness controlled gonad response rather than the duration of light. This aroused much controversy but was thought to have been disposed of by experiments [23] in which night activity was stimulated by raising the ambient temperature (rather than, as earlier, by crude mechanical methods). This was ineffective in producing gonad response in the absence of a long photoperiod. It still seems to be the case that an effective photoperiod must be available to the bird whether or not activity-wakefulness plays a part in the decision-making process.

It would have seemed "obvious" that the light information would be received by the eyes and transmitted by the optic nerves to the decision-making centre in the brain. Yet pre-war experiments with Mallards [4] had suggested that light "piped" direct to the brain of birds deprived of the optic apparatus was just as effective. Most experimenters are properly reluctant to blind birds to satisfy their curiosity, so little more work has been done in this respect. Recently [51] blinded House Sparrows have been shown to be capable of entraining their activity cycles on to light signals.

In any event a decision is reached somewhere in the brain that a given circadian cycle contains a long or a short photoperiod. It may be that the pineal organ plays some part in the system, since its removal [24] destroys those circadian activity rhythms that are maintained in darkness by intact birds. Yet the operated birds can be entrained again to light signals.

The decision "long photoperiod" sets going a chain of events in the neuroendocrine system [22]. The median eminence of the hypothalmus is known to secrete neurohormones which travel via the blood in the hypophysial portal vessels to the anterior pituitary gland. This in turn produces gonadotropic hormones. One group, luteinising/interstitial cell stimulating hormones (LH/ICSH), activates the cells in the gonads producing oestrogen or testosterone which influence the appearance of secondary sex characters and associated behaviour, including migration. The other, follicle stimulating/tubule-ripening hormones (FSH/TRH), activates ovarian follicles or stimulates spermatogenesis. Recently it has been demonstrated [53, 54] that these two groups of hormones are released by different photoperiods. Greenfinches *Chloris chloris* and House Sparrows were held on cycles of LD 6 : 18 (non-stimulating) and then given one extra hour of light each day as an interruption of the dark phase. By direct assay of LH concentrations and by inferential estimation of FSH concentrations it was apparent that LH was produced at a maximum when the interruption occurred early in the dark phase, FSH when it occurred later (Fig. 8.5). The one-hour light pulses were thus impinging on different phases of a circadian oscillator (or possibly on two distinct oscillators).

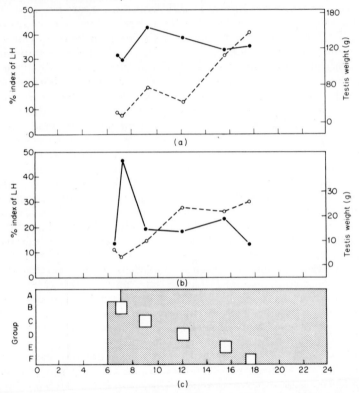

Figure 8.5. Variations in LH activity (solid lines) and FSH activity (broken lines) in House Sparrows *Passer domesticus* (a) and Greenfinch *Chloris chloris* (b) subjected to days of LD 6:18 with 1 h light interruptions at different parts of the dark period, as in (c). LH titres determined by radio-immunoassay, FSH activity inferred from testis weight. (From Murton *et. al.* [54])

8.4 CONTROL OF THE AUTUMN MIGRATION BY PHOTOPERIOD

Controversial though some of the opinions have been, and still are, regarding the control of vernal migration and breeding, they are nothing to the battles that have raged over the method of control in the post-breeding phase and autumn migration. No attempt can be made to review all this in historic detail, particularly as this has been essayed by all the main protagonists [19, 30, 42, 78]. We can only try and summarize the present understanding.

On the completion of breeding the birds become "refractive" in that long photoperiods fail to induce recrudescence of the gonads. Whether the termination of refractoriness is spontaneous or is brought about by photostimulation has been one of the main points of controversy. Probably there is an absolute refractory period whose specific length is quite independent of photoperiod and whose initiation and termination are under endogenous control. Then comes a period of relative refractoriness which can be terminated by exposure to "short" days. We have seen that the longer "long" photoperiods are the more effective they are in stimulating gonad growth. By contrast there appears little difference in effectiveness of "short" photoperiods between 6 and 14 h. However most experimenters have reported that LD 12 : 12 is the maximum for a short-photoperiod decision to be made. We have already seen that this is also the minimum for a long-photoperiod decision. So birds experiencing this combination could both complete the period of relative refractoriness and hasten the recrudescence of the gonads. Such an overlap of the acceptable extremes of short and long photoperiods is well known in plants [9]. Again as in many plants, experience of short photoperiods seems to be a prerequisite for reaction to the stimulating effects of long photoperiods. It is clear that birds wintering on the Equator will be able to react in both ways. It might even be that inaccuracies in the measuring-mechanism could on one day cause the decision to be "long" and on another "short". Indeed an indication of the time-measuring accuracy might be obtained, e.g. if LD 12.1 : 11.9 never resulted in the relatively refractory phase being terminated, measurement to within ± 6 min would be indicated.

At one time it was thought that termination of the relatively refractory period was brought about by a critical length of dark period rather than of photoperiod. This was because LD 8 : 16 was effective whereas LD 4 : 8 was not; nor was LD 6 : 6. However this is now explicable on variation in sensitivity within a circadian rhythm [30]. Thus LD 6 : 42 was as effective as LD 6 : 18 in terminating refractoriness. However half the birds exposed to the regime LD 6 : 30 also terminated even though the second light period fell in the middle of the subsequent 24 h period and should have been interpreted as a long (12+6) photoperiod. So interaction with the circadian rhythm would appear to be facilitatory but not absolutely necessary as for the long-photoperiod gonad-growth response. But it is clear that long dark periods are not effective in dissipating photorefractoriness.

It has been suggested [30] that as winter approaches there is a readjustment in the bird's timing mechanism such that progressively shorter photoperiods become stimulatory. The readjustment is brought about by changing the relative duration of the two phases (active/inactive) of the circadian clock which is also used for timing photoperiod length in the spring. Interestingly, this would suggest endogenous control, by a *circennial* clock, of the circadian control mechanism. Some support for this hypothesis was obtained using the interrupted night technique. A one hour light signal at 12 h from "dawn" had no effect in October but was highly stimulatory in January.

8.5 ADAPTIVENESS OF RESPONSE TO PHOTOPERIOD STIMULATION

The preceding account of responses to stimulation by varying photoperiods has inevitably been a generalized one. It is the subtle difference in response between species that enables one basic control system to be used in widely varying circumstances. The length of the endogenously determined absolute refractory period, the photoperiod lengths that terminate the partially refractory period and those which accelerate gonad development all appear to differ markedly and specifically. By combining in various ways they permit the control of the early-breeding single-brooded Rook *Corvus frugilegus;* the multi-brooded sedentary House Sparrow; the late-breeding migratory Whimbrel *Numenius phaeopus;* the transequatorial Bobolink [41].

The adjustments continue down to a very fine level. Thus Greenfinches and House Sparrows are closely related and have similar multi-brooded seasons. Yet there are distinct differences in their responses to "long" photoperiods mediated through the circadian cycle and particularly with regard the release of LH (Fig. 8.4). These in turn reflect differences in territorial behaviour. The House Sparrow is semi-colonial and territorial throughout the breeding season; the LH reaches a high level early and is maintained. The Greenfinch has a brief period of intensive territorial behaviour prior to mating and thereafter is much less obviously territorial; its early peak of LH is short-lived [53].

While measurement of photoperiod length through a circadian mechanism is undoubtedly a widespread control, it is by no means the only one. Even basically photoperiodic birds have their activities influenced by many proximate factors such as temperature, humidity, snow cover, and social pressures. In many others, particularly in the Tropics, photoperiods are of little concern even though the ability to respond can be shown experimentally. One of the late A. J. Marshall's provocative dicta was "Daylength is important only to species for which it is important that it should be important".

By the same token it should be emphasized that gonad development is not the only factor in migration. Birds must be in a suitable somatic state as well, in such respects as the deposition of fat reserves for the journey and the completion of at least parts of the moult. There is argument as to the extent to which sex hormones are responsible for the "total physiological state" required before migration can start. The post-breeding migration would seem less likely to be under the control of sex hormones, yet it has been shown [40, 43] that some migrants develop a potentially secretory interstitium in the gonads before the migratory journey.

8.6 TIMING OF THE MIGRATORY JOURNEY'S DURATION

The first autumnal migration is undoubtedly the critical one for the young migrant, hatched only a few months previously and with no experience of its winter quarters. It is much easier to postulate "instinctive knowledge" of the latter than to produce evidence of such a magical attribute. Some young birds, such as geese and swans, remain with their parents throughout the winter and so can benefit from their guidance. But in the great majority the family breaks up and the young migrate independently, sometimes to different areas. The extreme of independence of parental tuition is represented by the young Cuckoo *Cuculus canorus*.

A minimal requirement to reach an unknown goal is "distance-and-bearing navigation". If the bird were equipped with the innate information of how far it is to fly and in which direction, then it would stand a good chance of arriving in the general area of the target. Distance is a function of speed and time. If the young bird's migration was limited in time, and there was not a great variation in its average speed, then the distance parameter would be roughly fixed.

Birds prevented from migrating by being restrained in cages become intensively active at the migration season. This migratory restlessness or *Zugunruhe* is particularly obvious in normally diurnal birds which migrate by night (presumably to leave the day free for food gathering) (Fig. 8.3). It is easily monitored automatically, and its duration, several hours per night, coincides rather precisely with the time for which the free-flying migrants are airborne. We now have good estimates for the latter from observations in the field, most recently by the use of radar equipment [14]. The migration restlessness continues in phase with other indications of circadian rhythm even when the bird is in constant light or dark. Its duration is thus presumably measurable by the bird on a circadian basis just as is the normal activity.

The seasonal duration of the restlessness also fits well with what is known of the extent of the migration undertaken by different species. This has been most clearly shown when comparing the closely-related and sympatric Willow Warbler and Chiff Chaff [25, 26, 27]. As mentioned earlier, after breeding in N. Europe the former migrates transequatorially to S. Africa, the latter no further than N. Africa. Hand-reared Willow Warblers maintained a regular phasing of their physiological events under constant light conditions much better than did the Chiff Chaffs. Further, the migration restlessness of the Willow Warblers in autumn continued considerably longer (Fig. 8.6). Another observation was that individuals hatched later in the year come into migratory condition later. This all suggests a rather rigid programming of migration duration in the long distance migrant. As a further check Willow Warblers at the height of their migration restlessness were transferred across the Equator to the Congo. The intensity and duration of their restlessness continued to be on a par with birds left behind in Germany. So the stimuli of the southern wintering area did not stop the programme from running its full course.

This agrees with earlier field experiments [59, 60] in which Starlings trapped while on migration in the Netherlands were flown to Spain, a much favoured wintering area for the species. Despite this the birds

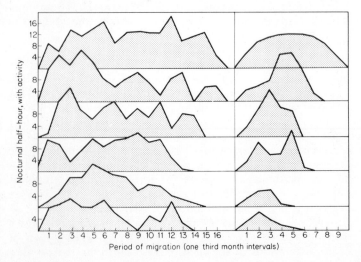

Figure 8.6. Duration of migration activity in a transequatorial migrant, Willow Warbler *Phylloscopus trochilus* (left), and in a shorter distance migrant, Chiff Chaff *P. collybita* (right). Six individuals of each species shown. (From Gwinner [27])

moved across the width of the Iberian peninsula, for a distance roughly equivalent to that which they would have covered from the Netherlands to their normal wintering grounds in England and France.

8.7 TIME-COMPENSATED SUN-COMPASS ORIENTATION

These, and earlier [58], Starling translocation experiments from the Netherlands to Switzerland (Fig. 8.7) confirmed that the (young) birds

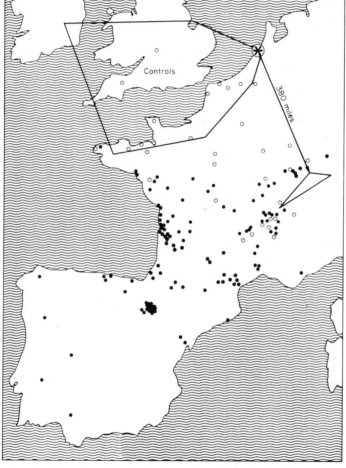

Figure 8.7. Recoveries of Starlings *Sturnus vulgaris* displaced from Holland to Switzerland while on autumn migration. Normal winter area marked "Controls". ○ Adults. ● Juveniles. (From Matthews [48] after Perdeck [58])

were not only programmed for distance but also for direction. Caught on migration and transferred to one side of the normal migration track they continued on a parallel course, so finishing up in an area entirely abnormal for their population. In Western Europe the general migration axis is NE/SW so there was a possibility that the young birds were merely joining up with others passing through the release point. This doubt was eliminated when Blue-winged Teal *Anas discors* were held back where trapped until migration had ceased or when Hooded Crows *Corvus cornix,* in addition, were transported to a new area. The programmed bearing was maintained. The latter experiments involved birds returning on their first vernal migration. Indeed, the imposed displacement, whether in autumn or spring, appeared to be learned and maintained in subsequent years.

These experiments involved the catching, marking with numbered leg rings, and transportation of thousands of birds. Results were then gradually built up as a small percentage were found dead or shot; a slow, laborious and uncertain business. On release the birds generally landed a short distance away so that orientation data could not be collected. A most important advance, therefore, was that made by Kramer [38] who found that the migration restlessness of caged birds was clearly oriented, generally in the appropriate migratory direction. This opened up the possibility of determining what stimuli were being used by the bird to determine its direction. Various automatic devices were devised for recording and measuring the orientation. If necessary, experiments with caged birds could be continued at seasons other than those of migration, by training them to go in one particular direction for food. Even better, tame pigeons could be trained to fly in one direction, so avoiding the inevitable restrictions imposed on caged birds which may distort their behaviour. Another form of free-flying one-direction orientation, not necessarily coincident with directions of migration nor with that of the home, has been found innate in a number of species of birds, such as Mallard. Because there seems to be no sensible explanation for its existence such orientation has been called "nonsense" orientation for interim convenience [46].

All these types of "compass" orientation would seem to have common sensory bases. By day they are based on the sun position. When the sun is occluded artificially or by heavy cloud, orientation is not achieved. If the sun position is distorted by mirrors the direction taken up is likewise distorted. There is suggestive evidence [77] that the Earth's magnetic field might in some way be involved but the consensus of experimental work favours sun compass orientation [48].

It is, of course, logical that there should be a time-measuring element since the birds can take up the appropriate direction at any time of day. They must therefore be able to compensate for the apparent movement of the sun round the sky of approximately 15° per h. Indeed, if they are presented with an artificial sun fixed in one position, they take up varying angles to it during the course of the day. In most cases it was actually more difficult to train a bird to go at a fixed angle to a light source that to go in a fixed compass direction with reference to a moving source.

The clock element is easily demonstrated by keeping birds on a light/dark cycle shifted out of phase. Their clocks then lock on to the new time within a few days. A bird with its clock "advanced" six hours takes up a direction approximately 90° left of the expected direction; with a clock retarded six hours, 90° right (Fig. 8.8). The circadian nature of the clock involved was shown by keeping direction-trained Starlings in constant light conditions for 11-12 days [34]. Records of their circadian activity pattern indicated that their clocks had a period of about 23½ h, i.e. activity began half an hour earlier each day (Fig. 8.9). When again shown the sun they took up an orientation to the left of the training direction, slightly less than the 90° that would be expected if their clocks had slipped six hours out of phase. When again exposed to the normal light/dark sequence the direction taken swung back to the original as their clocks were brought back into phase.

In lower latitudes, with night and day more nearly equal in length, massive time shifts, of 12 h, introduce a further complication. If artificial day and the true day overlap then the change in orientation is simply 180°. If the bird is tested during the day, but its shifted clock indicates that it is night, a variety of directions are taken up. Thus Mallard which had a NW "nonsense" orientation and were time switched 12 h near the equinoxes oriented SE around 0600, SW around 0900, NW at noon, NE at 1500 and SE again at 1800 [48]. This is understandable if we

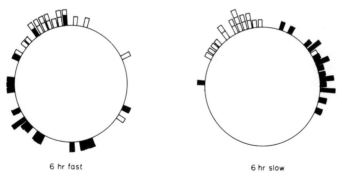

6 hr fast   6 hr slow

Figure 8.8. Bearings at which Mallard *Anas platyrhynchos* showing "nonsense" orientation were lost to sight by day. □ Controls. ■ Subjected to 6 h time shifts. (From Matthews [47])

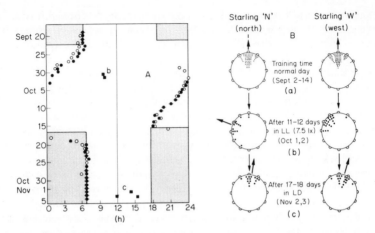

Figure 8.9. Two Starlings *Sturnus vulgaris* held in constant light conditions showed an anti-clockwise shift in sun-compass choice of training direction (B) which matches their increasingly early onset of locomotory activity (A). Mean choice directions shown by black arrows in B; onset of activity shown by open circles for "N", solid circles for "W" in A. In A, ■ mark the corresponding test times (b and c) in B.

postulate a rhythmic oscillation of the angle-correcting mechanism. Thus a bird flying NW requires to make a progressively smaller angle to the sun position through the day. The small angle at sunset returns to the large sunrise angle during the night by a reversal of the daytime process. If the partially unwound mechanism is suddenly exposed, like the Walrus and the Carpenter, to the sun shining in the middle of the night, it "locks on" and produces an orientation. There is, of course, a further complication around mid-summer and mid-winter when the days and nights are of very unequal lengths. This has yet to be investigated experimentally.

A number of invertebrates and fish apparently have a similar "unwinding" mechanism; others "wind-on" through the night, and yet others may do either. It has been suggested [68] that the winding-on mechanisms evolved in high latitudes, where the sun is seen to make a complete circuit, and the un-winding mechanisms in tropical zones.

Birds migrating into high or low latitudes would, of course, encounter unusual sun-position relationships. This is especially so of northern birds migrating south across the equator. There they would find the midday sun in the north and apparently moving, to an observer facing it, anticlockwise instead of clockwise. However, during its migration under, as it were, the sun's arch, the bird would have time to make adjustments.

If moved suddenly and out of contact with external clues, a bird might be expected to show confusion or actual false orientation. Thus a bird transported suddenly south across the Equator and seeking to go NW would do so correctly at sunrise and sunset but go SW at 0900, SE at noon and NE at 1500. Such work as has been done on sudden major latitude shifts [33, 65, 66, 56] have given results that suggest the birds continued to make corrections appropriate to the home situation with consequent predictable errors. Where continued observation was possible the birds may have been able to adjust to the new situation. Indeed learning may well be involved in the young bird's initial appreciation of the sun's direction and speed of movement [34].

Change in latitude also introduces changes in sun height (altitude), and this affects the horizontal component (azimuth) of the sun's "movement" round the sky. It is only viewed from the poles at midsummer that the sun's azimuth changes at a regular $15°$ per h. Elsewhere because of the inclination of the sun arc to the horizon the azimuth change is small early and late, large around midday. Moreover, this diurnal variation alters as the sun arc rises and falls with the seasons (Fig. 8.10). Experiments have yet to be done with birds at seasons and latitudes such that divergences from a regular $15°$ per h angle-correction could be detected against the experimental scatter. Yet varying rates of change of angle to the sun's azimuth according to time of day have been demonstrated in spiders and in fish. In the latter, unequivocal evidence was forthcoming that the animals were using the sun height (which was manipulated by a mirror) to decide how large a horizontal angle to

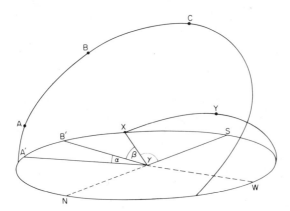

Figure 8.10. Perspective diagram of the sun arc at summer (upper) and winter (lower) solstices for $51°$N. The altitude of the sun affects the rate at which its downward projection (azimuth) moves round the horizon. (From Matthews [48])

select. Day length is also governed by seasonal changes in sun altitude and the velocity of angle correction was decreased during short photoperiods and increased during long [69, 7, 70]. It is quite likely that birds likewise adjust their orientation mechanism as well as the other physiological responses discussed earlier.

Birds migrating east or west enter different time zones, but could easily entrain their clocks to the new time. If transported suddenly then false orientation might be expected. Massive shifts are more economically achieved near the poles where the lines of longitude converge. Thus Adelie Penguins *Pygoscelis adeliae* had only to be moved 900 miles in Antarctic to experience a time shift of 8 h [55]. As would be expected south of the equator, the birds deviated anti-clockwise from the north-seeking nonsense orientation when transported east (internal clock slow) and clockwise when transported west. Surprisingly, birds held at the new site changed their clocks to local time within 3 weeks despite the 24 h daylight with but small fluctuations in sun height or illumination.

Other major longitudinal shifts have involved caged migrants moved over 114° from the Baltic across the Soviet landmass to Eastern Siberia [12]. These changed the directional trend of their migrational restlessness as would be expected if their clocks were behind the local time. Thus Barred Warblers *Sylvia nisoria* orientated SE at 21° E, S at 69° E, SW at 135° E. These birds winter in E. Africa so the clockwise swing of orientation gives the impression of a correction for displacement, to maintain a heading towards a geographical area. This would be a much greater feat than maintaining a simple compass heading. Similarly, Scandinavian migrants wind-drifted west to England showed more easterly headings (SE to SSE) than their normal migration directions (SSE to SW). As these, like the Barred Warblers, were mainly night migrants, it was argued [18] that they were less influenced by the passage of other migrants which might have carried day-flying starlings (p. 293) on a parallel course. The ambiguity could be resolved if, for example, birds were moved East in Spring when heading north. They would orientate NE if using a simple compass orientation, but NW if correcting towards the breeding quarters.

Among the Starlings there was a clear distinction between the parallel course behaviour of the young birds which did not know the area to which they were "programmed" and the orientation of the birds which were at least a year old which had been to the winter area before. They headed back towards it; transported SSE they turned NW (Fig. 8.7). Before considering the navigation mechanisms by which such corrective orientation of "homing" can be achieved, we must first consider compass orientation at night.

8.8 STAR-COMPASS ORIENTATION WITHOUT
TIME COMPENSATION

Many caged birds have demonstrated migration orientation throughout the night. Mallards also showed "nonsense" orientation just as well by night as by day, their flight being tracked by the use of small electric lamps attached to their legs. When the night sky was occluded, artificially or by heavy cloud, orientation was lacking, according to most investigators. The suggestion that the stars were being used as orientational clues received final confirmation when birds took up appropriate directions under the artificial stars of a planetarium [63].

The time relation of star "movement" is complicated. The rotation of the earth causes a given star to move across the sky, the rate of change of azimuth altering in accordance with the changing altitude between rising and setting. There are no seasonal changes in the arc altitude as there is with that of the sun. But there is a complication due to a difference between the length of solar and stellar days. Both are based on transits of an astronomical datum across the observer's meridian. Successive transits of a point on the star sphere, for practical purposes at infinity, occur at intervals of 23 h 56 min. Successive transits of the sun take 4 min longer since the earth is moving in orbit round it in the same direction as its own rotation. Because of this parallax effect the star sphere appears to slip westward through the year, a given star rising 4 min earlier each night, 2 h earlier after a month has elapsed. Thus at a given (solar) time a bird would have to fly at a different angle according to the calendar.

However, these complications are avoided because there is apparently no time-compensation element involved. When Mallard have their circadian clocks shifted by light/dark phase shift (Fig. 8.11) no appropriate change in direction occurs when they are showing "nonsense" orientation in free flight [47]. In the case of migration orientation in a planetarium the reverse test can be applied by presenting birds with skies appropriate to 3, 6 or 12 h earlier or later. None of these combinations produced any deflection in the orientation of Indigo Buntings *Passerina cyanea* [15].

It would appear, therefore, that the birds are getting directional information from the *patterns* of the constellations, much as we ourselves, for example, determine north from the pointers of the Great Bear. Because of the apparent western slippage of the star sphere many constellations are only visible at night during one season of the year. Only those close to the Pole Star never pass below the horizon, and it does seem that these circumpolar constellations are essential for

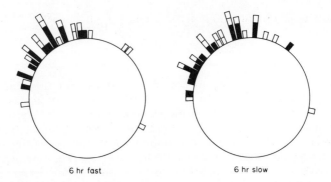

6 hr fast   6 hr slow

Figure 8.11. Bearings at which Mallard showing "nonsense" orientation were lost to sight at night under the stars. □ Controls. ■ Subjected to 6 h time shifts. Compare Fig. 8.8. (From Matthews [47])

obtaining direction information. The others may be "switched off" in the planetarium without preventing the birds' orientation.

Claims [63, 64] that the autumn or spring direction of orientation was selected according to which non-circumpolar constellations were visible would not appear to be substantiated. Indigo Buntings were exposed to long photoperiods in winter so that by May they were in the physiological state appropriate to the autumnal migration. Under spring stars in the planetarium they orientated south. Other birds held to normal photoperiods and so in the physiological state for vernal migration orientated north. Orientation was determined by internal state rather than from the constellation information [16].

The seasons follow in sequence as the earth, its axis inclined at an angle of 23½° to its orbit, moves round the sun. The stars visible at night are those opposite the dark side of the spinning earth, at a given point on the orbit. A complication is that the axis, rather like that of an off-balance top, gyrates slowly in a cone while still maintaining the same inclination to the orbit. This axial precession has the effect that one hemisphere is tilted towards or away from the sun by a given amount slightly earlier each solar year. The sidereal year can be said to be 50 sec shorter. After 13,000 years the constellations visible in spring will be visible in autumn and vice versa. This is a short time for evolutionary change to allow a new programming of the genetic code [2], especially as the whole cycle must have been gone through 40 times since modern birds were extant. However, the circumpolar constellations are not affected and the complication can be ignored in the present context.

In fact it seems that young birds must learn the configurational relationships of the stars and the directional information that they afford, even if the direction selected is innate [17]. Indeed Mallard can be made to learn wholly unnatural star patterns [76].

8.9 TIME-COMPENSATED MOON-COMPASS ORIENTATION

There is no sharp correlation in the intensity of nocturnal migration with the presence or absence of the moon. Nevertheless such an obvious feature of the night sky is a possible candidate as an orientation cue. In many studies of caged birds showing migration restlessness the moon has been reported as merely disruptive in its influence. Sometimes a simple phototactic response resulted, or the birds held a constant angle to the moon azimuth [8]. In most cases stars were also visible whereas the only critical test in the field can be when the sky is overcast with cloud sufficiently thick to obscure the stars yet thin enough to allow the moon's disc to be visible.

Such "moon-only" conditions are somewhat infrequent, of short duration, difficult to forecast and often accompanied by strong winds. Relatively few results with birds are thus available, but Mallard show "nonsense" orientation quite markedly and have done so with the moon in a variety of azimuths (Fig. 8.12) strongly suggesting time-compensation. The moon bears constant relations to the sun position below the horizon, opposite when full, 90° right at first quarter, 90° left at last quarter. If the phases can be recognized and the appropriate corrections made solar time-keeping could be applied and there would

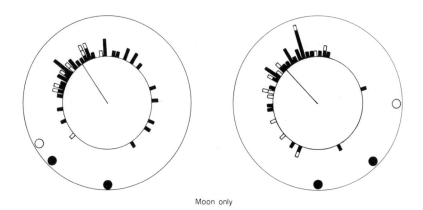

Moon only

Figure 8.12. Bearings at which Mallard showing "nonsense" orientation were lost to sight under moon-only conditions. The moon's azimuths varied, open and black circles, but the matched bearings remain NW indication time-compensation.

be no need to postulate a separate lunar clock [35]. However, we have seen that the sun-angle-compensation operates as if the sun runs backward through the south, so this simple relation would not hold. Instead possibilities of confusion between a backward running "sun" and a forward running moon are conjured up. Good orientation would then only be achieved at full moon near midnight when the two coincide in the south. "Moon-only" orientation is by no means so limited.

If the bird simply allows 15° per h for the moon's movement, as for the sun, then having seen moonrise it could orientate quite well through the night. But as the moon is relatively close in orbit round the earth the parallax effect is more marked than in the case of the stars and each night it rises nearly an hour earlier, by solar time. If the bird did not allow for this and was prevented from seeing the moon for a week, it would be around 90° in error. The limited information available for the Mallard suggests that they do have sufficient of an almanac-in-their-heads to carry them through such periods without viewing the moon and still make appropriate orientation when they next see it [48]. Time-shifting experiments have so far given confusing results. Birds time-shifted 6 h early gave the expected anti-clockwise shift. Birds shifted 6 h late and released when, according to their clocks, the moon should not yet have risen, produced results suggesting that the moon angle-correcting mechanism may oscillate back after moonset rather like the sun angle-correcting mechanism after sunset. It would be surprising if such a complicated mechanism for moon-orientation had evolved in view of the relatively few occasions on which birds would have to rely exclusively on its information. More research is certainly needed in this field.

8.10 THE TIME ELEMENT IN BI-COORDINATE NAVIGATION

We have seen that some birds displaced on migration are able to correct for that displacement and regain the original goal (p. 298). Such homing ability has been extensively studied by experiments in which birds were removed from their nests. Apart from wild birds a mass of experimental data have been accumulated with the homing pigeon. Such investigations have recently been reviewed [48] and here we need do no more than indicate the main lines of evidence that birds can actually determine and correct for displacement.

Firstly there is the return of birds from areas with which they would not be familiar in times which did not allow for their having wandered at random until they reached known territory. Examples are a Manx Shearwater *Procellaria puffinus* homing some 3,000 miles from Boston,

Mass. to Wales in 12½ days; a Leach's Petrel *Oceanodroma leucorrhea* doing the reverse trip, from Sussex to Maine in 14 days; a Laysan Albatross *Diomedea immutabilis* homing 3,200 miles to Midway Island across the Pacific from Washington State in 10 days. These are extreme cases but they are buttressed by a mass of shorter distance releases in which birds gave essentially straight-line homing times. Some observations en route confirmed that birds were on the direct line home. Recently developments in biotelemetry have enabled their actual tracks to be followed and these are found to have remarkably few deviations. Finally it has repeatedly been observed that birds are lost to sight in the general direction of home within a few minutes of release. Such homeward orientation has indeed been one of the most investigated of homing phenomena. Care must, of course, be taken that the bird cannot be using known landmarks. In many cases "nonsense" orientations have given a false appearance of homeward orientations. Releases must therefore be from several directions before an acceptable verdict is reached (Fig. 8.13).

A large number of hypotheses have been advanced to account for the homing phenomenon. One heterogeneous group suggests that the bird in some way maintains sensory contact with home. "Inertial navigation" has been suggested [3] and in passing we would note that this would require accurate measurement of time elapsed as well as the time element in velocity. However, the group of hypotheses as a whole is discredited by the demonstration that homeward orientation is not effective below about 50 miles [67]. Sometimes the gap is bridged by known landmarks [49] but there may be a zone in which orientation is apparently impossible. This strongly suggests that the navigational process involves the comparison of stimuli at the release point with remembered values at home. If two or more stimuli vary quantitatively in regular ways but in different directions across the earth's surface, a "grid" of isolines would result. The bird could then "fix" its position, in two co-ordinates, in relation to home or to the last point on which it had values.

Grids based on the dynamic consequences of the earth's rotation and/or its magnetism have frequently been discussed but are not much in favour at present. Again, as an aside, we may note that velocity measurements, and hence a time element, are inescapable requirements in such theoretical mechanisms.

It has, however, become axiomatic that good homeward orientation and swift homing returns from previously unknown release points are not achieved under heavy overcast [75, 48]. Some recent evidence to the contrary [36] was both inadequate [49] and quickly contradicted [37]. There is thus a strong suggestion that some form of astronavigation is involved; that birds, like Man, derive their geographical position from the

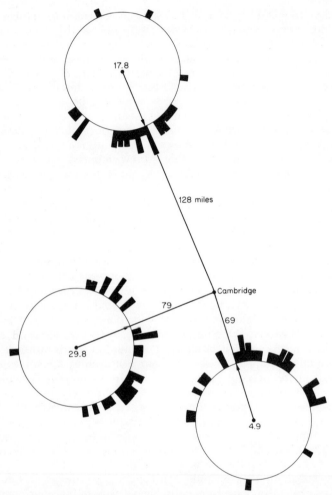

Figure 8.13. Bearings at which the same individual pigeons were lost to sight after being released successively (dates in circles) in three different directions from home. The first demonstration, in 1949, of true homeward orientation in birds. (From Matthews [48])

positions of the heavenly bodies in the sky. In this, measurement of time is an essential.

At any one time a star will be overhead at one point on the earth and have a decreasing altitude on a widening series of "circles of position" centred on that point. A certain altitude at a certain time means that the observation is made from somewhere on the appropriate circle.

Measuring the altitude of a second star places the observer on a second circle and if this overlaps with the first the observation point must be at either of the two points at which the circles cross. This ambiguity is resolved by taking a timed altitude for a third star, when the three circles are found to cross at a unique point or at worst form a small triangle of position within which the observer must be. The bird would obviously not proceed in this methodical way, using an almanac giving pre-calculated data on star altitude in relation to time. But it could well be comparing the tilt and degree of rotation of parts of the star sphere with those it would expect to see at home at the same time. It must remember what the time is at home with considerable accuracy.

Now, in fact we have little evidence that orientation and flight towards a goal, as opposed to a simple compass direction, is achieved at night by birds. Some early planetarium results were interpreted as showing that they could detect both longitude and latitude from the position of the star sphere [63, 64] but the interpretation has been challenged [74] and later workers have not obtained more than compass orientation [15]. By day, when clear evidence of bi-coordinate navigation has been obtained there is only one star available, the sun.

The human navigator proceeds in this case by watching the sun until it reaches the highest point on its arc across the sky. Its altitude at this point is inversely proportional to the observer's latitude. It is also reached at local noon and the difference between this and noon at Greenwich (recorded on a chronometer) is proportional to the degrees of longitude the observer is east (early) or west (late). The noon altitude rises and falls with the seasons (p. 297) and this must be allowed for if there is a substantial time lapse since the last noon-altitude measurement at home.

Since homeward orientation in birds has been observed at many times of day, there can be no question of their waiting to make the noon observation. Two hypotheses have been put forward. One postulates that the bird observes the movement along a small portion of the arc and then by extrapolation estimates the highest point and when it will be reached; comparisons are then made as above [48]. The other hypothesis suggests that the bird observes the slope of the segment of the arc, in effect by measuring the altitude and rate of change of altitude, and compares this segment of arc with that expected at the same time at home [57]. Both versions require very fine analysis of the sun's movement and the comparison of observed with remembered values. The first adds the complication of extrapolation, but reduces the memory requirement to one point; the second requires memory of the whole home arc. On present evidence it is not possible to decide between these hypotheses nor, indeed, to state categorically that sun bi-coordinate navigation is a proven fact. The experimental evidence has been reviewed in detail [48] along with anatomical and physiological investigations that suggest that the angle-measuring requirements should be within a bird's ability, likewise the necessary memory. Here we consider the time-measuring requirements.

The detection and measurement of short time intervals needed to estimate and extrapolate the sun's apparent movement would seem to be plausible. Precise estimates of time interval are implicit in rapid flight in the presence of obstacles, in the stereotyped nature of many behavioural movements, and in rhythmic performances elicited by Skinnerian training or in the precise duetting of song birds.

With regard to longitude measurement, the requirement is for birds to detect displacement in time (longitude is time, time is longitude) of the order of 4 mins. This equates with the minimal displacement of 50 miles in temperate latitudes found to be necessary for pigeons to show homeward orientation (p. 303). Biological clocks with this degree of constancy have been demonstrated. For instance Flying Squirrels *Glaucomys volens* in constant darkness maintained their activity cycles for up to 81 days and with standard deviations of as little as ±2 min [11]. The length of such free-running cycles can be changed by changing light intensity but this does not imply a modification of the underlying cycle, only an alteration of the threshold for emergence of activity. Similarly the clock's precision may be much greater than that manifest through the linking processes which terminate in the onset of activity.

But of course no matter how constant the clock may be in its period length, it will drift out of phase with solar time if it is circadian. In passing we may wonder whether there would not be advantage in clocks with periods of 23.9 h (sidereal day) and 24.8 h (lunar day) but clearly for solar bi-coordinate navigation there would be a premium on clocks with 24.0 h periods. Unfortunately when such clocks are demonstrated, as in two Starlings kept for 28 days in constant light and temperature [32], there is a tendency to suspect an overlooked Zeitgeber. Training birds to take food at fixed times can give some idea of the accuracy of the clocks, but will tend to give even more outside values than the onset of unlearned activity cycles [73, 1]. However tests with homing pigeons [52] did suggest that longitude judgement of between 50 and 60 miles was possible. Attempts in the field to impart false longitudinal information by resetting the biological clocks have given rather equivocal results [45, 48].

Even given that biological clocks constantly cycling at 24.0 h may well exist, it has been suggested that they would be little use for navigational purposes if they were as easily phase-shifted by light/dark cycles as those controlling sun-compass orientation. One could argue, of course, that one clock need not subserve both functions; there could be two, one flexible, one much more rigid. There is indeed some evidence for this [45]. On the other hand there seems no reason to insist on absolute, long-time rigidity, only one sufficient to last through the period of transportation (natural or artificial) in which the bird does not have access to astronomical information. Even a bird unable to leave the area at once, and adjusted to the new light/dark phase, could well remember the time difference first experienced, and hence the appropriate direction in which to fly when able to do so.

It would certainly seem an impossible requirement for a migrant, in journeys taking it through many time zones, to maintain a rigid chronometer set to one longitude, say that of the breeding area, where it may be for but a small fraction of the year. It would seem more likely that it would adjust to the time zone in which it was and be able to detect sudden changes in longitude if these were forced on it. The main function of bi-coordinate navigational ability would then be to detect and compensate for deflections from the programmed distance-and-bearing movement between breeding grounds and wintering area (p. 291). It would be no great surprise if even young birds on their first migration used bi-coordinate navigation to correct displacement. They could be comparing the situation as it is with that which they last experienced before losing touch (through a storm, say, or experimental transportation) with astronomical information. They could correct by heading back towards their last actual "fix" or, more likely, this corrective heading would vectorize with the programmed migratory orientation to point again at the unknown goal. What does seem pretty certain is that young birds could not have innate knowledge of the geographical location of that goal when one of the co-ordinates is a time-zone which the bird has never experienced.

REFERENCES

1. H. E. Adler, Psychophysical limits of celestial navigation hypotheses. *Ergbn. Biol.,* **26**, 235-252 (1963).
2. S. L. Agron, Evolution of bird navigation and the earth's axial precession. *Evolution,* **16**, 524-527 (1963).
3. J. S. Barlow, Inertial navigation as a basis for animal navigation. *J. Theor. Biol.,* **6**, 76-117 (1964).
4. J. Benoit and R. Kehl, Nouvelles recherches sur les voies nerveuses photoréceptrices et hypophysostimulantes chez le canard domestique. *C.R. Soc. Biol.,* **131**, 89-96 (1939).
5. J. Benoit, I. Assenmacher and E. Brard, Apparition et maintien de cycles sexuels non saisonniers chez le Canard domestique placé pendant plus de trois ans à l'obscurité totale. *J. Physiol. Paris,* **48**, 388-391 (1956).

308   G. V. T. MATTHEWS

6. J. Benoit, I. Assenmacher and E. Brard, Etude de l'évolution testiculaire du Canard domestique soumis tres jeune à un éclairement artificiel permanent pendant deux ans. *C.r. hebd. Séanc. Acad. Sçi. Paris,* 242, 3113-3115 (1956).

7. W. Braemar and H. O. Schwassmann, Von Rhythmus der Sonnenorientierung am Äquator (bei Fischen). *Ergebn. Biol.,* 26, 182-201 (1963).

8. I. L. Brown and L. R. Mewaldt, Behaviour of Sparrows of the Genus *Zonotrichia,* in orientation cages during the lunar cycle. *Z. Tierpsychol.,* 25, 668-700 (1968).

9. E. Bünning, *The Physiological Clock,* Springer-Verlag, Berlin (1964).

10. J. P. Chapin, The calendar of Wideawake Fair. *Auk,* 71, 1-15 (1954).

11. P. J. De Coursey, Effect of light on the circadian activity rhythm of the Flying Squirrel, *Glaucomys volans. Z. vergl. Physiol.,* 44, 331-354.

12. V. R. Dolnik and M. E. Shumakov, Testing the navigational abilities of birds. (In Russian.) Bionica, Moscow, 500-507 (1967).

13. D. F. Dorward, Comparative biology of the White Booby and the Brown Booby *Sula* spp. at Ascension. *Ibis,* 103b, 174-220 (1962).

14. E. Eastwood, *Radar Ornithology,* Methuen, London (1967).

15. S. T. Emlen, Migratory orientation in the Indigo Bunting *Passerina cyanea.* Part II: Mechanism of celestial orientation. *Auk,* 84, 463-489 (1967).

16. S. T. Emlen, Bird migration: influence of physiological state upon celestial orientation. *Science,* 165, 716-718 (1969).

17. S. T. Emlen, The development of migratory orientation in young Indigo Buntings. *Living Bird,* 8, 113-124 (1969).

18. P. R. Evans, Re-orientation of small passerine night migrants after displacement by the wind. *Brit. Birds,* 61, 281-303 (1968).

19. D. S. Farner, The photoperiodic control of reproductive cycles in birds. *Am. Scient.,* 52, 137-156 (1964).

20. D. S. Farner, Circadian systems in the photoperiodic responses of vertebrates. In *Circadian Clocks,* J. Aschoff (ed.), North Holland Publishing Co., Amsterdam 357-369 (1965).

21. D. S. Farner, The control of avian reproductive cycles.Proc. XIV Int. Orn. Cong. Oxford, 107-133 (1967).

22. D. S. Farner and B. K. Follett, Light and other environmental factors affecting avian reproduction. *J. Anim. Sci.,* 25, 90-115 (1966).

23. D. S. Farner and L. R. Mewaldt, Is increased activity or wakefulness an essential element in the mechanism of the photoperiodic responses of avian gonads? *Northwest Science,* 29, 53-65 (1955).

24. S. Gaston and M. Menaker, Pineal function: the biological clock in the sparrow? *Science,* 160, 1125-1127 (1968).

25. E. Gwinner, Circannuale Periodik als Grundlage des jahreszeitlichen Funktionswandels bei Zugvögeln. Untersuchungen am Fitis (*Phylloscopus trochilus*) und am Waldlaubsänger (*P. sibilatrix*). *J. Orn.,* 109, 70-95 (1968).

26. E. Gwinner, Untersuchungen zur Jahresperiodik von Laubsängern. *J. Orn.,* 110, 1-21 (1969a).

27. E. Gwinner, Artspezifische Muster der Zugunruhe bei Laubsängern und ihre mögliche Bedeutung für die Beendigung des Zuges im Winterquartier. *Z. Tierpsychol.,* 25, 843-853 (1969b).

28. E. Gwinner, The adaptive nature of circannual rhythms in birds. Abstracts XV Cong. Int. Ornith., The Hague, 15-16 (1970).

29. W. M. Hamner, Circadian control of photoperiodism in the House Finch demonstrated by interrupted-night experiments. *Nature, Lond.,* 203, 1400-1401 (1964).

30. W. M. Hamner, The photorefractory period of the House Finch. *Ecology*, 49, 211-227 (1968).
31. R. A. Hinde, Aspects of the control of avian reproductive development within the breeding season. Proc. XIV Int. Orn. Cong. Oxford, 133-153 (1967).
32. K. Hoffmann, Versuche zu der im Richtungsfinden der Vögel enthaltenen Zeitschätzung. *Z. Tierpsychol.*, 11, 453-475 (1954).
33. K. Hoffmann, Die Richtungsorientierung von Staren unter der Mitternachtssonne. *Z. vergl. Physiol.*, 41, 471-480 (1959).
34. K. Hoffmann, Experimental manipulation of the orientational clock in birds. *Cold Spring Harbour Symp.*, 25, 379-387 (1960).
35. K. Hoffmann, Clock-mechanisms in celestial orientation of animals. In *Circadian Clocks*, Ed. J. Aschoff, North-Holland Publishing Co., Amsterdam, 426-441 (1965).
36. W. T. Keeton, Orientation by pigeons: is the sun necessary? *Science*, 165, 922-928 (1969).
37. W. T. Keeton and A. Gobert, Orientation by untrained pigeons requires the sun. *Proc. Nat. Acad. Sci.*, 65, 853-856 (1970).
38. G. Kramer, Eine neue Methode zur Erforschung der Zugorientierung und die bisher damit erzielten Ergebnisse. Proc. X Int. Orn. Cong. Uppsala, 271-280 (1951).
39. D. S. Lehrman, Gonadal hormones and parental behaviour in birds and infrahuman mammals. In *Sex Internal Secretions*, Ed. W. C. Young, Williams & Wilkins, Baltimore (1961).
40. B. Lofts and A. J. Marshall, The interstitial and spermatogenetic tissue of autumn migrants in Southern England. *Ibis*, 99, 621-627 (1957).
41. B. Lofts and R. K. Murton, Photoperiodic and physiological adaptations regulating avian breeding cycles and their ecological significance. *J. Zool. Lond.*, 155, 327-394 (1968).
42. A. J. Marshall, Breeding seasons and migration. In *Biology and Comparative Physiology of Birds*, Ed. A. J. Marshall, Academic Press, New York and London (1961).
43. A. J. Marshall and C. J. F. Coombs, The interaction of environmental, internal and behavioural factors in the Rook *Corvus frugilegus* Linnaeus. *Proc. Zool. Soc. Lond.*, 128, 545-589 (1957).
44. A. J. Marshall and D. L. Serventy, Experimental demonstration of an internal rhythm of reproduction in a trans-equatorial migrant (the Short-tailed Shearwater *Puffinus tenuirostris*). *Nature*, 184, 1704-1705 (1959).
45. G. V. T. Matthews, An investigation of the "chronometer" factor in bird navigation. *J. Exp. Biol.*, 32, 39-58 (1955).
46. G. V. T. Matthews, "Nonsense" orientation in Mallard, *Anas platyrhynchos*, and its relation to experiments in bird navigation. *Ibis*, 103a, 211-230 (1961).
47. G. V. T. Matthews, The astronomical bases of "nonsense" orientation. Proc. XIII Int. Orn. Cong., Ithaca, 415-429 (1963).
48. G. V. T. Matthews, *Bird Navigation*, Cambridge University Press, Cambridge (1968).
49. G. V. T. Matthews, Comment. *Nature*, 227, 627 (1970).
50. M. Menaker, Circadian rhythms and photoperiodism in *Passer domesticus*. In *Circadian Clocks*, ed. J. Aschoff, North Holland Publishing Co., Amsterdam, 385-395 (1965).
51. M. Menaker, Extraretinal light perception in the Sparrow. I. Entrainment of the biological clock. *Proc. Nat. Acad. Sci. USA*, 95, 414-421 (1968).
52. M. E. Meyer, The internal clock hypothesis for astronavigation in homing pigeons. *Psychon. Sci.*, 5, 259-260 (1966).

53. R. K. Murton, B. Lofts and A. H. Orr, The significance of circadian-based photosensitivity in the House Sparrow *Passer domesticus*. *Ibis*, 112, 448-456 (1970).

54. R. K. Murton, B. Lofts and N. J. Westwood, The circadian basis of periodically controlled spermatogenesis in the Greenfinch *Chloris chloris*. *J. Zool. Lond.*, 161, 125-136 (1970).

55. R. L. Penney and J. T. Emlen, Further experiments on distance navigation in the Adelie Penguin. *Ibis*, 109, 99-109 (1967).

56. R. L. Penney and D. K. Riker, Adelie Penguin orientation under the northern sun. *Antarctic J. US*, 4, 116-117 (1969).

57. C. J. Pennycuick, The physical basis of astronavigation in birds: theoretical considerations. *J. Exp. Biol.*, 37, 573-593 (1960).

58. A. C. Perdeck, Two types of orientation in migrating starlings *Sturnus vulgaris* L. and chaffinches *Fringilla coelebs* L., as revealed by displacement experiments. *Ardea*, 46, 1-37 (1958).

59. A. C. Perdeck, An experiment on the ending of autumn migration in starlings. *Ardea*, 52, 133-139 (1964).

60. A. C. Perdeck, Orientation of starlings after displacement to Spain. *Ardea*, 55, 194-202 (1967).

61. W. Rowan, On photoperiodism, reproductive periodicity and the annual migration of birds and certain fishes. *Proc. Boston Soc. nat. Hist.*, 38, 147-189 (1926).

62. W. Rowan, *The Riddle of Migration*, Williams & Wilkins, Baltimore (1931).

63. E. G. F. Sauer, Die Sternenorientierung nächtlich ziehender Grasmücken (*Sylvia atricapilla, borin* und *curruca*). *Z. Tierpsychol.*, 14, 29-70 (1957).

64. E. F. G. Sauer, Further studies of the stellar orientation of nocturnally migrating birds. *Psychol. Forschung.*, 26, 224-244 (1961).

65. K. Schmidt–Koenig, Sun compass orientation of pigeons upon equatorial and trans-equatorial displacement. *Biol. Bull.*, 124, 311-321 (1963).

66. K. Schmidt-Koenig, Sun compass orientation of pigeons upon displacement north of the arctic circle. *Biol. Bull.*, 127, 154-158 (1963).

67. K. Schmidt-Koenig, Über die Entfernung als Parameter bei der Anfangs-orientierung der Brieftaube. *Z. vergl. Physiol.*, 52, 33-55 (1966).

68. H. O. Schwassmann, Environmental cues in the orientation rhythm of fish. *Cold Spring Harbour Symp.*, 25, 443-450 (1960).

69. H. O. Schwassmann and W. Braemer, The effect of experimentally changed photoperiod on sun-orientation in fish. *Physiol. Zool.*, 34, 273-286 (1961).

70. H. O. Schwassmann and A. D. Hasler, The role of the sun's altitude in sun orientation of fish. *Physiol. Zool.*, 37, 163-178 (1963).

71. D. L. Serventy, Egg-laying timetable of the Slender-billed Shearwater *Puffinus tenuirostris*. Proc. XIII Int. Ornith. Cong., Ithaca, 338-343 (1963).

72. D. W. Snow, The breeding of Audubon's Shearwater (*Puffinus herminieri*) in the Galapagos. *Auk*, 82, 591-597 (1965).

73. H. Stein, Untersuchungen über dem Zeitsinn bei Vögeln. *Z. vergl. Physiol.*, 33, 387-403 (1951).

74. H. G. Wallraff, Können Grasmücken mit Hilfe des Sternenhimmels navigieren. *Z. Tierpsychol.*, 17, 165-177 (1960).

75. H. G. Wallraff, Über die Anfangsorientierung von Brieftauben unter geschlossener Wolkendecke. *J. Orn.*, 107, 326-336 (1966).

76. H. G. Wallraff, Über das Orientierungsvermögen von Vögeln unter natürlichen und künstlichen Sternenmustern. Dressurversuche mit Stockenten. *Verh. Deutsch Zool. Gesell.*, 348-357 (1968).

77. W. Wiltschko, Über den Einfluss statischer Magnetfelder auf die Zugorientierung der Rotkehlchen *Erithacus rubecula. Z. vergl. Physiol.,* 25, 537-558 (1968).
78. A. Wolfson, Environmental and neuroendocrine regulation of annual gonadal cycles and migratory behaviour in birds. In *Recent Progress in Hormone Research,* Academic Press, New York, 22, 177-244 (1966).

Index